STATISTICS I

Descriptive Statistics and Probability

HARCOURT BRACE COLLEGE OUTLINE SERIES

STATISTICS I

Descriptive Statistics and Probability

Elliot A. Tanis

Department of Mathematics
Hope College
Holland, Michigan

Harcourt Brace College Publishers
Fort Worth Philadelphia San Diego
New York Orlando Austin San Antonio
Toronto Montreal London Sydney Tokyo

Permissions Department
Harcourt Brace & Company
Orlando, Florida 32887

Printed in the United States of America

ISBN 0-15-601616-8

5678901234 074 121110987654

PREFACE

Do not *read* this Outline—**use** it. You can't learn probability and statistics simply by reading about it: You have to *do* it. Solving specific, practical problems is a good way to master—and to demonstrate your mastery of—the definitions and theories upon which probability and statistics are based. Outside the classroom, you need three tools to solve problems and analyze data: a pencil, paper, and a calculator or computer. Add a fourth tool, this Outline, and you're all set.

This HBJ College Outline has been designed as a tool to help you sharpen your problem-solving skills in probability and statistics. Each chapter covers a topic (or related topics), whose fundamental principles are broken down in outline form for easy reference. This outline text is heavily interspersed with worked-out examples, so you can see immediately how each new idea is applied in problem form. Each chapter also contains a Summary and a Raise Your Grades section, which (taken together) give you an opportunity to review the primary principles of a topic and the problem-solving techniques implicit in those principles.

Most important, this Outline gives you plenty of problems to practice on. Work the Solved Problems on your own and then check yourself against the step-by-step solutions provided. Test your mastery of the material in each chapter by doing the Supplementary Problems. (In the Supplementary Problems, you're given the answers only—the details of the solution are up to you.) Finally, you can review all the topics in a unit (five or six chapters) by working the problems in the Unit Exams. (The solution to each exam question is explained and the section that gives the necessary background for the solution is given, so you can diagnose your own strengths and weaknesses.)

Having the tools is one thing; knowing how to use them is another. The solution to any problem in probability or statistics requires six procedures: (1) UNDERSTANDING, (2) ANALYZING, (3) PLANNING, (4) EXECUTING, (5) CHECKING, (6) REPORTING. Let's look at each of these procedures in more detail.

1. **UNDERSTANDING:** Read over the problem carefully and be sure you understand every part of it. If you have difficulty with any of the terms or ideas in the problem, reread the text material on which the problem is based. (In this Outline, important ideas, principles, and terms are printed in boldface type, so they will be easy to find.) Make certain that you understand what kind of answer will be required. If the problem is quantitative, make an estimate of the magnitude of the answer.

2. **ANALYZING:** Break the problem down into its components. Ask yourself

    - What, if any, are the data?
    - What is (are) the unknown(s)?
    - What question is really being asked?
    - What equation(s) or definition(s) can be used to help answer the question?

3. **PLANNING:** Outline the steps that will be used to solve the problem in a series of operations.

4. **EXECUTION:** Follow your plan and execute any mathematical operations. It helps to work with symbols whenever possible: Substituting data for variables should be the *last* thing you do. Make sure you've used the correct signs, exponents, and units.

5. **CHECKING:** Never consider a problem solved until you have checked your work. Does your answer

    - make sense?
    - have the right units?
    - answer the question?

    Is your math right?

6. **REPORTING:** Make sure you have shown your reasoning and method clearly, and that your answer is readable. (It can't hurt to write the word "Answer" in front of your answer. That way, you—and your instructor—can find it at a glance, saving time and trouble all round.)

This Outline is written in two volumes—Volume I covers descriptive statistics and probability; Volume II covers inferential statistics. Together these two volumes cover most of the topics that are taught in an elementary or introductory statistics course that has algebra as its only mathematics prerequisite.

If you are a student who simply wants to know "How do I work this problem" and are not so much interested in "What are the basic ideas and principles that are needed for an understanding of this problem," my hope for you is that, as you become more confident with the manipulative aspect of the solution of a problem, you will also learn to appreciate the theoretical underpinnings of the solution.

I would like to thank the administration of Hope College for their encouragement during the preparation of these educational materials. I am especially grateful for a grant from Hope College that was funded from an Exxon Educational Foundation grant secured by Neal Sobania, Director of International Education at Hope College. The purpose of this Exxon grant was to internationalize our curriculum. Thus several examples have been developed for this Outline that have an international dimension.

Special thanks for the completion of this Outline go to Joyce Chandler Zandee, who worked every problem and offered helpful suggestions for making the text, problems and the solutions more understandable for you, the reader.

Hope College
Holland, Michigan

ELLIOT A. TANIS

CONTENTS

CHAPTER 1	**Graphical Presentation of Data**	**1**
	1-1: Frequency Distributions	1
	1-2: Histograms	4
	1-3: Stem-and-Leaf Diagrams	6
	1-4: Empirical Distribution Function and the Ogive	7
	Solved Problems	*11*
CHAPTER 2	**Measures of Central Tendency**	**22**
	2-1: The Sample Mean for Ungrouped Data	22
	2-2: The Sample Mean for Grouped Data	23
	2-3: The Median	26
	2-4: The Mode	27
	Solved Problems	*28*
CHAPTER 3	**Measures of Dispersion**	**33**
	3-1: The Variance and Standard Deviation for Ungrouped Sample Data	33
	3-2: The Variance and Standard Deviation for Grouped Sample Data	35
	3-3: Chebyshev's Inequality	39
	3-4: The Range	40
	Solved Problems	*41*
CHAPTER 4	**Measures of Relative Standing**	**48**
	4-1: Quartiles, Percentiles, and Deciles	48
	4-2: z-Score	52
	4-3: Box-and-Whisker Diagrams	53
	Solved Problems	*55*
CHAPTER 5	**Bivariate Data**	**64**
	5-1: Scatter Diagrams	64
	5-2: Correlation Coefficient	67
	5-3: Linear Regression	70
	Solved Problems	*75*
EXAM 1		**88**
CHAPTER 6	**Probability**	**93**
	6-1: The Concept of Probability	93
	6-2: Properties of Probability	97
	6-3: Methods of Enumeration	100
	6-4: Probabilities for Intersections of Events	104
	Solved Problems	*109*
CHAPTER 7	**Discrete Probability Distributions**	**121**
	7-1: Discrete Random Variables	121
	7-2: Probability Functions	122
	7-3: Distribution Functions	124

	7-4: Mean, Variance, and Standard Deviation of a Random Variable	126
	Solved Problems	*129*
CHAPTER 8	**The Binomial Distribution**	**140**
	8-1: Bernoulli Experiments	140
	8-2: Binomial Probabilities	140
	8-3: The Mean, Variance, and Standard Deviation of a Binomial Random Variable	143
	Solved Problems	*145*
CHAPTER 9	**Some Other Discrete Distributions**	**153**
	9-1: Hypergeometric Distribution	153
	9-2: Poisson Distribution	156
	9-3: Multinomial Distribution	161
	Solved Problems	*164*
CHAPTER 10	**The Normal Distribution**	**171**
	10-1: Continuous Distributions	171
	10-2: The Standard Normal Distribution	175
	10-3: Other Normal Distributions	180
	Solved Problems	*185*
CHAPTER 11	**Sampling Distributions**	**195**
	11-1: Random Samples: The Mean, Variance, and Standard Deviation of $\overline{X}$	195
	11-2: The Distribution of the Sample Mean	198
	11-3: Normal Approximation of Binomial Probabilities	204
	11-4: The Chi-Square Distribution	208
	11-5: Student's t Distribution	211
	11-6: The F Distribution	214
	Solved Problems	*219*
EXAM 2		**230**
APPENDIX	**Table 1: Binomial Coefficients**	**239**
	Table 2: Binomial Probabilities	**240**
	Table 3: Poisson Probabilities	**244**
	Table 4: Probabilities for the Standard Normal Distribution	**249**
	Table 5: Chi-Square Critical Values	**250**
	Table 6: Student's t Critical Values	**251**
	Table 7: F Distribution Critical Values	**252**
	Table 8: Critical Values for the Correlation Coefficient	**255**
	Table 9: Random Numbers on the Interval (0, 1)	**256**
INDEX		**257**

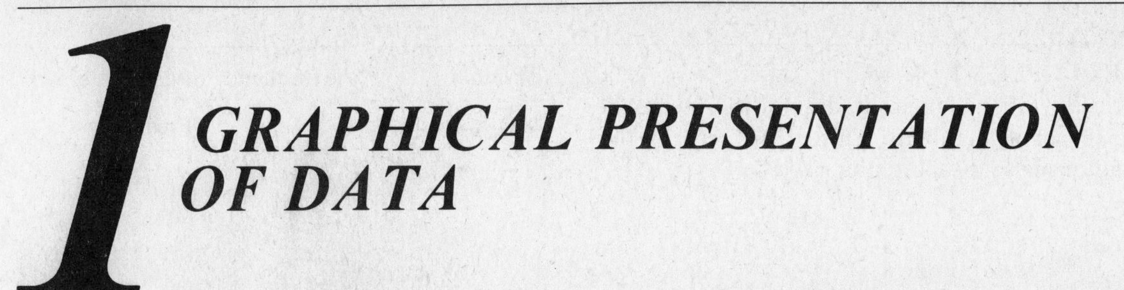

1 GRAPHICAL PRESENTATION OF DATA

THIS CHAPTER IS ABOUT

☑ **Frequency Distributions**
☑ **Histograms**
☑ **Stem-and-Leaf Diagrams**
☑ **Empirical Distribution Function and the Ogive**

In statistics we observe random experiments for which the outcome can't be predicted with certainty. The observations that we collect or code as numbers are called **data**.

1-1. Frequency Distributions

A. Discrete data

If in an experiment we *count*, for example, the number of successes in 10 trials or the number of telephone calls per hour, each count is an observation of a **discrete variable** (or discrete-type variable) and the data produced are **discrete data** (or discrete-type data).

note: Discrete data are observed by counting, NOT by measuring; so these data can only have values that are integers or specified fractions. Thus, you would never observe 5.6 successes in 10 trials. Likewise, if you were observing the hourly rates earned by assembly-line workers, you would never record a value of $7.3564, although you could expect $7.35.

But numbers by themselves are often difficult to interpret. In order to be useful, data must be *grouped* in some way. A frequency distribution is a way of grouping observations to summarize numeric findings.

- A **frequency distribution of discrete data** is a *tally* of the number of times each possible outcome occurs.

EXAMPLE 1-1 To make some quick-and-easy spending money, you decide to raise rabbits. You buy 30 pairs of domestic rabbits, breed them, and then await the blessed events. It's not long before baby bunnies abound. You cleverly observe that the litter sizes aren't all the same. Use the following litter-size data to create a frequency distribution.

10	12	8	9	7	11	12	8	5	9
10	11	6	7	8	12	11	12	8	10
9	6	11	9	10	6	7	8	9	14

Solution To group the data into a frequency distribution, count the number of times each possible outcome occurs.

<table>
<tr><th>Litter size</th><th>Tally</th><th>Frequency</th></tr>
<tr><td>5</td><td>|</td><td>1</td></tr>
<tr><td>6</td><td>|||</td><td>3</td></tr>
<tr><td>7</td><td>|||</td><td>3</td></tr>
<tr><td>8</td><td>卌</td><td>5</td></tr>
<tr><td>9</td><td>卌</td><td>5</td></tr>
<tr><td>10</td><td>||||</td><td>4</td></tr>
<tr><td>11</td><td>||||</td><td>4</td></tr>
<tr><td>12</td><td>||||</td><td>4</td></tr>
<tr><td>13</td><td></td><td>0</td></tr>
<tr><td>14</td><td>|</td><td>1</td></tr>
<tr><td></td><td>Total</td><td>30</td></tr>
</table>

EXAMPLE 1-2 A wildlife expert, who isn't raising rabbits for profit, decides to observe 30 pairs of wild rabbits. This expert records the following wild-rabbit litter sizes:

1	7	5	3	4	2	8	6	7	9
6	5	2	7	4	8	9	10	1	6
3	6	5	2	8	7	4	3	7	2

Group these data into a frequency distribution.

Solution Tally the frequency of each litter size.

Litter size	Tally	Frequency
1	ǀǀ	2
2	ǀǀǀǀ	4
3	ǀǀǀ	3
4	ǀǀǀ	3
5	ǀǀǀ	3
6	ǀǀǀǀ	4
7	~~ǀǀǀǀ~~	5
8	ǀǀǀ	3
9	ǀǀ	2
10	ǀ	1
	Total	30

B. Continuous data

Experiments can also result in data that are measurements of a continuous variable. In general, any values found by measurement are called **continuous data** (or continuous-type data), because they can assume any value within a range or interval of real-number values. For example, height, weight, and age are continuous variables which, when measured, yield continuous data.

note: Continuous data can *look* like discrete data. For example, heights are often recorded as integers. But when we say that the height of a person is 183 cm ($\sim$6 ft) we really mean that the actual height is *between* 182.5 and 183.5 cm.

To make a frequency distribution with a set of continuous data, we must first group the observations. We do this by dividing the scale of possible outcomes into intervals called **classes**, using the following terminology and suggestions.

Terminology:

(1) **Class limits** give the smallest and largest possible observed values in a given class (interval).
(2) **Class boundaries** give the smallest and largest actual values in a given class. (Remember that 183 cm is actually between 182.5 and 183.5.) Class boundaries fall halfway between the upper class limit for the smaller class and the lower class limit for the larger class.
(3) The **class width** is the difference between the class boundaries for a given class.
(4) The **class mark** is the midpoint of a class.

Suggestions:

(1) The number of classes (or intervals) is usually between 6 and 20 for best results—but the actual number chosen is arbitrary.
(2) Usually classes are chosen so that they all have the same class width.
(3) It is advantageous (but not necessary) to use an odd class width so that the class mark (middle value) is easy to determine and is the same accuracy as the data.

Now we can make a frequency distribution.

- A **frequency distribution of continuous data** is a tally of outcomes of a continuous variable grouped in (arbitrary) intervals or classes.

EXAMPLE 1-3 You have measured the heights of 50 women and recorded the following data (in cm):

171	157	163	170	173	170	168	167	164	170
169	157	168	[178]	163	161	165	160	166	155
170	173	171	174	171	173	175	167	164	168
169	170	157	169	172	156	160	158	172	165
171	[152]	163	165	170	177	159	157	158	165

(a) Divide the heights into classes to create an appropriate frequency distribution for your data.
(b) Determine the class marks.

Solution

(**a**) First, find the *range*, which is the difference between the largest and smallest values. Scanning through the data, you find that the smallest height is 152 and the largest height is 178, so the range is 178 − 152 = 26. Now, determine the class width and the number of classes, so that *at least* the range is covered; i.e., (number of classes) × (class width) > range. With a range of 26, you can pick any of the tabulated combinations here (plus many more).

Number of classes	Class width	Range covered
3	9	27 > 26
9	3	27 > 26
5	6	30 > 26
6	5	30 > 26
4	7	28 > 26
7	4	28 > 26

There is no single correct solution, although some choices may be better than others, depending on the number of observations in the set. Trial-and-error is the best way to approach these problems: If your choice of class width and number of classes clumps all the observations together—or spreads them out too far—you may want to choose another combination. The choice of 6 classes with a class width of 5 is made because it's an advantage to have an *odd* class width.

Next, pick 150 as the first class limit, so that 152—the lowest observed value—will fall in the first class. Then, determine the lower class boundary for the first class and add the class width to get the upper class boundary. The lower boundary of the first class is 149.5; the upper class boundary is 149.5 + 5 = 154.5, which is also the lower class boundary for the second class. Continue this procedure until all class boundaries are set. After you determine all the class boundaries, fill in the class limits; for example, the third class has boundaries 159.5–164.5, so the class limits are 160–164. Finally, tally the observations and total the frequency of occurrences in each class.

Class boundaries	Class limits	Tally	Frequency	Class marks
149.5–154.5	150–154	|	1	152
154.5–159.5	155–159	~~||||~~ ||||	9	157
159.5–164.5	160–164	~~||||~~ |||	8	162
164.5–169.5	165–169	~~||||~~ ~~||||~~ |||	13	167
169.5–174.5	170–174	~~||||~~ ~~||||~~ ~~||||~~ |	16	172
174.5–179.5	175–179	|||	3	177
		Total	50	

(**b**) You can easily determine the class mark by adding the upper and lower limits for each class and dividing by 2. The class marks turn out to be integers because you had the foresight to choose an odd class width. You have now completed the grouped frequency distribution.

EXAMPLE 1-4 The weights of 24 "three-pound" (1.36 kg) bags of apples were as follows:

3.26	[3.62]	3.39	3.12	3.53	3.30	3.10	3.26	3.19	3.22	3.14	3.39
3.31	3.21	3.49	3.41	[3.02]	3.17	3.20	3.12	3.42	3.36	3.21	3.26

Use these data to make a grouped frequency distribution. Include class limits, class boundaries, tallies, frequencies, and class marks.

Solution Here you see that the largest weight is 3.62 and the smallest weight is 3.02, so the range is 3.62 − 3.02 = .60. One choice you can make is to use 6 classes with a class width of .11, since 6(.11) = .66 > .60. (Also, remember that it's advantageous to have an odd class width.) Note that the class boundaries are halfway between two possible weight measurements.

Class boundaries	Class limits	Tally	Frequency	Class marks
3.005–3.115	3.01–3.11	||	2	3.06
3.115–3.225	3.12–3.22	~~||||~~ |||	8	3.17
3.225–3.335	3.23–3.33	~~||||~~ |	6	3.28
3.335–3.445	3.34–3.44	~~||||~~	5	3.39
3.445–3.555	3.45–3.55	||	2	3.50
3.555–3.665	3.56–3.66	|	1	3.61
		Total	24	

EXAMPLE 1-5 A random sample of 50 textbooks was selected, yielding the following costs:

22.95	46.90	32.45	34.45	28.95	12.95	18.95	11.55	45.90	25.95
37.55	42.90	19.20	22.95	11.95	32.65	25.30	19.95	4.95	28.95
33.05	33.95	16.65	17.70	44.85	35.90	50.45	22.85	22.95	23.95
26.95	15.95	18.95	20.40	18.55	23.95	33.25	33.55	30.90	23.45
26.95	13.95	36.85	8.90	21.95	25.95	22.95	33.90	16.90	12.55

Use this sampling to form a grouped frequency distribution that uses 10 classes with a class width of 5. Use 4.50 as the lower class limit in the first class.

Solution

Class limits	Tally	Frequency
4.50–9.49	||	2
9.50–14.49	||||	5
14.50–19.49	卌 |||	8
19.50–24.49	卌 卌 |	11
24.50–29.49	卌 ||	7
29.50–34.49	卌 ||||	9
34.50–39.49	|||	3
39.50–44.49	|	1
44.50–49.49	|||	3
49.50–54.49	|	1
	Total	50

1-2. Histograms

A. Histograms for discrete data

To obtain a better understanding of a set of data, it's often useful to construct a **histogram**, which depicts data as a series of rectangles. For discrete data, we can construct a rectangle that has a base of length one, centered at each observed integer value, and a height equal to the frequency of that observed integer value.

EXAMPLE 1-6 Construct a histogram that depicts the data on litter sizes of domestic rabbits in Example 1-1.

Solution The litter sizes ranged from 5 to 14 and had frequencies from 0 to 5. Putting the frequency on the y-axis and the litter size on the x-axis, you construct the histogram shown in Fig. 1-1. Notice that the base of each rectangle is one unit long and is centered at the observed litter size.

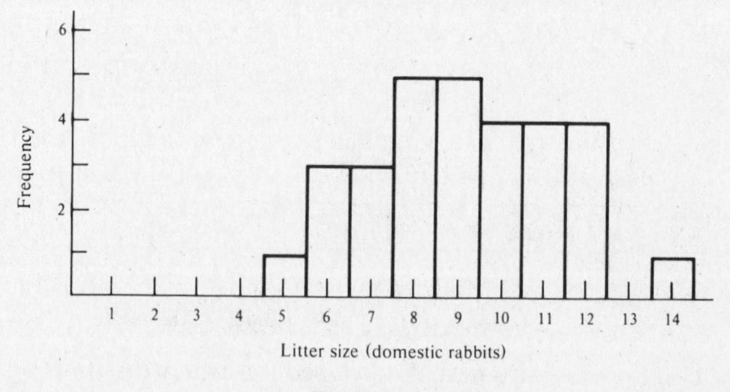

Figure 1-1

EXAMPLE 1-7 Construct a histogram of the data on wild-rabbit litter sizes in Example 1-2.

Solution The litter sizes ranged from 1 to 10. A histogram of these data is shown in Fig. 1-2.

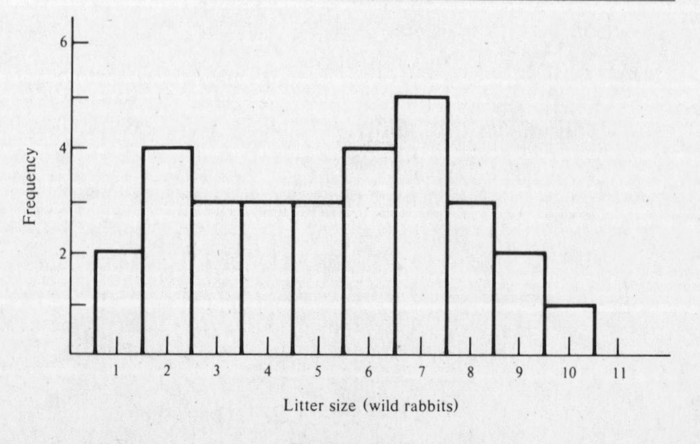

Figure 1-2

B. Histograms for continuous data

To construct a frequency histogram for continuous-type data, for each class we construct a rectangle whose height is equal to the frequency of the class and whose base is given by the class boundaries. There are several ways to label the horizontal axis: We can label the class boundaries, the class limits, the class marks, or a combination of any or all of these, depending on what we're interested in.

EXAMPLE 1-8 Construct a histogram for the height data grouped in Example 1-3.

Solution On a height vs. frequency graph, construct a series of rectangles so that the bases of the rectangles are the class boundaries (149.5–154.5, 154.5–159.5, etc.) and the heights of the rectangles are the frequencies (1, 9, etc.), as shown in Fig. 1-3. On this histogram, label the x-axis by specifying the class boundaries.

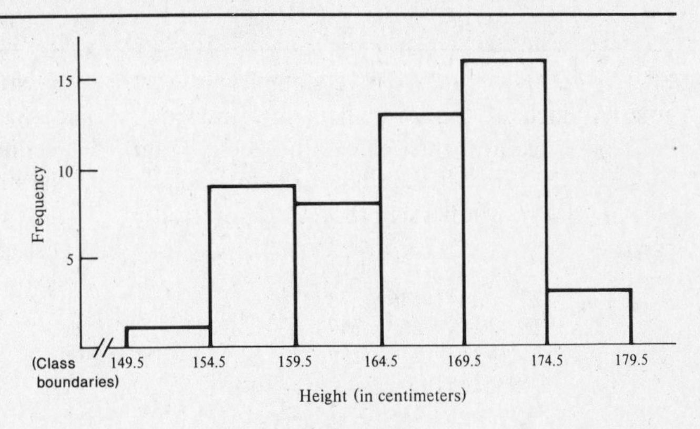

Figure 1-3

EXAMPLE 1-9 Construct a histogram for the weight data grouped in Example 1-4.

Solution Construct the histogram, using the class boundaries as bases for the rectangles. Label the class marks as illustrated in Fig. 1-4. (Notice that the class marks are indeed at the midpoint between the class boundaries.)

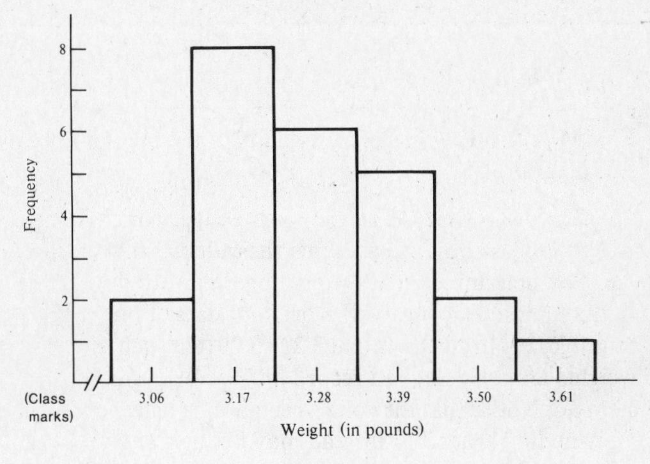

Figure 1-4

EXAMPLE 1-10 Use the textbook costs grouped in Example 1-5 to construct a histogram that shows the class limits.

Solution Figure 1-5 shows a histogram for the textbook costs given in Example 1-5. The class limits are given on the horizontal axis, between the class boundaries (which you can figure out yourself).

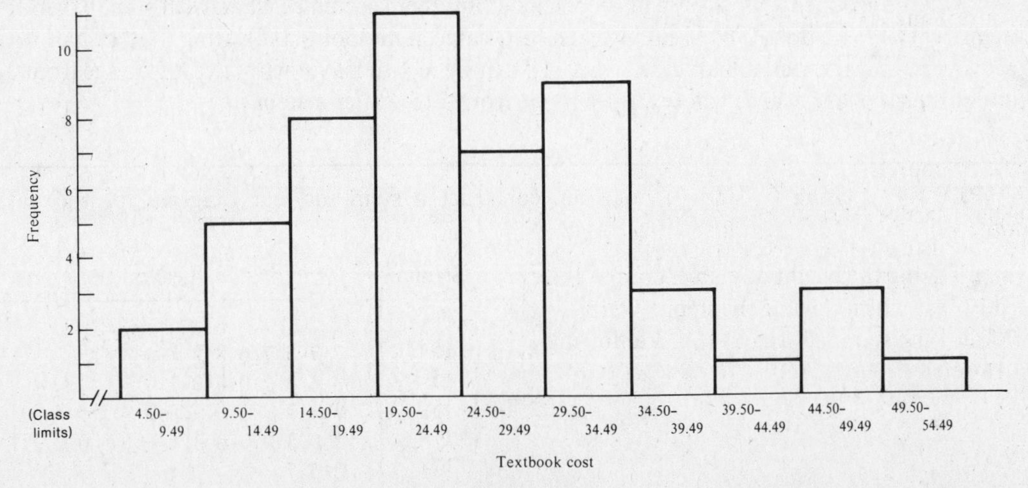

Figure 1-5

1-3. Stem-and-Leaf Diagrams

A. One-sided stem-and-leaf diagrams

A **stem-and-leaf diagram** is an exploratory data-analysis technique that allows us to group data without losing the original data. We use the leading digit(s) as the "stems" and the trailing digit(s) as the "leaves," so that the numbers themselves become a graph of the data.

EXAMPLE 1-11 Twenty-five statisticians went golfing. Each of the 25 golfers hit three golf balls off a tee and then calculated the average length of the three drives. (This is a statistician's idea of fun.) They tabulated the following data (in yards):

227	244	246	278	262
252	269	260	247	277
250	235	274	257	282
269	255	263	236	289
258	231	255	241	261

Construct a stem-and-leaf diagram using as stems the integers 22, 23, 24, 25, 26, 27, and 28.

Solution The first of our data values, 227, contains 22 as a stem and 7 as a leaf, while the second value, 244, is recorded as leaf 4 following the stem 24. The resulting chart gives us a feel for the spread and frequency of the data—as does a histogram—but the stem-and-leaf approach saves the original data values.

Stems	Leaves
22	7
23	5 6 1
24	4 6 7 1
25	2 0 7 5 8 5
26	2 9 0 9 3 1
27	8 7 4
28	2 9

EXAMPLE 1-12 Using the weights of the bags of apples given in Example 1-4, construct a stem-and-leaf diagram.

Solution The weights of the bags range from 3.02 to 3.62, so you use as stems the values 3.0–3.6. You determine the leaves by the digit found in the hundred's place of the original data. For example, any bag with weight 3.20–3.29 (ten such weights are possible) you record in the row with stem 3.2. Notice that the values need not be listed in ascending order, but instead may be recorded as they are encountered in the data set.

Stems	Leaves
3.0	2
3.1	2 0 9 4 7 2
3.2	6 6 2 1 0 1 6
3.3	9 0 9 1 6
3.4	9 1 2
3.5	3
3.6	2

---

note: Compare the shape of the stem-and-leaf diagram in Example 1-12 with the shape of the histogram in Example 1-9, Fig. 1-4. Although the grouping isn't the same, the shapes of the histogram and the stem-and-leaf diagram are similar.

So far, we've grouped ten possible items per stem. But there are times when we'll want to group five items per stem. We do this by breaking each stem value at midpoint, indicating the first half with an asterisk (*) and the second half with a dot (·). Then we record leaves whose values range from 0 to 4 after a stem (*), and leaves whose values range from 5 to 9 after a stem (·).

EXAMPLE 1-13 Using (*) and (·) notation, construct a stem-and-leaf diagram for the data in Example 1-3.

Solution Using the height data in Example 1-3, record 171 as leaf 1 following the stem 17*, 157 as leaf 7 following stem 15·, 163 as leaf 3 following stem 16·, etc.

Stems	Leaves
15*	2
15·	7 7 5 7 6 8 9 7 8
16*	3 4 3 1 0 4 0 3
16·	8 7 9 8 5 6 7 8 9 9 5 5 5
17*	1 0 3 0 0 0 3 1 4 1 3 0 2 2 1 0
17·	8 5 7

B. Two-sided stem-and-leaf diagrams

Stem-and-leaf diagrams can be used to depict a sample from a population composed of two types. We depict such a sample with a two-sided diagram.

EXAMPLE 1-14 Some birdwatchers deep in marshland caught 32 common moorhens (gallinules) in order to measure the lengths of their culmens (frontal shields)—a task much easier said than done, but they probably had their reasons. They obtained the following data (in mm) for 16 males and 16 females, respectively:

Males:	45.5	42.3	44.4	41.8	44.0	47.3	43.2	45.8
	45.5	46.2	45.5	45.7	46.0	42.8	44.2	43.9
Females:	39.8	41.2	38.9	38.3	41.8	41.5	41.5	43.3
	41.5	40.9	40.9	42.9	40.7	40.2	39.2	40.3

Construct a stem-and-leaf diagram that fully reflects these data.

Solution If we construct a single-sided stem-and-leaf diagram, we lose the division into male and female:

Stems	Leaves
38	9 3
39	8 2
40	9 9 7 2 3
41	8 2 8 5 5 5
42	3 8 9
43	2 9 3
44	4 0 2
45	5 8 5 5 7
46	2 0
47	3

But we can show the separation between female and male moorhens by constructing a double-sided stem-and-leaf diagram. This two-sided diagram reflects all the data and gives a better picture of how culmen lengths are distributed, thus making comparison easier. We see at a glance that male moorhens are more likely to have longer frontal shields than female moorhens:

Females	Stems	Males
3 9	38	
2 8	39	
3 2 7 9 9	40	
5 5 5 8 2	41	8
9	42	3 8
3	43	2 9
	44	4 0 2
	45	5 8 5 5 7
	46	2 0
	47	3

1-4. Empirical Distribution Function and the Ogive

A. For discrete data: The empirical distribution function

For a given value x the **empirical distribution function** gives the cumulative relative frequency of observations up to and including x. This function can easily be calculated for discrete observations by adding two columns to the frequency distribution table. First, add a column that gives the **cumulative frequency**, which is the number of observations less than or equal to x. Then, add a **cumulative relative frequency** column by dividing the cumulative frequencies by the total number of observations in the data set.

EXAMPLE 1-15 **(a)** Using the data from Example 1-1, create a table that gives frequencies, cumulative frequencies, and cumulative relative frequencies. **(b)** Give the value of the empirical distribution function at $x = 8$ and at $x = 11$.

Solution

(a) In the first column, list the litter sizes observed, which range from 5 to 14. Next, tally the number of times each of these litter sizes appears in the sample data. To check your count, make sure that the sum of the frequencies equals the total number of litter observations. Then, determine the cumulative frequency up to and including each observed x value (litter size). There is one litter in the sample with 5 or fewer baby bunnies in it. There are three litters of 6 bunnies, so there are $3 + 1 = 4$ observations with 6 or fewer bunnies per litter. In the same way, you add each succeeding frequency to the current total to compute cumulative frequency. Finally, find the cumulative relative frequency by dividing each of the cumulative frequencies by the total number of observations in the data set.

Litter size (x)	Frequency		Cumulative frequency	Cumulative relative frequency	
5	1	....	1	1/30	Empirical distribution function
6	3	4	4	4/30	
7	3	7...	7	7/30	
8	5		12	12/30	
9	5	etc.	17	17/30	
10	4		21	21/30	
11	4		25	25/30	
12	4		29	29/30	
13	0		29	29/30	
14	1		30	30/30	
Total	30				

(b) The value of the empirical distribution function at $x = 8$ is 12/30; at $x = 11$ it's 25/30.

The graph of the empirical distribution function depicts the cumulative relative frequency of a set of data, i.e., the relative frequency of the number of observations less than or equal to a given value of x.

EXAMPLE 1-16 Use the information obtained in Example 1-15 to graph the empirical distribution function.

Solution We locate the litter sizes on the x-axis and the cumulative relative frequency (the values of the empirical distribution function) on the y-axis. Then we draw a straight line from each dot representing the cumulative relative frequency of a litter size straight across to the next increase in litter size, as in Fig. 1-6.

note: Because the data are discrete, the empirical distribution function has breaks in it—it's not a continuous function.

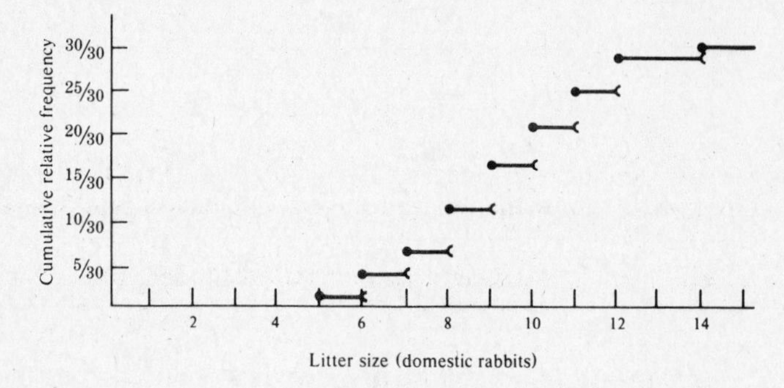

Figure 1-6

EXAMPLE 1-17 Using the litter sizes for wild rabbits given in Example 1-2, graph the empirical distribution function for these data.

Solution Add the cumulative frequency and cumulative relative frequency columns to the frequency distribution table in Example 1-2.

Litter size	Frequency	Cumulative frequency	Cumulative relative frequency
1	2	2	2/30
2	4	6	6/30
3	3	9	9/30
4	3	12	12/30
5	3	15	15/30
6	4	19	19/30
7	5	24	24/30
8	3	27	27/30
9	2	29	29/30
10	1	30	30/30
	30		

Plot the cumulative relative frequency against litter size, as in Fig. 1-7 (facing page).

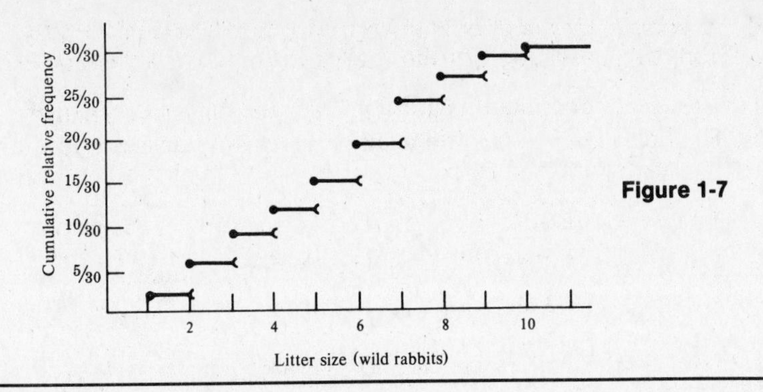

Figure 1-7

B. For continuous data: The ogive

To describe the cumulative distribution of *continuous-type* data, we have to add a cumulative frequency and a cumulative relative frequency column to the *grouped* frequency distribution table. A cumulative frequency column for continuous data gives the number of observations up to and including a particular *class.* For the cumulative relative frequency, we divide the cumulative frequencies by the total number of observations in the data set.

EXAMPLE 1-18 Find the cumulative frequency and the cumulative relative frequency for the heights of 50 women given in Example 1-3.

Solution We construct the following table from the frequency distribution table:

Class boundaries	Class limits	Frequency	Cumulative frequency	Cumulative relative frequency
149.5–154.5	150–154	1	1	1/50 = .02
154.5–159.5	155–159	9	10	10/50 = .20
159.5–164.5	160–164	8	18	18/50 = .36
164.5–169.5	165–169	13	31	31/50 = .62
169.5–174.5	170–174	16	47	47/50 = .94
174.5–179.5	175–179	3	50	50/50 = 1.00
		50		

The graph of the cumulative relative frequency for a set of continuous data is the relative frequency **ogive** curve. At each upper class boundary, we plot a point that corresponds to the cumulative relative frequency. Then we connect these points with straight-line segments to construct the relative frequency ogive curve.

EXAMPLE 1-19 Use the height data tabulated in Example 1-18 to construct a relative frequency ogive curve.

Solution Along the x-axis, label the class boundaries, and along the y-axis give the cumulative relative frequencies. Connect the points representing each cumulative relative frequency to form the ogive curve, as in Fig. 1-8.

note: Because the data here are continuous, the graph of the cumulative relative frequency ogive has no breaks in it—it's a continuous function.

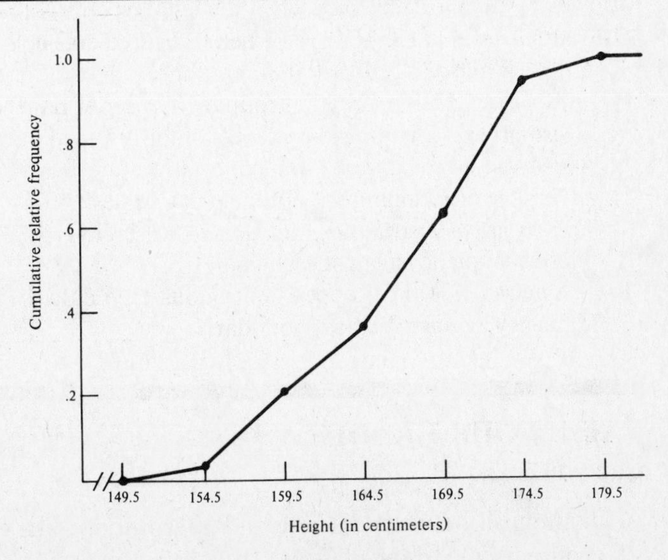

Figure 1-8

EXAMPLE 1-20 Graph the ogive curve for the data in Example 1-4.

Solution Expand the frequency distribution table to include the cumulative frequencies and cumulative relative frequencies. Then plot the last column against the class boundaries to draw the ogive curve shown in Fig. 1-9.

Class boundaries	Class limits	Frequency	Cumulative frequency	Cumulative relative frequency
3.005–3.115	3.01–3.11	2	2	2/24
3.115–3.225	3.12–3.22	8	10	10/24
3.225–3.335	3.23–3.33	6	16	16/24
3.335–3.445	3.34–3.44	5	21	21/24
3.445–3.555	3.45–3.55	2	23	23/24
3.555–3.665	3.65–3.66	1	24	24/24
		24		

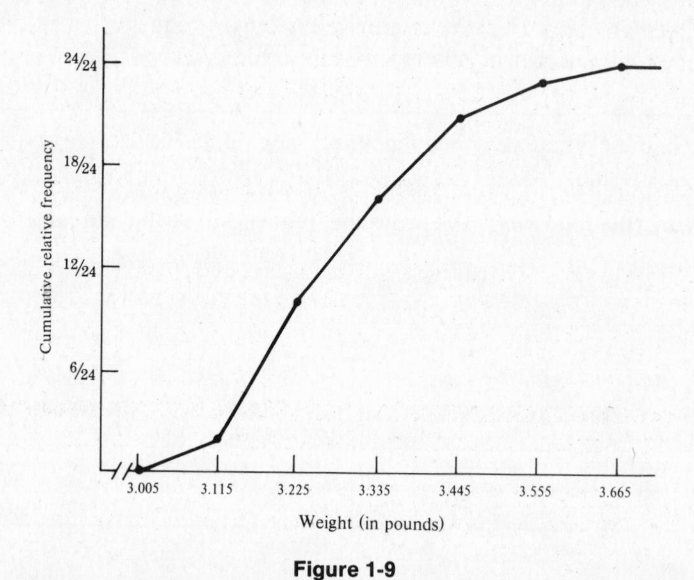

Figure 1-9

SUMMARY

1. *Discrete data* are observed when the number of possible outcomes is finite.
2. *Continuous data* are observed when measurements come from a continuous scale.
3. A *class* is a group or interval of continuous data.
4. A *frequency distribution* is a grouping of a set of data.
5. A *histogram* is a graphical presentation of grouped data in rectangular form.
6. A *stem-and-leaf diagram* is an exploratory data-analysis technique for grouping data that has the advantage of retaining the original data values.
7. The *empirical distribution function* for discrete-type data is defined by plotting the cumulative relative frequency against each value x.
8. The *ogive curve* for grouped continuous-type data is defined by plotting the cumulative relative frequency against the class boundaries.

RAISE YOUR GRADES

Can you . . . ?

☑ distinguish between discrete and continuous data
☑ group discrete data
☑ set up class boundaries for continuous data

- ☑ find the class marks for continuous data
- ☑ draw histograms
- ☑ construct a stem-and-leaf diagram
- ☑ graph an empirical distribution function
- ☑ define and graph an ogive curve
- ☑ explain the similarities and differences between an empirical distribution function and an ogive curve

RAPID REVIEW

1. The weight of a bag of candy that contains several pieces is an example of ______________ data.
2. The number of pieces of candy in a bag is an example of ______________ data.
3. A graphical presentation of grouped data is called a ______________.
4. The midpoint of a class interval for grouped data is called the ______________.
5. The smallest and largest possible observed values in a class are called ______________.
6. It's possible that an observation is equal to the class boundary. (*True* or *False*)
7. An exploratory data-analysis technique that gives information similar to that shown by a histogram is called a ______________ diagram.
8. The empirical distribution function is defined using ______________ relative frequency.
9. The graph of the cumulative relative frequency versus class boundaries of grouped continuous data is called the ______________.

Answers (1) continuous (2) discrete (3) histogram (4) class mark (5) class limits (6) False (7) stem-and-leaf (8) cumulative (9) ogive

SOLVED PROBLEMS

PROBLEM 1-1 The number of cars per household for 40 homes was observed, yielding the following data:

4 2 3 2 1 1 3 2 2 0 2 2 3 4 1 1 1 1 2 3
2 2 1 1 3 2 2 2 2 2 2 2 1 2 0 3 2 2 2 3

(a) Group these data into a frequency distribution.
(b) Calculate the cumulative frequency and cumulative relative frequency.
(c) Draw the histogram for these data.
(d) Draw the graph of the empirical distribution function for these data.

Solution

(a) Construct the frequency distribution table.

Number of cars	Tally	Frequency
0	||	2
1	~~||||~~ ||||	9
2	~~||||~~ ~~||||~~ ~~||||~~ ~~||||~~	20
3	~~||||~~ ||	7
4	||	2
	Total	40

(b) Add the frequencies cumulatively; then divide each cumulative frequency by the total to get the cumulative relative frequency:

Number of cars	Frequency	Cum. freq.	Cum. rel. freq.
0	2	2	.050
1	9	11	.275
2	20	31	.775
3	7	38	.950
4	2	40	1.000
	40		

(c) Putting the number of cars on the x-axis and the frequency on the y-axis, draw rectangles—each having a base equal to one and height equal to frequency. See Fig. 1-10a.

(d) Plot the cumulative relative frequency against the number of cars. See Fig. 1-10b. Notice that you're plotting discrete-type data.

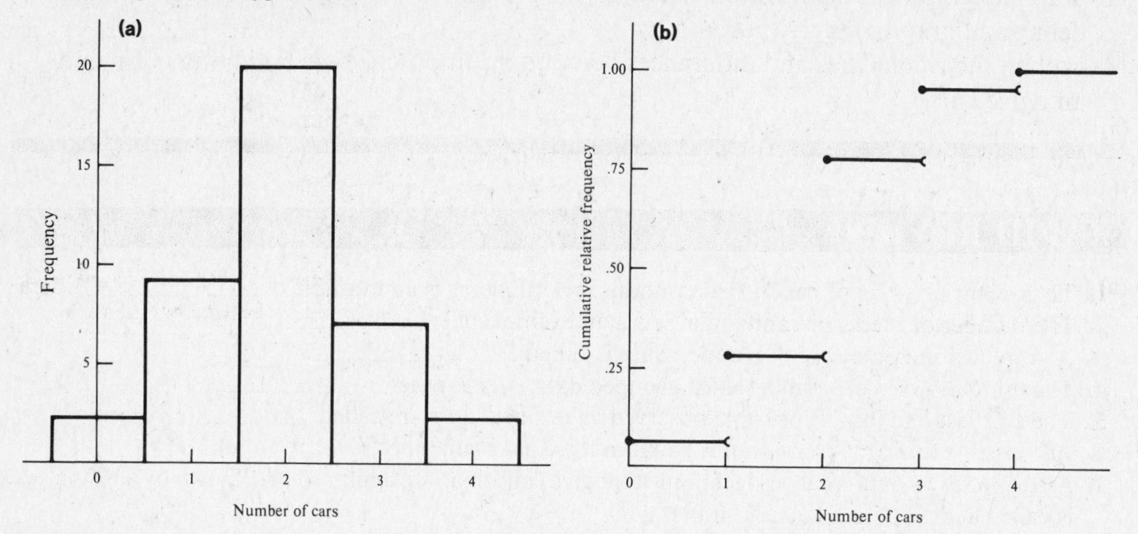

Figure 1-10

PROBLEM 1-2 If you roll a 4-sided die, it can come up 1, 2, 3, or 4. If, for purposes of statistical experiment only, you roll a pair of 4-sided dice 100 times and observe the sums of their outcomes, you might record the following data:

4 3 6 6 4 8 7 2 6 7 5 5 7 7 7 5 4 3 5 6
5 5 4 2 3 6 4 8 3 8 6 8 4 3 6 7 3 6 2 5
3 6 6 5 7 5 7 7 3 5 7 5 5 7 8 5 4 4 4 4
4 7 6 6 4 5 3 3 7 4 4 5 2 5 4 8 8 4 2 7
5 3 6 5 5 2 4 2 3 7 4 3 6 5 5 5 5 5 6 5

(a) Group these data and calculate the cumulative relative frequency. **(b)** Construct a histogram for these data. **(c)** Draw the graph of the empirical distribution function.

Solution

(a)

Outcome (sum)	Frequency	Cumulative frequency	Cumulative relative frequency
2	7	7	.07
3	13	20	.20
4	18	38	.38
5	25	63	.63
6	15	78	.78
7	15	93	.93
8	7	100	1.00
	100		

(b) See Fig. 1-11a.
(c) See Fig. 1-11b.

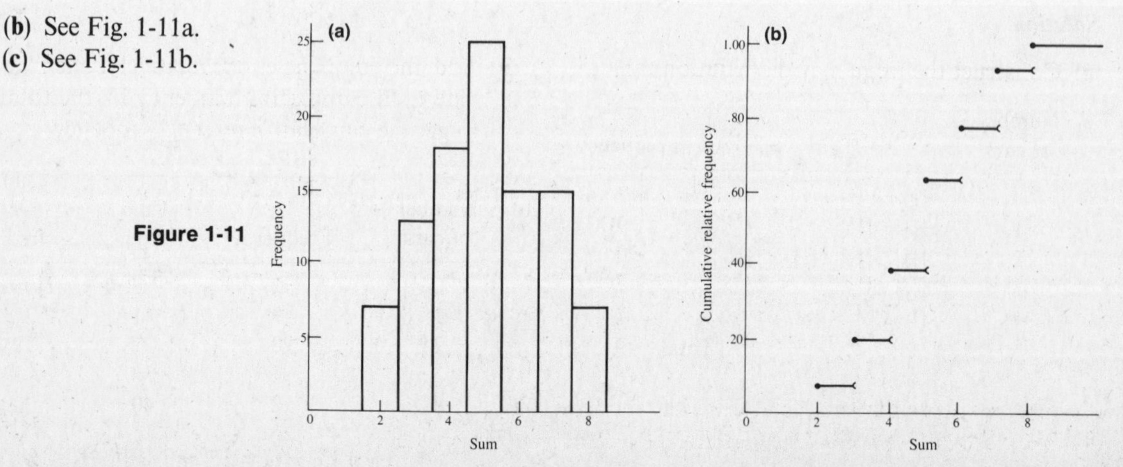

Figure 1-11

PROBLEM 1-3 A farmer weighed the milk produced in a day by each of 50 cows. He recorded the following weights (in lb):

42	53	38	40	48	44	44	47	20	38
50	41	32	37	67	14	43	38	36	42
29	21	51	45	73	15	50	71	49	29
25	37	39	28	17	64	76	42	52	50
40	38	48	57	21	49	46	32	28	24

(a) Group these data using 7 classes having a class width of 9. Include in your frequency distribution the class boundaries, class limits, frequencies, and class marks.
(b) Draw a histogram of the grouped data. Label the x-axis with the class marks.

Solution

(a) The range is $76 - 14 = 62$, so set the lower class limit at 14. Set the class boundaries to give the smallest and largest actual values in each class. The class mark is the midpoint of each class.

Class boundaries	Class limits	Tally	Frequency	Class marks
13.5–22.5	14–22	~~IIII~~ I	6	18
22.5–31.5	23–31	~~IIII~~ I	6	27
31.5–40.5	32–40	~~IIII~~ ~~IIII~~ II	12	36
40.5–49.5	41–49	~~IIII~~ ~~IIII~~ IIII	14	45
49.5–58.5	50–58	~~IIII~~ II	7	54
58.5–67.5	59–67	II	2	63
67.5–76.5	68–76	III	3	72
			50	

(b) Since these data are continuous-type data, each rectangle in the histogram has a base whose length is equal to the class boundaries—the base length is bisected by the class mark. See Fig. 1-12.

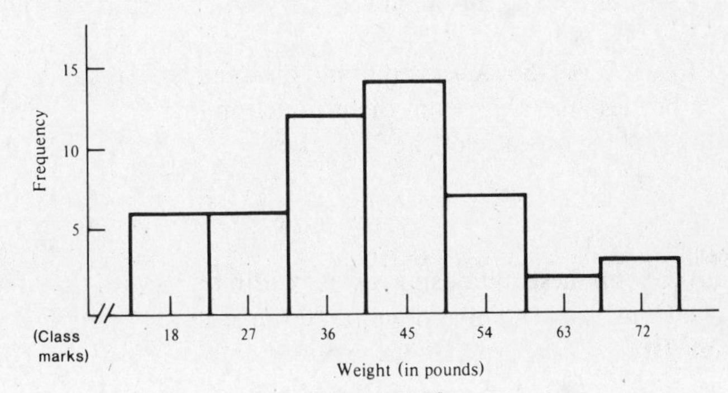

Figure 1-12

PROBLEM 1-4 Two hundred participants in a banking survey were asked to give their ages. The following distribution was constructed from the resulting data:

Age	Frequency	Cumulative frequency	Cumulative relative frequency
10–19	10	10	.05
20–29	30	40	.20
30–39	54	94	.47
40–49	50	144	.72
50–59	24	168	.84
60–69	18	186	.93
70–79	12	198	.99
80–89	2	200	1.00
	200		

(a) What are the class boundaries for the first class? **(b)** What is the class mark for the first class? **(c)** Draw a histogram for these data. **(d)** Draw the ogive curve.

Solution

(a) The class boundaries for the first class are 9.5–19.5.
(b) The class mark for the first class is $(10 + 19)/2 = 29/2 = 14.5$.

(c) See Fig. 1-13a. (d) See Fig. 1-13b.

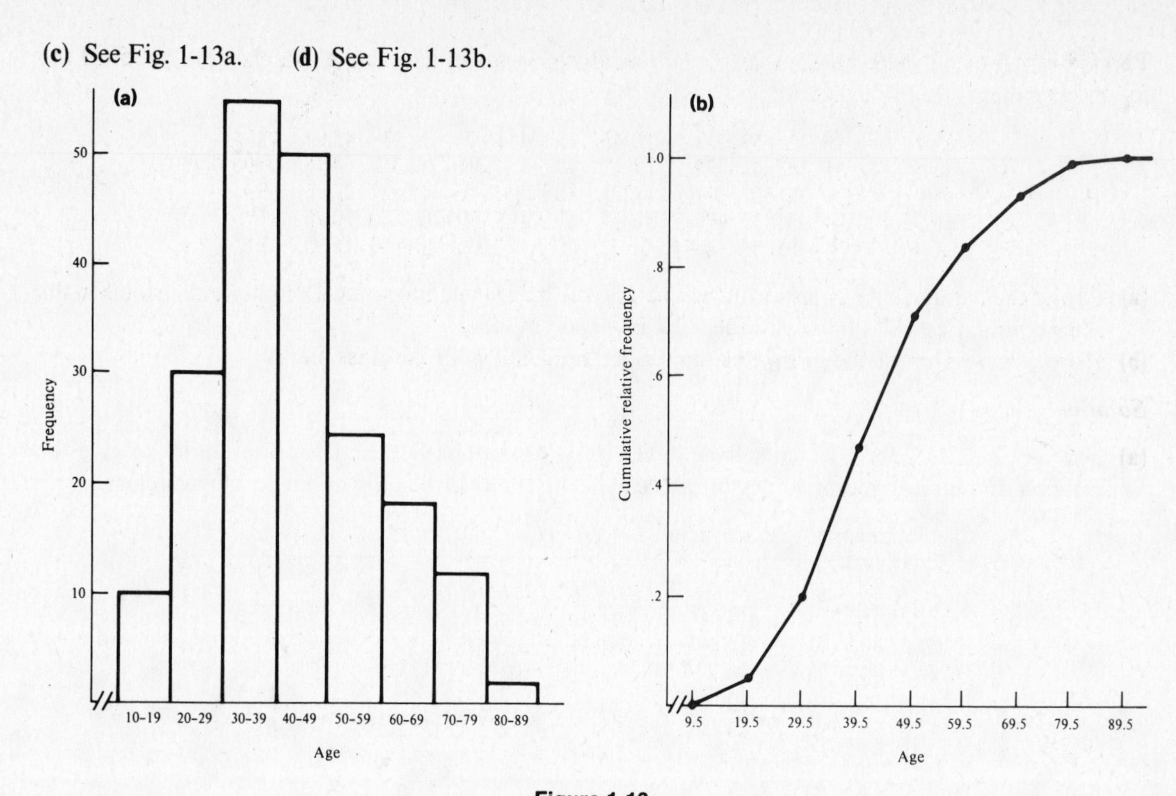

Figure 1-13

PROBLEM 1-5 A certain brand of liquid soap is supposedly packaged in 1000-g bottles. A sample of 25 bottles was taken from the production line and weighed. The following data give the excess weight over 1000 g for each of the 25 bottles:

60	31	19	43	47	48	47	13	28	35	51	24	24
27	33	57	38	24	39	40	63	51	25	43	26	

(a) Group these data using a class width of 10 with the first class limits of 10–19.
(b) Construct a stem-and-leaf diagram using as stems 1, 2, 3, 4, 5, and 6.
(c) Draw a histogram of the grouped data.

Solution

(a)

Weight	Tally	Frequency	Class mark
10–19	||	2	14.5
20–29	~~||||~~ ||	7	24.5
30–39	~~||||~~	5	34.5
40–49	~~||||~~ |	6	44.5
50–59	|||	3	54.5
60–69	||	2	64.5
		25	

(b)

Stems	Leaves
1	9 3
2	8 4 4 7 4 5 6
3	1 5 3 8 9
4	3 7 8 7 0 3
5	1 7 1
6	0 3

(c) See Fig. 1-14. Notice that the length of the base of each rectangle is given by the difference of the class boundaries.

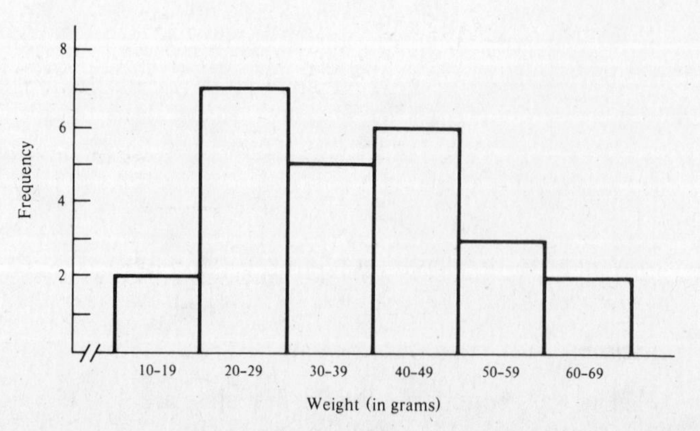

Figure 1-14

PROBLEM 1-6 A random sample of 50 textbooks was selected (see Example 1-5), yielding the following costs:

22.95	46.90	32.45	34.45	28.95	12.95	18.95	11.55	45.90	25.95
37.55	42.90	19.20	22.95	11.95	32.65	25.30	19.95	4.95	28.95
33.05	33.95	16.65	17.70	44.85	35.90	50.45	22.85	22.95	23.95
26.95	15.95	18.95	20.40	18.55	23.95	33.25	33.55	30.90	23.45
26.95	13.95	36.85	8.90	21.95	25.95	22.95	33.90	16.90	12.55

(a) Round the textbook costs off to the nearest dollar and make a stem-and-leaf diagram, using 5 items per stem.
(b) Draw a histogram of the (unrounded) grouped data, showing the class limits.

Solution

(a)

Stems	Leaves
0·	5 9
1*	3 2 2 4 3
1·	9 9 7 8 6 9 9 7
2*	3 3 0 3 3 4 0 4 3 2 3
2·	9 6 5 9 7 7 6
3*	2 4 3 3 4 3 4 1 4
3·	8 6 7
4*	3
4·	7 6 5
5*	0

(b) Group the data using 10 classes having a class width of 5 (as in Example 1-5), and draw the histogram as shown in Fig. 1-15. Notice the correspondence between the shapes of the stem-and-leaf diagram and the histogram.

Class limits	Class boundaries	Frequency
4.50– 9.49	4.495– 9.495	2
9.50–14.49	9.495–14.495	5
14.50–19.49	14.495–19.495	8
19.50–24.49	19.495–24.495	11
24.50–29.49	24.495–29.495	7
29.50–34.49	29.495–34.495	9
34.50–39.49	34.495–39.495	3
39.50–44.49	39.495–44.495	1
44.50–49.49	44.495–49.495	3
49.50–54.49	49.495–54.495	1
		50

Now draw the histogram:

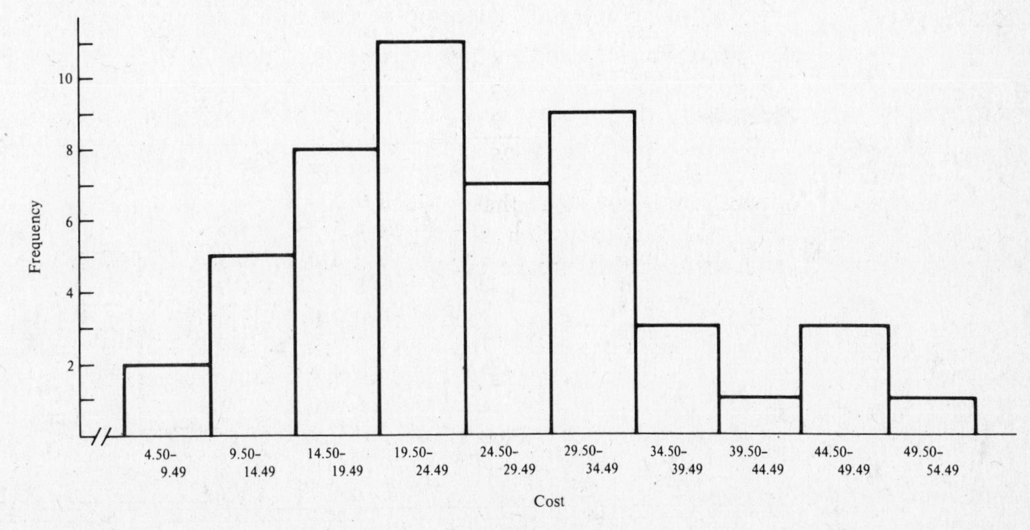

Figure 1-15

PROBLEM 1-7 The following data give the maximum lung capacity (in liters) for 50 college students: 25 first-year females and 25 first-year males:

Females					Males				
4.2	5.3	3.5	4.3	3.7	6.8	5.7	5.6	4.8	3.7
3.2	3.5	2.8	3.5	3.7	5.0	3.8	5.6	5.4	5.2
3.5	3.7	2.7	2.8	3.3	4.9	4.7	7.6	4.0	5.6
2.7	3.0	3.1	3.0	3.7	4.2	4.0	5.5	5.1	4.9
3.3	3.4	2.3	2.6	3.2	5.4	5.9	3.4	6.1	4.1

(a) Construct a stem-and-leaf diagram with 5 items per stem.
(b) Construct a two-sided stem-and-leaf diagram with 5 items per stem.

Solution

(a)

Stems	Leaves
2*	3
2·	8 7 8 7 6
3*	2 3 0 1 0 3 4 2 4
3·	5 7 5 5 7 5 7 7 7 8
4*	2 3 0 2 0 1
4·	8 9 7 9
5*	3 0 4 2 1 4
5·	7 6 6 6 5 9
6*	1
6·	8
7*	
7·	6

(b)

Females	Stems	Males
3	2*	
8 7 8 7 6	2·	
2 3 0 1 0 3 4 2	3*	4
5 7 5 5 7 5 7 7	3·	7 8
2 3	4*	0 2 0 1
	4·	8 9 7 9
3	5*	0 4 2 1 4
	5·	7 6 6 6 5 9
	6*	1
	6·	8
	7*	
	7·	6

PROBLEM 1-8 The following heights (in inches) for 82 students in a statistics class have already been put in order:

60	61	61	61	62	62	63	63	63	64	64	65	65	65	65	65	65
66	66	66	66	67	67	67	67	67	67	67	67	67	68	68	68	68
68	68	69	69	70	70	70	70	70	70	70	71	71	71	71	71	71
71	71	71	71	71	72	72	72	72	72	72	72	72	72	72	72	73
73	73	73	74	74	74	74	75	75	75	76	76	76	76			

Use these data to construct a stem-and-leaf diagram that has 2 items per stem. *Hint:* You'll need to set up a system for breaking the stems. Try the following rules:

Record leaves 0 and 1 with a stem having an asterisk.
Record leaves 2 and 3 (*t*wos and *t*hrees) with a stem having a *t*.
Record leaves 4 and 5 (*f*ours and *f*ives) with a stem having an *f*.
Record leaves 6 and 7 (*s*ixes and *s*evens) with a stem having an *s*.
Record leaves 8 and 9 with a stem having a dot.

Solution

Stems	Leaves
6*	0 1 1 1
6*t*	2 2 3 3 3
6*f*	4 4 5 5 5 5 5 5
6*s*	6 6 6 6 7 7 7 7 7 7 7 7 7
6·	8 8 8 8 8 8 9 9
7*	0 0 0 0 0 0 0 1 1 1 1 1 1 1 1 1 1 1
7*t*	2 2 2 2 2 2 2 2 2 2 2 3 3 3 3
7*f*	4 4 4 4 5 5 5
7*s*	6 6 6 6

Supplementary Exercises

PROBLEM 1-9 To determine the distribution of clutch sizes for the common moorhen (gallinule), a birdwatcher counted the numbers of eggs in 100 nests and found the following clutch sizes:

7 9 9 9 9 10 7 8 9 10 8 10 7 7 8 9 8 8 8 9
9 10 9 9 9 8 7 7 8 7 7 9 7 6 9 9 10 8 13 13
8 8 9 9 9 10 10 7 8 7 7 7 11 8 8 7 7 5 6 12
9 10 8 8 10 9 10 11 9 13 9 7 8 9 10 10 9 8 10 8
9 8 8 7 6 9 9 8 8 7 5 7 9 9 5 9 9 7 6 11

(a) Group these data, giving both frequencies and cumulative relative frequencies. **(b)** Draw the histogram for these data. **(c)** Draw the graph of the empirical distribution function for these data.

Answer **(a)**

Clutch size	Frequency	Cumulative relative frequency
5	3	.03
6	4	.07
7	20	.27
8	23	.50
9	30	.80
10	13	.93
11	3	.96
12	1	.97
13	3	1.00
	100	

(b) See Fig. 1-16a.

(c) See Fig. 1-16b.

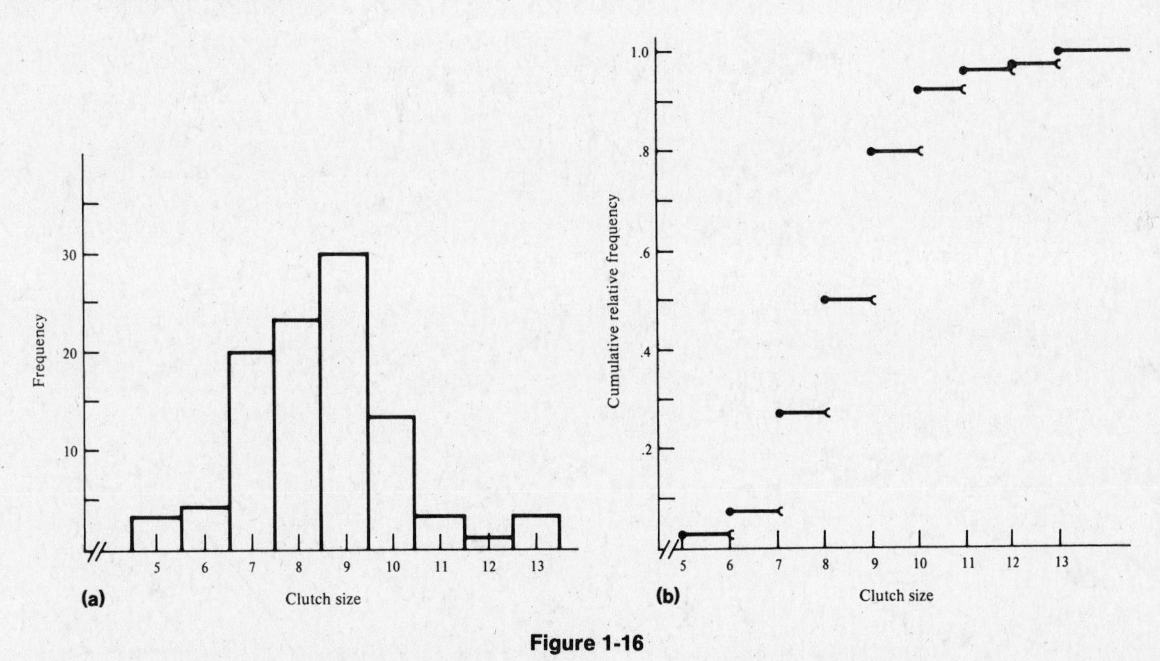

Figure 1-16

PROBLEM 1-10 The number of sit-ups that each member of a 20-player volleyball team could do at the beginning of the season was compared with the number each player could do at the end of the season. The differences, post-season minus pre-season, were as follows:

0 −1 7 1 13 3 5 3 1 −2
3 6 −6 3 5 1 0 6 2 2

Construct the graph of the empirical distribution function for these data.

Answer See Fig. 1-17 at the top of p. 18. [Note that there is a step of size 1/20 at each observation.]

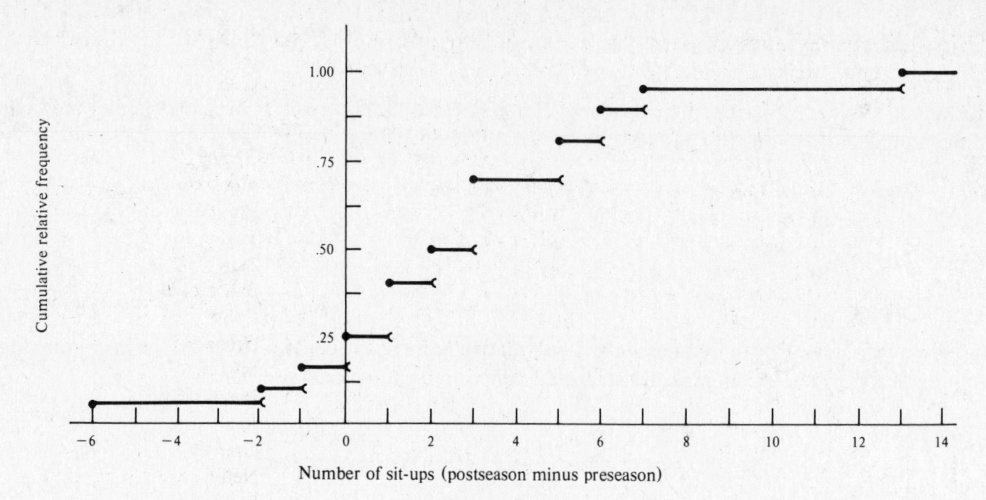

Figure 1-17

PROBLEM 1-11 The weights (in kg) of 50 male college freshmen were recorded:

76.0	64.6	80.6	84.8	78.9	84.3	68.3	72.2	63.7	92.6
66.4	72.6	62.6	74.7	60.4	72.0	77.2	59.0	75.0	54.1
90.8	66.1	68.1	69.1	79.7	65.3	81.3	56.9	71.6	74.9
73.0	88.0	59.5	62.8	75.0	71.3	74.4	70.8	77.8	71.9
63.8	70.3	72.0	66.0	74.0	70.7	69.6	75.3	63.4	69.4

(**a**) Group these data into 6 classes using a class width of 7. (**b**) Construct a histogram for these data. (**c**) Draw the ogive curve for the grouped data.

Answer (**a**)

Class boundaries	Frequency	Cumulative relative frequency
52.25–59.25	3	.06
59.25–66.25	11	.28
66.25–73.25	17	.62
73.25–80.25	12	.86
80.25–87.25	4	.94
87.25–94.25	3	1.00
	50	

(**b**) See Fig. 1-18a.

(**c**) See Fig. 1-18b.

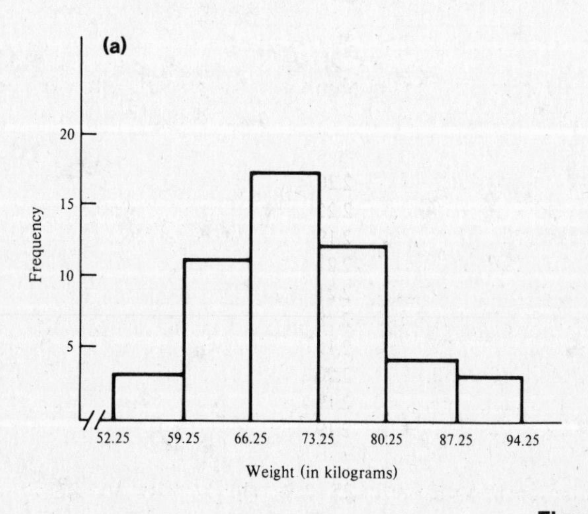

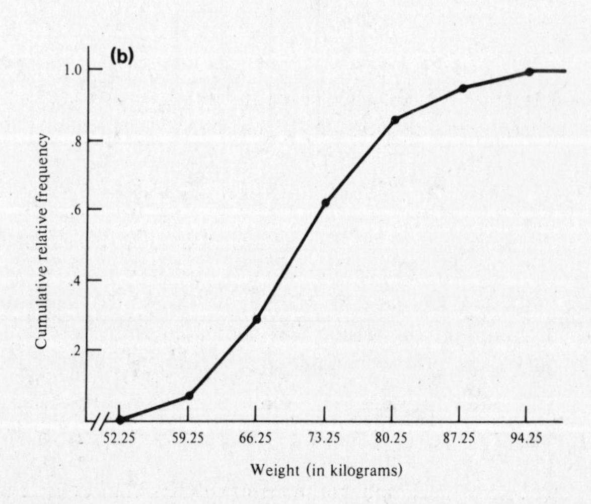

Figure 1-18

PROBLEM 1-12 The following data from the United Nations give the infant mortality (death within the first year) per 1000 infants for 50 countries from Africa, North America, and Europe.

125.3	Algeria	126.6	Burundi	154.3	Chad
134.5	Congo	150.0	Ethiopia	107.3	Ghana
92.0	Kenya	120.9	Mali	114.4	Morocco
140.5	Nigeria	127.0	Rwanda	152.6	Senegal
100.6	South Africa	131.1	Sudan	114.7	Togo
106.5	Tunisia	100.5	Uganda	116.6	Zaire
140.0	Zanzibar	78.8	Zimbabwe	22.4	Bahamas
9.6	Canada	18.0	Costa Rica	17.3	Cuba
73.1	Dominican Republic	44.0	El Salvador	32.4	Greenland
15.4	Grenada	16.2	Jamaica	59.8	Mexico
96.5	Nicaragua	36.2	Panama	18.6	Puerto Rico
11.2	United States	22.5	Virgin Islands	12.8	Austria
11.7	Belgium	9.3	France	14.3	Greece
10.6	Ireland	12.7	Italy	8.1	Netherlands
7.5	Norway	26.0	Portugal	10.3	Spain
6.8	Sweden	7.6	Switzerland	11.4	Scotland
13.5	Northern Ireland	33.5	Yugoslavia		

(a) Make a frequency distribution for these data using a class width of 25. For the first class use the class limits 0–24.9. **(b)** Draw a histogram of your grouped data.

Answer

(a)

Class limits	Frequency
0– 24.9	22
25– 49.9	6
50– 74.9	1
75– 99.9	3
100–124.9	8
125–149.9	7
150–174.9	3
	50

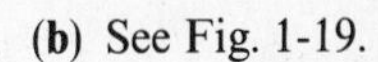

(b) See Fig. 1-19.

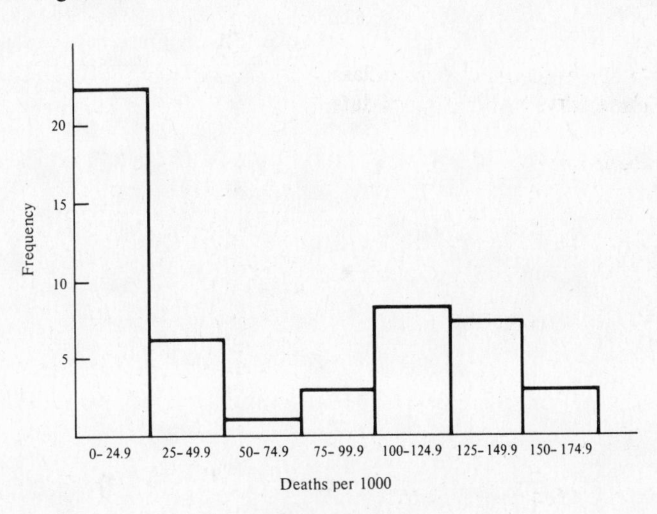

Figure 1-19

PROBLEM 1-13 You've been hired to test a sample of one hundred 2200-ohm resistors. You find that the resistors give the following measurements in hundreds of ohms:

2.17	2.18	2.17	2.19	2.23	2.18	2.16	2.23	2.29	2.20
2.24	2.13	2.18	2.22	2.21	2.22	2.23	2.18	2.23	2.25
2.24	2.20	2.21	2.20	2.25	2.15	2.20	2.25	2.16	2.19
2.20	2.18	2.18	2.19	2.22	2.19	2.20	2.19	2.22	2.21
2.22	2.18	2.18	2.28	2.19	2.23	2.21	2.19	2.21	2.21
2.19	2.18	2.21	2.17	2.20	2.18	2.18	2.21	2.22	2.18
2.22	2.23	2.19	2.23	2.18	2.19	2.18	2.21	2.18	2.15
2.20	2.23	2.20	2.20	2.23	2.20	2.19	2.22	2.17	2.20
2.20	2.17	2.19	2.19	2.25	2.19	2.19	2.20	2.20	2.19
2.21	2.14	2.24	2.21	2.19	2.23	2.18	2.22	2.23	2.19

(a) Group these data into 9 classes with the class boundaries of the first class at 2.125–2.145. **(b)** Draw the histogram for the grouped data.

Answer

(a)

Class boundaries	Frequency
2.125–2.145	2
2.145–2.165	4
2.165–2.185	21
2.185–2.205	33
2.205–2.225	20
2.225–2.245	14
2.245–2.265	4
2.265–2.285	1
2.285–2.305	1
	100

(b) See Fig. 1-20.

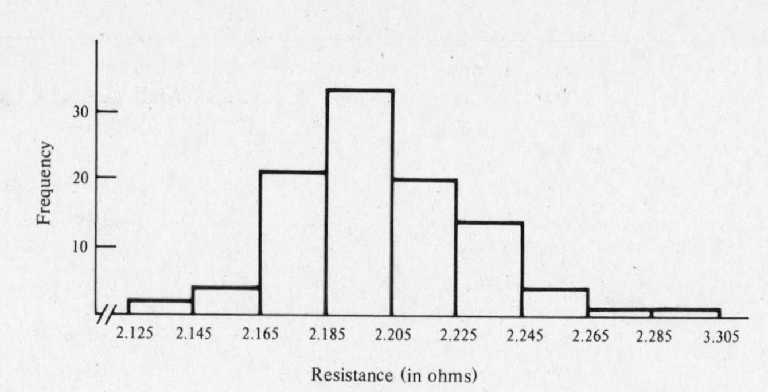

Figure 1-20

PROBLEM 1-14 The lung capacity (in liters) of each of the 20 members of a volleyball team was measured, yielding the following data:

3.4 3.7 3.6 3.4 3.9 3.4 3.3 3.7 3.4 3.6
4.1 3.5 4.1 3.7 3.6 4.3 3.7 4.2 3.8 4.7

(a) Group these data into 5 classes with the class limits for the first class at 3.3–3.5. **(b)** Draw a histogram of the grouped data. **(c)** Draw the ogive curve for the grouped data.

Answer **(a)**

Class limits	Frequency	Cumulative relative frequency
3.3–3.5	6	.30
3.6–3.8	8	.70
3.9–4.1	3	.85
4.2–4.4	2	.95
4.5–4.7	1	1.00
	20	

(b) See Fig. 1-21a.

(c) See Fig. 1-21b.

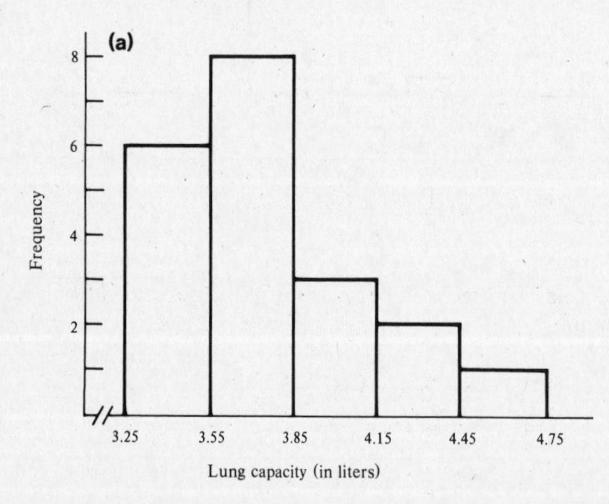

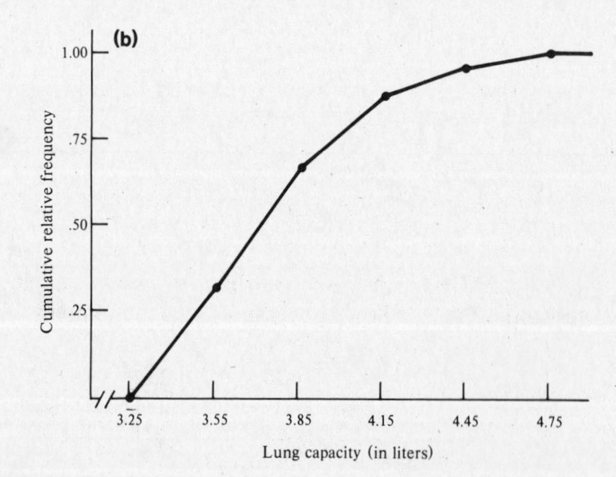

Figure 1-21

PROBLEM 1-15 At the beginning of a health/fitness program, the grip strengths of 40 female college students were measured. The following data were recorded:

43 43 49 55 45 33 56 43 45 40 42 53 50 39 43 33
47 35 21 35 37 30 36 44 37 35 29 39 49 27 24 29
19 34 28 22 34 40 29 36

Use these data to construct a stem-and-leaf diagram using 5 items per stem.

Answer

Stems	Leaves
1·	9
2*	1 4 2
2·	9 7 9 8 9
3*	3 3 0 4 4
3·	9 5 5 7 6 7 5 9 6
4*	3 3 3 0 2 3 4 0
4·	9 5 5 7 9
5*	3 0
5·	5 6

PROBLEM 1-16 At the beginning of a health/fitness program, the grip strengths of 40 male college students were measured. The following data were recorded:

68 57 47 66 52 67 53 70 54 53 43 61 65 59 54 55
58 50 71 61 42 57 45 48 41 38 62 46 67 69 58 58
51 52 61 68 71 62 50 47

(**a**) Use the data in Problems 1-15 and 1-16 to construct a two-sided stem-and-leaf diagram. (**b**) Combine the data in Problems 1-15 and 1-16 to construct a one-sided stem-and-leaf diagram.

Answer

(**a**)

Women	Stems	Men
9	1·	
2 4 1	2*	
9 8 9 7 9	2·	
4 4 0 3 3	3*	
6 9 5 7 6 7 5 5 9	3·	8
0 4 3 2 0 3 3 3	4*	3 2 1
9 7 5 5 9	4·	7 5 8 6 7
0 3	5*	2 3 4 3 4 0 1 2 0
6 5	5·	7 9 5 8 7 8 8
	6*	1 1 2 1 2
	6·	8 6 7 5 7 9 8
	7*	0 1 1

(**b**)

Stems	Leaves
1·	9
2*	1 4 2
2·	9 7 9 8 9
3*	3 3 0 4 4
3·	8 9 5 5 7 6 7 5 9 6
4*	3 2 1 3 3 3 0 2 3 4 0
4·	7 5 8 6 7 9 5 5 7 9
5*	2 3 4 3 4 0 1 2 0 3 0
5·	7 9 5 8 7 8 8 5 6
6*	1 1 2 1 2
6·	8 6 7 5 7 9 8
7*	0 1 1

2 MEASURES OF CENTRAL TENDENCY

THIS CHAPTER IS ABOUT

☑ The Sample Mean for Ungrouped Data
☑ The Sample Mean for Grouped Data
☑ The Median
☑ The Mode

Measures of central tendency are sometimes referred to as measures of location. They provide numbers that give the *average* or *middle* or *most typical value* of a set of data.

2-1. The Sample Mean for Ungrouped Data

When observing a random experiment or when collecting data, we often denote the outcomes as $x_1, x_2, \ldots, x_n$, where n is the sample size or number of observations. The **sample mean** $\bar{x}$ (read "*x-bar*") is the *arithmetic average* of a set of measurements; i.e., the sample mean is equal to the sum of the measurements divided by the number of measurements. In symbols, the sample mean is defined by

SAMPLE MEAN

$$\bar{x} = \frac{1}{n}\sum_{i=1}^{n} x_i = \frac{1}{n}\sum x_i = \frac{1}{n}\sum x \tag{2.1}$$

$$= \frac{x_1 + x_2 + \cdots + x_n}{n}$$

$$= \frac{\text{sum of the observations}}{\text{number of observations}}$$

note: In statistics, a **sample** is a representative part of an entire group, called a *population* or *universe*. Instead of examining a population, which is often large and difficult to count, we can examine a sample of the population in order to draw our conclusions.

EXAMPLE 2-1 Three cars of the same model, each having a 41-liter gas tank, were chosen as a sample. Each car was driven on an oval track until it ran out of gas. The first car traveled 931 kilometers (km), the second car traveled 972 km, and the third car traveled 935 km. Find the average number of kilometers traveled on a tankful of gas.

Solution We use the definition of the sample mean (formula 2.1) to find the average distance traveled on a tank of gas. If you say that $x_1 = 931$, $x_2 = 972$, and $x_3 = 935$, you can write

$$\bar{x} = \frac{1}{n}\sum x = \frac{x_1 + x_2 + x_3}{n} = \frac{931 + 972 + 935}{3} = \frac{2838}{3} = 946$$

EXAMPLE 2-2 Using the data in Example 2-1, find the average number of kilometers per liter (kpl) traveled for the sample of the 3 cars.

Solution First you determine the kpl for each car:

$$x_1 = \frac{931}{41} = 22.7 \qquad x_2 = \frac{972}{41} = 23.7 \qquad x_3 = \frac{935}{41} = 22.8$$

Then the sample mean of the numbers of kpl is the arithmetic average of these values:

$$\bar{x}_1 = \frac{22.7 + 23.7 + 22.8}{3} = \frac{69.2}{3} = 23.07$$

remark: The quantity 23.07 kpl is equivalent to 54.26 miles per gallon—impressive either way you say it.

EXAMPLE 2-3 A 12-sided die was rolled 7 times. The sample results were $x_1 = 8$, $x_2 = 1$, $x_3 = 5$, $x_4 = 11$, $x_5 = 3$, $x_6 = 5$, and $x_7 = 9$. What's the sample mean of these 7 rolls?

Solution

$$\bar{x} = \frac{8 + 1 + 5 + 11 + 3 + 5 + 9}{7} = \frac{42}{7} = 6$$

EXAMPLE 2-4 Suppose you know a poverty-stricken gambler whose funds are limited to \$5.00. Having decided to place \$1 bets on red in roulette until the \$5 and any accumulated winnings are lost, your friend asks you to tell her the number of bets she can place before going broke. You simulate nine times on a computer the number of bets that can be placed before losing the original \$5, and come up with the following data:

$x_1 = 43$ $x_2 = 95$ $x_3 = 7$ $x_4 = 29$ $x_5 = 701$ $x_6 = 97$ $x_7 = 13$ $x_8 = 15$ $x_9 = 17$

What can you tell your friend?

Solution You can tell your friend the *average* number of bets that you observed during your nine simulations to give her some indication of the number of bets she can place before she loses the \$5. For these data

$$\bar{x} = \frac{43 + 95 + 7 + 29 + 701 + 97 + 13 + 15 + 17}{9} = \frac{1017}{9} = 113$$

(Also see Example 2-12.)

2-2. The Sample Mean for Grouped Data

A. Sample mean for grouped discrete data

If certain values are repeated in a data set, we can modify the formula for the sample mean to simplify our calculations. Suppose that x_1 is observed with frequency f_1, x_2 is observed with frequency $f_2, \ldots,$ and x_k is observed with frequency f_k. Then the sample mean is defined by

SAMPLE MEAN FOR GROUPED DATA

$$\bar{x} = \frac{1}{n}\sum_{i=1}^{k} f_i x_i = \frac{1}{n}\sum fx \qquad \textbf{(2.2)}$$

where n is the sum of the frequencies

$$n = \sum_{i=1}^{k} f_i$$

That is, n is the total number of observations.

EXAMPLE 2-5 A 4-sided die was rolled 20 times, and the following sample observations, which have been ordered, were recorded:

1 1 1 1 1 2 2 2 2 3 3 3 3 3 3 3 4 4 4 4

Find the sample mean.

Solution For these data the sample mean is

$$\bar{x} = \frac{1 + 1 + 1 + 1 + 1 + 2 + 2 + 2 + 2 + 3 + 3 + 3 + 3 + 3 + 3 + 3 + 4 + 4 + 4 + 4}{20}$$

or, more simply, using formula (2.2),

$$\bar{x} = \frac{1}{n}\sum fx = \frac{5 \cdot 1 + 4 \cdot 2 + 7 \cdot 3 + 4 \cdot 4}{20} = \frac{5 + 8 + 21 + 16}{20} = \frac{50}{20} = 2.5$$

EXAMPLE 2-6 The frequency distribution for the litter sizes of 30 sample litters of domestic rabbits is

Litter size:	5	6	7	8	9	10	11	12	13	14
Frequency:	1	3	3	5	5	4	4	4	0	1

Calculate the sample mean.

Solution Using these data, you can let x represent the litter size and f represent the frequency with which litter size x was observed. To calculate the sample mean, set up a frequency distribution table that lists f and x and then add a third column (fx) that gives the products of f and x.

If you use the table to find $\sum fx$, you can use the simplified formula (2.2) to find the sample mean

$$\bar{x} = \frac{1}{n}\sum fx = \frac{275}{30} = 9.17$$

Litter size (x)	Frequency (f)	fx
5	1	5
6	3	18
7	3	21
8	5	40
9	5	45
10	4	40
11	4	44
12	4	48
13	0	0
14	1	14
	$n = \sum f = 30$	$\sum fx = 275$

EXAMPLE 2-7 The following data for the litter sizes of wild rabbits are given:

Litter size:	1	2	3	4	5	6	7	8	9	10
Frequency:	2	4	3	3	3	4	5	3	2	1

Calculate the sample mean for these litter sizes.

Solution Set up the frequency distribution table, where x is the litter size and f is the frequency with which x was observed. Calculate fx for each value of x; then find n and $\sum fx$.

The sample mean is

$$\bar{x} = \frac{1}{n}\sum fx = \frac{157}{30} = 5.23$$

x	f	fx
1	2	2
2	4	8
3	3	9
4	3	12
5	3	15
6	4	24
7	5	35
8	3	24
9	2	18
10	1	10
	$n = 30$	$\sum fx = 157$

B. Sample mean for grouped continuous data

When we're dealing with a large number of continuous-type measurements, we don't have exactly repeated measures very often. But we can use the grouped frequency distribution to obtain a *close approximation* of $\bar{x}$. We do this by assuming that all of the observations in a particular class are equal to the *class mark* of that class. Then we denote the class marks of k classes by $u_1, u_2, \ldots, u_k$, which are observed with frequencies $f_1, f_2, \ldots, f_k$. Thus, setting n equal to the sum of the frequencies, we can write a formula for $\bar{u}$:

APPROXIMATE SAMPLE MEAN FOR CONTINUOUS DATA

$$\bar{u} = \frac{1}{n}\sum fu \qquad \textbf{(2.3)}$$

note: This formula gives a value that's usually close to $\bar{x} = \frac{1}{n}\sum x$, but is often easier to calculate than $\bar{x}$.

EXAMPLE 2-8 The heights (in cm) of a sample of 50 women (x) were measured, so that the following data and grouped frequency distribution table were obtained:

171	157	163	170	173	170	168	167	164	170
169	157	168	178	163	161	165	160	166	155
170	173	171	174	171	173	175	167	164	168
169	170	157	169	172	156	160	158	172	165
171	152	163	165	170	177	159	157	158	165

Class limits	Frequency (f)	Class mark
150–154	1	152
155–159	9	157
160–164	8	162
165–169	13	167
170–174	16	172
175–179	3	177
	50	

(a) Find the approximate sample mean ($\bar{u}$) of these grouped data. **(b)** How does $\bar{u}$ compare with the (exact) sample mean $\bar{x}$?

Solution

(a) To find the approximate sample mean, set up the table as shown, calling the class marks u and calculating fu. Then you can calculate the approximate sample mean of the grouped data by substituting the class marks into formula (2.3):

$$\bar{u} = \frac{1}{n}\sum fu = \frac{8315}{50} = 166.3$$

Class mark (u)	Frequency (f)	fu
152	1	152
157	9	1413
162	8	1296
167	13	2171
172	16	2752
177	3	531
	$n = 50$	$\sum fu = 8315$

(b) For the original data, $\sum x = 8306$. Thus by formula (2.1),

$$\bar{x} = \frac{8306}{50} = 166.12$$

So you see that $\bar{u}$ is very close to $\bar{x}$.

EXAMPLE 2-9 Each of a sample of 24 "three-pound" bags of apples was weighed. The following data and grouped frequency distribution table were obtained:

3.26	3.62	3.39	3.12	3.53	3.30
3.10	3.26	3.19	3.22	3.14	3.39
3.31	3.21	3.49	3.41	3.02	3.17
3.20	3.12	3.42	3.36	3.21	3.26

Class limits	Frequency (f)	Class mark
3.01–3.11	2	3.06
3.12–3.22	8	3.17
3.23–3.33	6	3.28
3.34–3.44	5	3.39
3.45–3.55	2	3.50
3.56–3.66	1	3.61
	24	

Calculate the approximate sample mean $\bar{u}$, and compare this value with the exact value of the sample mean $\bar{x}$.

Solution Using the apple weights, you obtain the table shown to the right.

The approximate sample mean for the grouped data is

$$\bar{u} = \frac{1}{n}\sum fu = \frac{78.72}{24} = 3.28$$

Class mark (u)	Frequency (f)	fu
3.06	2	6.12
3.17	8	25.36
3.28	6	19.68
3.39	5	16.95
3.50	2	7.00
3.61	1	3.61
	24	78.72

For the ungrouped data, $\sum x = 78.70$. Thus

$$\bar{x} = 78.70/24 = 3.279$$

Again, you see that $\bar{u}$ is a close approximate to $\bar{x}$.

2-3. The Median

The **median** m of a set of observations is the midpoint of a set of observations after the observations have been ordered from small to large or from large to small.

- When the number of observations is *odd*, the median is the *middle observation.*
- When the number of observations is *even*, the median is the *average of the two middle observations.*

A. Finding the median for a small number of data

For a small number of data, all we have to do is order the data and find the middle.

EXAMPLE 2-10 In Example 2-1, the numbers of km traveled per tankful of gas for 3 cars were 931, 972, and 935. Find the median value.

Solution If you order these observations from small to large, you have $931 < 935 < 972$. The number of observations is odd, so the median is the middle observation, $m = 935$.

EXAMPLE 2-11 Ordering the 7 rolls of a 12-sided die as given in Example 2-3, you obtain 1, 3, 5, 5, 8, 9, 11. What's the median value?

Solution With 7 observations, the median is the fourth value in the ordered arrangement, so that $m = 5$.

EXAMPLE 2-12 Find the median number of bets it took to lose $5 in Example 2-4.

Solution If you order the outcomes of the 9 simulations, you have $7 < 13 < 15 < 17 < 29 < 43 < 95 < 97 < 701$. The median is the fifth observation in this ordered arrangement, so $m = 29$.

note: You also could have told your friend (Example 2-4) the median number of bets she could place before losing $5.

EXAMPLE 2-13 A pair of 6-sided dice was rolled 8 times, and the following sums were recorded for each of the 8 rolls:

$$x_1 = 3, \quad x_2 = 6, \quad x_3 = 7, \quad x_4 = 10, \quad x_5 = 7, \quad x_6 = 5, \quad x_7 = 8, \; x_8 = 6$$

What's the median value for these data?

Solution Ordering these observations, you have 3, 5, 6, 6, 7, 7, 8, 10. Because the number of observations is even, the median is the average of the fourth and fifth observations in the ordered arrangement, $m = (6 + 7)/2 = 6.5$.

B. Finding the median for a large number of data

It's not always a simple operation to order a large number of data, but we can use a stem-and-leaf diagram, which gives a visual representation of data, to help in locating the median. The rows of the stem-and-leaf diagram allow you to count observations at a glance (or at least very quickly).

EXAMPLE 2-14 The golf drives of 25 statisticians were measured, and the lengths (in yd) of the drives were recorded:

227	244	246	278	262	252	269	260	247	277	250	235	274
257	282	269	255	263	236	289	258	231	255	241	261	

Find the median value of these data.

Solution There are 25 observations, so the median must be the 13th observation after the observations have been ordered. By organizing the data in a stem-and-leaf diagram, you partially order the golfing data. The diagram shows that there are 8 observations after 3 rows and 14 observations after 4 rows. The median value will be in the fourth row, since 13 lies between 8 and 14. So all you have to do is order the observations in the fourth row from smallest to largest. The median is the fifth largest observation in the fourth row, so $m = 257$.

	Stems	Leaves
Row 1	22	7
Row 2	23	5 6 1
Row 3	24	4 6 7 1
Row 4	25	2 0 7 5 8 5
Row 5	26	2 9 0 9 3 1
Row 6	27	8 7 4
Row 7	28	2 9

It is even easier to find the median if the leaves in the stem-and-leaf diagram are already ordered from small to large within each row.

EXAMPLE 2-15 The stem-and-leaf diagram shown here gives the weights of apples in 24 three-pound bags. The leaves in each row are ordered from small to large. Find the median of these ordered values.

	Stems	Leaves
Row 1	3.0	2
Row 2	3.1	0 2 2 4 7 9
Row 3	3.2	0 1 1 2 6 6 6
Row 4	3.3	0 1 6 9 9
Row 5	3.4	1 2 9
Row 6	3.5	3
Row 7	3.6	2

Solution There are 24 observations, so the median is the average of the 12th and 13th observations. You find the 12th and 13th observations simply by counting them (starting with Row 1). Because the 12th and 13th observations both equal 3.26, the median is

$$m = \frac{3.26 + 3.26}{2} = 3.26$$

2-4. The Mode

The **mode** is the observation that occurs most frequently in a set of observations.

EXAMPLE 2-16 The number of cars per household for 40 homes was counted. There were 2 households with 0 cars, 9 with 1 car, 20 with 2 cars, 7 with 3 cars, and 2 with 4 cars. What's the modal number of cars per household?

Solution Households with 2 cars were observed most frequently (20 times), so the mode is 2.

SUMMARY

1. A set of n observations is denoted by $x_1, x_2, \ldots, x_n$.
2. The *sample mean* of a *set of observations* is $\bar{x} = \frac{1}{n}\sum x$.
   - The sample mean for a set of grouped *discrete data* is $\bar{x} = \frac{1}{n}\sum fx$, where n is the sum of the frequencies.
3. If $u_1, u_2, \ldots, u_k$ are observed with frequencies $f_1, f_2, \ldots, f_k$, then $\bar{u} = \frac{1}{n}\sum fu$, where $n = \sum f$ is the sum of the frequencies.
   - The approximate sample mean $\bar{u}$ for a set of grouped *continuous data* is the weighted arithmetic average of the class marks, where the weights are the class frequencies.
4. When a set of observations has been ordered, the *median* is the middle observation if the number of observations is odd and the average of the two middle observations if the number of observations is even.
5. The *mode* is the outcome that is observed most frequently.

RAISE YOUR GRADES

Can you . . . ?

☑ define the sample mean
☑ calculate the sample mean $\bar{x}$ for n observations
☑ calculate the sample mean $\bar{x}$ for grouped discrete data
☑ calculate the approximate sample mean $\bar{u}$ for grouped continuous data
☑ find the median
☑ find the mode

RAPID REVIEW

1. The arithmetic average of n measurements is called the ______________.
2. For n observations $x_1, x_2, \ldots, x_n$, $\sum x =$ ______________.
3. If outcomes $u_1, u_2, \ldots, u_k$ are observed with frequencies $f_1, f_2, \ldots, f_k$, then $\sum fu =$ ______________.
4. If $x_1 = 89$, $x_2 = 73$, and $x_3 = 81$, then $\bar{x} =$ ______________.
5. If $u_1 = 1, u_2 = 2, u_3 = 3, u_4 = 4$ and $f_1 = 3, f_2 = 2, f_3 = 4, f_4 = 1$, then $\bar{u} =$ ______________.
6. If $x_1 = 89, x_2 = 73, x_3 = 81$, then the median is $m =$ ______________.
7. If $x_1 = 9, x_2 = 3, x_3 = 1, x_4 = 6$, then the median is $m =$ ______________.
8. The outcome that is observed most frequently is called the ______________.

Answers: **(1)** sample mean **(2)** $x_1 + x_2 + \cdots + x_n$ **(3)** $f_1u_1 + f_2u_2 + \cdots + f_ku_k$ **(4)** $\frac{243}{3} = 81$ **(5)** $\frac{3 + 4 + 12 + 4}{10} = 2.3$ **(6)** 81 **(7)** $\frac{3 + 6}{2} = 4.5$ **(8)** mode

SOLVED PROBLEMS

PROBLEM 2-1 A fair 8-sided die was rolled 5 times, and the following sample outcomes were observed: $x_1 = 8, x_2 = 1, x_3 = 7, x_4 = 3, x_5 = 6$. **(a)** Find the sample mean $\bar{x}$. **(b)** Find the median m.

Solution

(a) Use formula (2.1) to calculate the sample mean:

$$\bar{x} = \frac{1}{n}\sum x = \frac{8 + 1 + 7 + 3 + 6}{5} = \frac{25}{5} = 5$$

(b) The median is the middle observation in the ordered arrangement: $1 < 3 < 6 < 7 < 8$, so $m = 6$.

PROBLEM 2-2 A sample of 10 mathematically minded students who applied for admission to a college had the following ACT verbal scores: 25, 21, 19, 30, 23, 25, 24, 25, 19, 23. **(a)** Find the sample mean of these ACT scores. **(b)** Find the median of these ACT scores.

Solution

(a) By the formula for the sample mean:

$$\bar{x} = \frac{1}{n}\sum x = \frac{25 + 21 + 19 + 30 + 23 + 25 + 24 + 25 + 19 + 23}{10}$$

$$= \frac{234}{10} = 23.4$$

(b) The ordered ACT scores are

19 19 21 23 23 24 25 25 25 30

The number of observations is even, so the median is the average of the fifth and sixth scores: $m = (23 + 24)/2 = 23.5$.

PROBLEM 2-3 As part of their first laboratory exercise, some budding young alchemists weighed a sample of eleven 50-cc retorts on an analytical balance. The following weights (in g) were recorded:

26.85 25.45 26.79 25.38 25.47 29.15 24.86 26.98 29.93 24.40 26.90

Find **(a)** the sample mean and **(b)** the median of these weights.

Solution

$$\textbf{(a)}\ \bar{x} = \frac{26.85 + 25.45 + 26.79 + 25.38 + 25.47 + 29.15 + 24.86 + 26.98 + 29.93 + 24.40 + 26.90}{11}$$

$$= \frac{292.16}{11} = 26.56$$

(b) The ordered weights are $24.40 < 24.86 < 25.38 < 25.45 < 25.47 < 26.79 < 26.85 < 26.90 < 26.98 < 29.15 < 29.93$. The median is the sixth observation so $m = 26.79$.

PROBLEM 2-4 According to the United Nations, the population estimates in thousands for a sample of six Asian countries are as follows:

Afghanistan	16,786	India	676,218
Indonesia	151,720	Japan	118,449
Philippines	50,740	Singapore	2,472

Find **(a)** the sample mean and **(b)** the median of these populations.

Solution

$$\textbf{(a)}\ \bar{x} = \frac{16{,}786 + 151{,}720 + 50{,}740 + 676{,}218 + 118{,}449 + 2{,}472}{6} = \frac{1{,}016{,}385}{6} = 169{,}397.5$$

(b) The ordered populations are $2{,}472 < 16{,}786 < 50{,}740 < 118{,}449 < 151{,}720 < 676{,}218$, so $m = (50{,}740 + 118{,}449)/2 = 84{,}594.5$.

PROBLEM 2-5 The number of cars per household was counted in a survey of a sample of 40 homes. The results of the survey showed that 2 households had 0 cars, 9 had 1 car, 20 had 2 cars, 7 had 3 cars, and 2 had 4 cars. Find the sample mean number of cars per household.

Solution Set up a frequency distribution table for discrete data, adding an fx product column. Then sum the f and fx columns to get $n = \sum f = 40$ and $\sum fx = 78$. Finally, use the simplified formula (2.2) to calculate the sample mean for grouped discrete data:

$$\bar{x} = \frac{1}{n}\sum fx = \frac{78}{40} = 1.95$$

Numbers of cars (x)	Frequency (f)	fx
0	2	0
1	9	9
2	20	40
3	7	21
4	2	8
	$n = \sum f = 40$	$\sum fx = 78$

PROBLEM 2-6 The sums of a sample of 100 rolls of a pair of 4-sided dice are as follows:

```
4 3 6 6 4 8 7 2 6 7 5 5 7 7 7 5 4 3 5 6
5 5 4 2 3 6 4 8 3 8 6 8 4 3 6 7 3 6 2 5
3 6 6 5 7 5 7 7 3 5 7 5 5 7 8 5 4 4 4 4
4 7 6 6 4 5 3 3 7 4 4 5 2 5 4 8 8 4 2 7
5 3 6 5 5 2 4 2 3 7 4 3 6 5 5 5 5 5 6 5
```

Find the sample mean of these sums.

Solution Set up the frequency distribution table. Then, using the table and formula (2.2) for grouped discrete data, the sample mean is

$$\bar{x} = \frac{501}{100} = 5.01$$

x	f	fx
2	7	14
3	13	39
4	18	72
5	25	125
6	15	90
7	15	105
8	7	56
	$n = 100$	$\sum fx = 501$

PROBLEM 2-7 The following sample milk weights (in lb) from 50 cows were recorded:

42	50	29	25	40	53	41	21	37	38
38	32	51	39	48	40	37	45	28	57
48	67	73	17	21	44	14	15	64	49
44	43	50	76	46	47	38	71	42	32
20	36	49	52	28	38	42	29	50	24

Find the sample mean for **(a)** the ungrouped data and **(b)** the grouped data. [*Hint:* Notice that the data are continuous.]

Solution

(a) For the 50 (ungrouped) weights, you just add, and find the arithmetic average: $\sum x = 2060$; thus $\bar{x} = 2060/50 = 41.2$.

(b) For the grouped data, you can construct a frequency distribution table to obtain the class marks u (see Problem 1-3). Then multiply each class by its frequency to obtain the fu product:

u	f	fu
18	6	108
27	6	162
36	12	432
45	14	630
54	7	378
63	2	126
72	3	216
	$n = \sum f = 50$	$\sum fu = 2052$

Summing the f and fu columns gives $n = \sum f = 50$ and $\sum fu = 2052$, respectively. Now you can use formula (2.3) for calculating the approximate sample mean for grouped continuous data: $\bar{u} = \frac{1}{n}\sum fu = 2052/50 = 41.04$.

PROBLEM 2-8 A sample of 25 bottles of liquid soap was taken from a production line and each bottle was weighed. The following data give the excess weight over 1000 g for each of the 25 bottles:

60	31	19	43	47	48	47	13	28	35	51	24	24
27	33	57	38	24	39	40	63	51	25	43	26	

Find the sample mean of the excess weight for **(a)** the ungrouped data and **(b)** the grouped data.

Solution

(a) For the ungrouped data, add and find the arithmetic average:
$\sum x = 936$; thus

$$\bar{x} = \frac{936}{25} = 37.44$$

(b) You can construct a frequency distribution table to calculate the mean for the grouped data (see Problem 1-5). The approximate sample mean is

$$\bar{u} = \frac{1}{n}\sum fu = \frac{932.5}{25} = 37.3$$

u	f	fu
14.5	2	29.0
24.5	7	171.5
34.5	5	172.5
44.5	6	267.0
54.5	3	163.5
64.5	2	129.0
	25	932.5

Supplementary Exercises

PROBLEM 2-9 A fair, 12-sided die was rolled 6 times with the following sample results: 4, 1, 8, 11, 3, 12. Find (**a**) the sample mean and (**b**) the median.

Answer (**a**) $\bar{x} = 6.5$ (**b**) $m = 6$

PROBLEM 2-10 A 12-sided die was rolled 18 times with the following sample results:

8 6 4 10 5 11 6 4 1 8 5 2 11 12 1 10 9 9

Find (**a**) the sample mean and (**b**) the median.

Answer (**a**) $\bar{x} = 6.78$ (**b**) $m = 7$

PROBLEM 2-11 Find (**a**) the sample mean, (**b**) the median, and (**c**) the mode for the following clutch sizes of a sample of 100 moorhens:

7	9	9	9	9	10	7	8	9	10	8	10	7	7	8	9	8	8	8	9
9	10	9	9	9	8	7	7	8	7	7	9	7	6	9	9	10	8	13	13
8	8	9	9	9	10	10	7	8	7	7	7	11	8	8	7	7	5	6	12
9	10	8	8	10	9	10	11	9	13	9	7	8	9	10	10	9	8	10	8
9	8	8	7	6	9	9	8	8	7	5	7	9	9	5	9	9	7	6	11

Answer (**a**) $\bar{x} = 8.47$ (**b**) $m = 8.5$ (**c**) mode $= 9$

PROBLEM 2-12 In one season, each of a sample of 7 hockey players scored the following points: 89, 93, 53, 46, 66, 47, 61. Find the sample mean number of points these 7 players scored.

Answer $\bar{x} = 65$

PROBLEM 2-13 The following data give the maximum lung capacity (in liters) for a sample of 25 first-year females and 25 first-year males:

Females					Males				
4.2	5.3	3.5	4.3	3.7	6.8	5.7	5.6	4.8	3.7
3.2	3.5	2.8	3.5	3.7	5.0	3.8	5.6	5.4	5.2
3.5	3.7	2.7	2.8	3.3	4.9	4.7	7.6	4.0	5.6
2.7	3.0	3.1	3.0	3.7	4.2	4.0	5.5	5.1	4.9
3.3	3.4	2.3	2.6	3.2	5.4	5.9	3.4	6.1	4.1

Find (**a**) the exact sample mean for the 25 females; (**b**) the exact sample mean for the 25 males; (**c**) the median for the 25 females; (**d**) the median for the 25 males.

Answer (**a**) $\bar{x} = 3.36$ (**b**) $\bar{x} = 5.08$ (**c**) $m = 3.3$ (**d**) $m = 5.1$

PROBLEM 2-14 The following numbers are the weights (in kg) of a sample of 50 male college freshmen:

76.0	64.6	80.6	84.8	78.9	84.3	68.3
72.2	63.7	92.6	66.4	72.6	62.6	74.7
60.4	72.0	77.2	59.0	75.0	54.1	90.8
66.1	68.1	69.1	79.7	65.3	81.3	56.9
71.6	74.9	73.0	88.0	59.5	62.8	75.0
71.3	74.4	70.8	77.8	71.9	63.8	70.3
72.0	66.0	74.0	70.7	69.6	75.3	63.4
69.4						

u	f
55.75	3
62.75	11
69.75	17
76.75	12
83.75	4
90.75	3
	50

Using the given frequency distribution table, find (**a**) the sample mean for the grouped data, and (**b**) the sample mean for the ungrouped data.

Answer (**a**) $\bar{u} = 71.43$ (**b**) $\bar{x} = 71.66$

PROBLEM 2-15 The resistance of a sample of one hundred 2200-ohm resistors was recorded as follows:

2.17	2.18	2.17	2.19	2.23	2.18	2.16	2.23	2.29	2.20
2.24	2.13	2.18	2.22	2.21	2.22	2.23	2.18	2.23	2.25
2.24	2.20	2.21	2.20	2.25	2.15	2.20	2.25	2.16	2.19
2.20	2.18	2.18	2.19	2.22	2.19	2.20	2.19	2.22	2.21
2.22	2.18	2.18	2.28	2.19	2.23	2.21	2.19	2.21	2.21
2.19	2.18	2.21	2.17	2.20	2.18	2.18	2.21	2.22	2.18
2.22	2.23	2.19	2.23	2.18	2.19	2.18	2.21	2.18	2.15
2.20	2.23	2.20	2.20	2.23	2.20	2.19	2.22	2.17	2.20
2.20	2.17	2.19	2.19	2.25	2.19	2.19	2.20	2.20	2.19
2.21	2.14	2.24	2.21	2.19	2.23	2.18	2.22	2.23	2.19

u	f
2.135	2
2.155	4
2.175	21
2.195	33
2.215	20
2.235	14
2.255	4
2.275	1
2.295	1
	100

(**a**) Find the sample mean for the data grouped according to the given table. (**b**) Find the mode.

Answer (**a**) $\bar{u} = 2.2018$ (**b**) mode = 2.19

PROBLEM 2-16 The lung capacity (in liters) for a sample of 20 members of a volleyball team was measured, with the following results:

3.4	3.7	3.6	3.4	3.9	3.4	3.3	3.7	3.4	3.6
4.1	3.5	4.1	3.7	3.6	4.3	3.7	4.2	3.8	4.7

u	f
3.4	6
3.7	8
4.0	3
4.3	2
4.6	1
	20

Using the given table, find (**a**) the sample mean for the grouped data, (**b**) the sample mean for the ungrouped data, and (**c**) the median.

Answer (**a**) $\bar{u} = 3.76$ (**b**) $\bar{x} = 3.755$ (**c**) $m = 3.7$

3 MEASURES OF DISPERSION

THIS CHAPTER IS ABOUT

☑ **The Variance and Standard Deviation for Ungrouped Sample Data**
☑ **The Variance and Standard Deviation for Grouped Sample Data**
☑ **Chebyshev's Inequality**
☑ **The Range**

Measures of dispersion give an indication of the amount of variation among data, e.g., a measure of how spread out the data are.

3-1. The Variance and Standard Deviation for Ungrouped Sample Data

The sample variance and the sample standard deviation give a measure of the spread of a set of measurements.

- The **sample variance** s^2 of a set of measurements $x_1, x_2, \ldots, x_n$ is defined by

SAMPLE VARIANCE

$$s^2 = \frac{\sum (x - \bar{x})^2}{n - 1} = \frac{(x_1 - \bar{x})^2 + (x_2 - \bar{x})^2 + \cdots + (x_n - \bar{x})^2}{n - 1} \tag{3.1}$$

where $\bar{x}$ is the sample mean $\bar{x} = \frac{1}{n}\sum x$.

- The **standard deviation** s is the square root of the variance; i.e.,

STANDARD DEVIATION

$$s = \sqrt{s^2} \tag{3.2}$$

EXAMPLE 3-1 The numbers of kilometers that a sample of three cars traveled were $x_1 = 931$, $x_2 = 972$, and $x_3 = 935$. Find the sample variance and the sample standard deviation for these data.

Solution First find the sample mean $\bar{x}$ by formula (2.1):

$$\bar{x} = \frac{1}{n}\sum x = \frac{931 + 972 + 935}{3} = 946$$

Now use formula (3.1) to find the sample variance:

$$s^2 = \frac{\sum (x - \bar{x})^2}{n - 1} = \frac{(x_1 - \bar{x})^2 + (x_2 - \bar{x})^2 + (x_3 - \bar{x})^2}{n - 1}$$

$$= \frac{(931 - 946)^2 + (972 - 946)^2 + (935 - 946)^2}{3 - 1}$$

$$= \frac{225 + 676 + 121}{2} = \frac{1022}{2} = 511$$

Then use the sample variance to find the sample standard deviation:

$$s = \sqrt{s^2} = \sqrt{511} = 22.6$$

If there are many data to manipulate, we can use a table to help with the calculations.

EXAMPLE 3-2 The sample outcomes of 7 rolls of a 12-sided die were 8, 1, 5, 11, 3, 5, 9. Find the sample variance and standard deviation of these data.

Solution To simplify your calculations, construct a table as shown.

From the table you can easily see that $\bar{x} = 42/7 = 6$, so

$$s^2 = \frac{\sum (x - \bar{x})^2}{n - 1} = \frac{74}{7 - 1} = 12.33$$

and

$$s = \sqrt{s^2} = \sqrt{12.33} = 3.51$$

x	$x - \bar{x}$	$(x - \bar{x})^2$
8	2	4
1	−5	25
5	−1	1
11	5	25
3	−3	9
5	−1	1
9	3	9
$\sum x = 42$	$\sum (x - \bar{x}) = 0$	$\sum (x - \bar{x})^2 = 74$

The second column of the table in Example 3-2 illustrates a fact that is always true, namely,

$$\sum (x - \bar{x}) = \sum x - \sum \bar{x} = \sum x - n\bar{x}$$

but $\bar{x} = \frac{1}{n}\sum x$, so $n\bar{x} = \sum x$ and

$$\sum x - n\bar{x} = \sum x - \sum x = 0$$

- The sum of the deviations of a set of measurements from their mean is always 0.

EXAMPLE 3-3 Nine simulations on a computer of the number of $1 bets on red in roulette before losing $5 were as follows: 43, 95, 7, 29, 701, 97, 13, 15, 17. Find the sample variance and standard deviation for these data.

Solution Construct a table using the given values, as shown. Since $\bar{x} = 1017/9 = 113$,

$$s^2 = \frac{398{,}336}{8} = 49{,}792$$

and $s = \sqrt{49{,}792} = 223.14$

x	$(x - \bar{x})$	$(x - \bar{x})^2$
43	−70	4,900
95	−18	324
7	−106	11,236
29	−84	7,056
701	588	345,744
97	−16	256
13	−100	10,000
15	−98	9,604
17	−96	9,216
$\sum x = 1017$	$\sum (x - \bar{x}) = 0$	$\sum (x - \bar{x})^2 = 398{,}336$

The formula for sample variance (3.1) can be written in a different form. Since

$$s^2 = \frac{\sum (x - \bar{x})^2}{n - 1}$$

we can rewrite this definition, after some algebraic manipulations, as

SAMPLE VARIANCE (alternate form)

$$s^2 = \frac{n\sum x^2 - (\sum x)^2}{n(n - 1)} \tag{3.3}$$

This alternate formula often simplifies our calculations.

note: Formulas (3.1) and (3.3) are equivalent. However, it's possible at times to obtain slightly different results due to round-off errors or the magnitudes of the observations.

EXAMPLE 3-4 Find the sample variance for the data given in Example 3-1 by using the alternate formula (3.3).

Solution Construct the table as shown. By (3.3), the sample variance is

$$s^2 = \frac{n\sum x^2 - (\sum x)^2}{n(n-1)}$$

$$= \frac{3(2{,}685{,}770) - (2838)^2}{3(3-1)} = \frac{3066}{6} = 511$$

x	x^2
931	866,761
972	944,784
935	874,225
$\sum x = 2838$	$\sum x^2 = 2{,}685{,}770$

EXAMPLE 3-5 Use the alternate formula (3.3) to find the sample variance for the 7 rolls of a 12-sided die given in Example 3-2.

Solution First construct a table:

x	x^2
8	64
1	1
5	25
11	121
3	9
5	25
9	81
42	326

The sample variance is

$$s^2 = \frac{7(326) - (42)^2}{7(7-1)}$$

$$= \frac{2282 - 1764}{42}$$

$$= \frac{518}{42}$$

$$= 12.33$$

3-2. The Variance and Standard Deviation for Grouped Sample Data

A. Variance and standard deviation for grouped discrete data

When we have discrete data with repeated measures, we can modify the formula for the sample variance to simplify our calculations. Suppose that x_1 is observed with frequency f_1, x_2 is observed with frequency $f_2, \ldots,$ and x_k is observed with frequency f_k. The sample variance is now defined by

SAMPLE VARIANCE FOR GROUPED DISCRETE DATA

$$s^2 = \frac{\sum_{i=1}^{k} f_i(x_i - \bar{x})^2}{n-1} = \frac{\sum f(x - \bar{x})^2}{n-1} \tag{3.4}$$

But it's often tedious and time-consuming to subtract the mean from each observation and square the result, as formula (3.4) requires. So we also have an equivalent, *working formula* for the sample variance:

SAMPLE VARIANCE FOR GROUPED DISCRETE DATA (alternate form)

$$s^2 = \frac{n\sum_{i=1}^{k} f_i x_i^2 - \left(\sum_{i=1}^{k} f_i x_i\right)^2}{n(n-1)} = \frac{n\sum fx^2 - (\sum fx)^2}{n(n-1)} \tag{3.5}$$

In both formulas (3.4) and (3.5), $\bar{x}$ is the sample mean and $n = \sum_{i=1}^{k} f_i$ is the sum of the frequencies.

note: Formulas (3.4) and (3.5) are exactly equivalent. You simply pick the easiest one for the data at hand. (Formula (3.5) is especially handy for large sample sizes.) It's possible, at times, to obtain slightly different results from formulas (3.4) and (3.5) due to round-off errors or the magnitudes of the observations.

EXAMPLE 3-6 A sample of 20 rolls of a 4-sided die resulted in $x_1 = 1, f_1 = 5; x_2 = 2, f_2 = 4; x_3 = 3, f_3 = 7; x_4 = 4, f_4 = 4$. For these results, $\bar{x} = 2.5$. Calculate the sample variance and standard deviation for these data.

Solution Construct the following table:

x	f	fx	fx^2	$f(x - \bar{x})^2$
1	5	5	5	11.25
2	4	8	16	1.00
3	7	21	63	1.75
4	4	16	64	9.00
	20	$\sum fx = 50$	$\sum fx^2 = 148$	$\sum f(x - \bar{x})^2 = 23.00$

Using formula (3.4),

$$s^2 = \frac{\sum f(x - \bar{x})^2}{n - 1} = \frac{23}{19} = 1.21$$

Using the working formula (3.5),

$$\begin{aligned} s^2 &= \frac{n\sum fx^2 - (\sum fx)^2}{n(n-1)} \\ &= \frac{20(148) - (50)^2}{20(20-1)} \\ &= \frac{2960 - 2500}{380} \\ &= \frac{460}{38} = 1.21 \end{aligned}$$

[Notice that using the working formula (3.5) saves you the trouble of calculating the $f(x - \bar{x})^2$ column in the table.]

The standard deviation is

$$s = \sqrt{s^2} = \sqrt{1.21} = 1.10$$

EXAMPLE 3-7 Find the sample variance and standard deviation for the following 30 sample litter sizes of domestic rabbits. (For these data, $\bar{x} = 275/30 = 9.17$.)

Litter size:	5	6	7	8	9	10	11	12	13	14
Frequency:	1	3	3	5	5	4	4	4	0	1

Solution

x	f	fx	fx^2	$f(x - x)^2$
5	1	5	25	17.39
6	3	18	108	30.15
7	3	21	147	14.13
8	5	40	320	6.84
9	5	45	405	.14
10	4	40	400	2.76
11	4	44	484	13.40
12	4	48	576	32.04
13	0	0	0	0.00
14	1	14	196	23.33
	30	275	2661	140.18

By formula (3.4),

$$s^2 = \frac{140.18}{29} = 4.83$$

By formula (3.5),

$$\begin{aligned} s^2 &= \frac{30(2661) - (275)^2}{30(30-1)} \\ &= \frac{79{,}830 - 75{,}625}{870} = \frac{4205}{870} \\ &= 4.83 \end{aligned}$$

The standard deviation is $s = \sqrt{4.83} = 2.20$.

B. Variance and standard deviation for grouped continuous data

When we're working with continuous data that have been grouped in a frequency distribution, we can use class marks and class frequencies to approximate the sample variance. We denote the class marks of k classes with $u_1, u_2, \ldots, u_k$ and the class frequencies of k classes with $f_1, f_2, \ldots, f_k$. The approximate sample variance of the grouped data is therefore

SAMPLE VARIANCE FOR GROUPED CONTINUOUS DATA

$$s_u^2 = \frac{\sum_{i=1}^{k} f_i(u_i - \bar{u})^2}{n-1} = \frac{\sum f(u - \bar{u})^2}{n-1} \tag{3.6}$$

A working formula for the sample variance is

SAMPLE VARIANCE FOR GROUPED CONTINUOUS DATA (alternate form)

$$s_u^2 = \frac{n\left(\sum_{i=1}^{k} f_i u_i^2\right) - \left(\sum_{i=1}^{k} f_i u_i\right)^2}{n(n-1)} = \frac{n\sum fu^2 - (\sum fu)^2}{n(n-1)} \tag{3.7}$$

In both formulas, n is the sum of the frequencies

$$n = \sum_{i=1}^{k} f_i$$

and in formula (3.6), $\bar{u}$ is the approximate sample mean for grouped continuous data (2.3):

$$\bar{u} = \frac{1}{n}\sum_{i=1}^{k} f_i u_i = \frac{1}{n}\sum fu$$

note: The sample variance s_u^2 calculated by formula (3.6) or (3.7) is an approximation based on the assumption that all the observations within a class are equal to the class mark. The value of s_u^2 is usually very close to s^2, as defined by formula (3.1).

EXAMPLE 3-8 Given the following frequency distribution table, find the approximate sample variance and standard deviation of the heights (in cm) of a sample of 50 women.

Class boundaries	Class limits	Frequency	Class mark
149.5–154.5	150–154	1	152
154.5–159.5	155–159	9	157
159.5–164.5	160–164	8	162
164.5–169.5	165–169	13	167
169.5–174.5	170–174	16	172
174.5–179.5	175–179	3	177
		50	

Solution First expand the frequency distribution table, denoting class marks u and frequency f:

u	f	fu	fu^2	$f(u-\bar{u})^2$	$f(u-\bar{u})$
152	1	152	23,104	204.49	−14.3
157	9	1413	221,841	778.41	−83.7
162	8	1296	209,952	147.92	−34.4
167	13	2171	362,557	6.37	9.1
172	16	2752	473,344	519.84	91.2
177	3	531	93,987	343.47	32.1
	50	8315	1,384,785	2000.50	0.0

Using formula (3.6),

$$s_u^2 = \frac{\sum f(u-\bar{u})^2}{n-1} = \frac{2000.50}{49} = 40.83$$

Using the working formula (3.7),

$$s_u^2 = \frac{n\sum fu^2 - (\sum fu)^2}{n(n-1)} = \frac{50(1{,}384{,}785) - (8315)^2}{50(49)} = \frac{100{,}025}{2450} = 40.83$$

The standard deviation for the grouped data is $s_u = \sqrt{40.83} = 6.39$.

note: As the last column in the table illustrates, the sum of the deviations from the mean is always 0.

EXAMPLE 3-9 The weights of a sample of 24 "three-pound" bags of apples were measured and the observations grouped in a frequency distribution table. Find (**a**) the approximate sample variance, (**b**) the exact sample variance, and (**c**) the standard deviation for each sample variance.

3.26	3.62	3.39	3.12	3.53	3.30
3.19	3.22	3.14	3.39	3.31	3.21
3.02	3.17	3.20	3.12	3.42	3.36
3.10	3.26	3.49	3.41	3.21	3.26

Class boundaries	Class limits	Frequency	Class mark
3.005–3.115	3.01–3.11	2	3.06
3.115–3.225	3.12–3.22	8	3.17
3.225–3.335	3.23–3.33	6	3.28
3.335–3.445	3.34–3.44	5	3.39
3.445–3.555	3.45–3.55	2	3.50
3.555–3.665	3.56–3.66	1	3.61
		24	

Solution

(**a**) Construct the table (as shown) to find the approximate sample variance for the grouped data. Then by formula (3.6),

$$s_u^2 = \frac{.4598}{23} = .0200$$

or, by formula (3.7),

$$s_u^2 = \frac{24(258.6614) - (78.72)^2}{24(23)}$$

$$= \frac{11.0352}{552} = .0200$$

u	f	fu	fu^2	$f(u - \bar{u})^2$
3.06	2	6.12	18.7272	.0968
3.17	8	25.36	80.3912	.0968
3.28	6	19.68	64.5504	.0000
3.39	5	16.95	47.4605	.0605
3.50	2	7.00	24.5000	.0968
3.61	1	3.61	13.0321	.1089
	24	78.72	258.6614	.4598

(**b**) To find the exact sample variance, you have to work with the ungrouped data. And because you have a lot of data here, you'll use formula (3.3), beginning with a table:

x	x^2	x	x^2	x	x^2	x	x^2
3.26	10.6276	3.19	10.1761	3.02	9.1204	3.10	9.6100
3.62	13.1044	3.22	10.3684	3.17	10.0489	3.26	10.6276
3.39	11.4921	3.14	9.8596	3.20	10.2400	3.49	12.1801
3.12	9.7344	3.39	11.4921	3.12	9.7344	3.41	11.6281
3.53	12.4609	3.31	10.9561	3.42	11.6964	3.21	10.3041
3.30	10.8900	3.21	10.3041	3.36	11.2896	3.26	10.6276
20.22	68.3094	19.46	63.1564	19.29	62.1297	19.73	64.9775

$$\sum x = 78.70$$
$$\sum x^2 = 258.5730$$

After all that, you can finally use the formula:

$$s^2 = \frac{n\sum x^2 - (\sum x)^2}{n(n-1)} = \frac{24(258.5730) - (78.70)^2}{24(24-1)}$$

$$= \frac{6205.7520 - 6193.690}{552} = \frac{12.0620}{552}$$

$$= .0219$$

(**c**) The sample standard deviation for the grouped data is $s_u = \sqrt{s_u^2} = \sqrt{.0200} = .14$. The sample standard deviation for the ungrouped data is $s = \sqrt{s^2} = \sqrt{.0219} = .15$. You can see that they're almost equal and s_u is easier to calculate than s.

3-3. Chebyshev's Inequality

If $x_1, x_2, \ldots, x_n$ is a set of n observations, the sample mean $\bar{x}$ gives a measure of the center of these data and the standard deviation s gives a measure of how spread out the data are from $\bar{x}$. (The smaller the value of s, the closer the x values are clustered around $\bar{x}$.) Now we need to find out what proportion of our observations will fall within a specific number (k) of standard deviations from the mean. We can do this exactly by counting the number of observations that fall within the interval from $\bar{x} - ks$ to $\bar{x} + ks$ and then taking a percentage. This percentage will have a lower limit, or *bound*, which we know by *Chebyshev's inequality*.

Chebyshev's inequality: For any number $k > 1$, the proportion of observations that are within k standard deviations of the sample mean is at least $1 - (1/k^2)$. That is, AT LEAST $1 - (1/k^2)$ of the data lie within the interval from $\bar{x} - ks$ to $\bar{x} + ks$.

TABLE 3-1: Values of $1 - (1/k^2)$ for Selected Values of k

k	$1 - \frac{1}{k^2}$
2	$1 - \frac{1}{2^2} = \frac{3}{4} = 75\%$
$\frac{5}{2}$	$1 - \frac{1}{(5/2)^2} = \frac{21}{25} = 84\%$
3	$1 - \frac{1}{3^2} = \frac{8}{9} = 89\%$
$\frac{7}{2}$	$1 - \frac{1}{(7/2)^2} = \frac{45}{49} = 92\%$

Table 3-1 gives some selected values of k and $1 - (1/k^2)$. For example, at least 75% of the data lie within 2 standard deviations of the mean; i.e., at least 75% of the data lie in the interval from $\bar{x} - 2s$ to $\bar{x} + 2s$. And at least 89% of the data lie within 3 standard deviations of the mean; i.e., at least 89% of the data lie between $\bar{x} - 3s$ and $\bar{x} + 3s$.

When the histogram of a set of data is *bell-shaped*, or the data come from a *normal population* (see Chapter 10), we can give an additional rule, sometimes called the empirical rule:

Empirical rule: Approximately 68% of the data lie within 1 standard deviation of the mean, approximately 95% of the data lie within 2 standard deviations of the mean, and approximately 99.7% of the data lie within 3 standard deviations of the mean when the data come from a normal population.

- Chebyshev's inequality is a guaranteed lower bound for all sets of data, while the empirical rule gives approximations for samples from normal populations only.

note: Chebyshev's inequality still applies, even when the empirical rule is used: 95% *is* greater than 75%, and 99.7% *is* greater than 89%.

EXAMPLE 3-10 The weights (in kg) of a sample of $n = 252$ male college freshmen have a sample mean $\bar{x} = 73.8$ and a sample standard deviation $s = 11.3$. How many of the 252 observed weights fall within **(a)** 2 standard deviations of $\bar{x}$, **(b)** 3 standard deviations of $\bar{x}$? Show the interval in each case.

Solution Use Chebyshev's inequality.

(a) When $k = 2$, at least

$$\left(1 - \frac{1}{k^2}\right)(252) = \left(1 - \frac{1}{2^2}\right)(252) = \left(\frac{3}{4}\right)(252) = 189$$

of the weights are in the interval

$$(73.8 - 2(11.3), 73.8 + 2(11.3)) = (73.8 - 22.6, 73.8 + 22.6) = (51.2, 96.4)$$

(b) According to Table 3-1, at least $\frac{8}{9}$ of the data fall within 3 standard deviations. So at least $(\frac{8}{9})(252) = 224$ of the weights are in the interval

$$(73.8 - 3(11.3), 73.8 + 3(11.3)) = (73.8 - 33.9, 73.8 + 33.9) = (39.9, 107.7)$$

note: Weights tend to be normally distributed, so assume that these data are normally distributed. Then you can use the empirical rule.

(a) Approximately 95% of the 252 weights or $.95 \times 252 \cong 239$ of the weights are in the interval $(73.8 - 2(11.3), 73.8 + 2(11.3)) = (51.2, 96.4)$.
(b) You can expect almost all ($\sim$99.7%) of these weights ($\sim .997 \times 252 \cong 251$) to lie in the interval $(73.8 - 3(11.3), 73.8 + 3(11.3)) = (39.9, 107.7)$.

EXAMPLE 3-11 For the heights (in cm) of a sample of 50 women the sample mean is $\bar{x} = 166.12$ and the sample standard deviation is $s = 6.26$.

(a) If the number of heights that lie within the interval $(166.12 - 2(6.26), 166.12 + 2(6.26)) = (153.6, 178.64)$ is 49, what proportion of these heights lie within 2 standard deviations of the mean?
(b) If the number of heights that lie within the interval $(166.12 - 6.26, 166.12 + 6.26) = (159.86, 172.38)$ is 35, what proportion of these heights lie within 1 standard deviation of the mean?
(c) What kind of distribution do these proportions suggest?

Solution

(a) If 49 of 50 measurements fall within the interval (153.6, 178.64), then $\frac{49}{50} = .98$, or 98%, of the data fall within 2 standard deviations of $\bar{x}$.
(b) If 35 of 50 measurements fall within the interval (159.86, 172.38), then $\frac{35}{50} = .70$, or 70%, of the data fall within 1 standard deviation.
(c) Since these proportions are nearly equal to the percentages predicted by the empirical rule, the height measurements are probably normally distributed.

3-4. The Range

The **range** of a set of data is the difference between the maximum and the minimum. That is, given a set of measurements $x_1, x_2, \ldots, x_n$,

$$\text{range} = \text{largest measurement} - \text{smallest measurement}$$

EXAMPLE 3-12 The numbers of kilometers traveled per liter of gas for 3 cars were $x_1 = 22.7$, $x_2 = 23.7$, $x_3 = 22.8$. What's the range?

Solution $23.7 - 22.7 = 1.0$

EXAMPLE 3-13 The numbers of $1 bets placed in roulette before losing $5 were 43, 95, 7, 29, 701, 97, 13, 15, 17. What's the range for these data?

Solution $701 - 7 = 694$

SUMMARY

1. The *sample variance* of $x_1, x_2, \ldots, x_n$ is

$$s^2 = \frac{\sum (x - \bar{x})^2}{n - 1} = \frac{n \sum x^2 - (\sum x)^2}{n(n - 1)}$$

where $\bar{x}$ is the sample mean.

2. The *sample standard deviation* is $s = \sqrt{s^2}$.
3. The sample variance for *grouped discrete data*, or repeated measures, is

$$s^2 = \frac{\sum f(x - \bar{x})^2}{n - 1} = \frac{n \sum fx^2 - (\sum fx)^2}{n(n - 1)}$$

4. The sample variance for *grouped continuous data* is

$$s_u^2 = \frac{\sum f(u - \bar{u})^2}{n - 1} = \frac{n \sum fu^2 - (\sum fu)^2}{n(n - 1)}$$

where u is the class mark and $\bar{u}$ is the approximate mean for continuous data.

5. *Chebyshev's inequality* gives a bound for a set of measurements: The proportion of data that fall within k standard deviations ($k > 1$) is at least $1 - (1/k^2)$.
6. The *empirical rule* gives an approximate bound for a set of measurements from a normal population.
7. The *range* is the difference between the maximum and minimum observations.

RAISE YOUR GRADES

Can you . . . ?

☑ define the sample variance
☑ calculate the sample variance for n observations
☑ calculate the sample variance for grouped discrete data
☑ calculate the sample variance for grouped continuous data
☑ define the sample standard deviation
☑ calculate the sample standard deviation
☑ apply Chebyshev's inequality
☑ state the empirical rule
☑ find the range

RAPID REVIEW

1. The sample standard deviation is a measure of ____________.
2. If $x_1 = 9$, $x_2 = 10$, and $x_3 = 11$, then **(a)** $\bar{x} =$ ____________, **(b)** $s^2 =$ ____________, and **(c)** $s =$ ____________.
3. If $x_1 = 8$, $x_2 = 10$, and $x_3 = 12$, then **(a)** $\bar{x} =$ ____________, **(b)** $s^2 =$ ____________, and **(c)** $s =$ ____________.
4. According to Chebyshev's inequality, out of 900 observations at least ____________ lie within 3 standard deviations of the sample mean.
5. According to the empirical rule, out of 100 observations **(a)** ____________ observations lie within 1 standard deviation of the sample mean; **(b)** ____________ observations lie within 2 standard deviations of the sample mean.
6. If $x_1 = 89$, $x_2 = 73$, and $x_3 = 81$, then the range equals ____________.
7. The sum of the deviations from the sample mean in a set of measurements is always ____________.

Answers (1) spread (dispersion) (2) **(a)** 10 **(b)** 1 **(c)** 1 (3) **(a)** 10 **(b)** 4 **(c)** 2 (4) 800 (5) **(a)** ~68 **(b)** ~95 (6) 16 (7) 0

SOLVED PROBLEMS

PROBLEM 3-1 A fair, 6-sided die was rolled 4 times, and the following sample outcomes were observed: $x_1 = 1, x_2 = 6, x_3 = 3, x_4 = 2$. Find **(a)** the sample mean $\bar{x}$, **(b)** the sample variance, and **(c)** the sample standard deviation.

Solution

(a) $\bar{x} = \dfrac{\sum x}{n} = \dfrac{1 + 6 + 3 + 2}{4} = \dfrac{12}{4} = 3$

(b) By formula (3.1) the sample variance is

$$s^2 = \frac{\sum (x - \bar{x})^2}{n - 1}$$

$$= \frac{(1 - 3)^2 + (6 - 3)^2 + (3 - 3)^2 + (2 - 3)^2}{4 - 1} = \frac{4 + 9 + 0 + 1}{3} = \frac{14}{3} = 4.67$$

(c) By formula (3.2), the standard deviation is

$$s = \sqrt{s^2} = \sqrt{4.67} = 2.16$$

PROBLEM 3-2 A fair, 8-sided die was rolled 5 times, and the following sample outcomes were observed: $x_1 = 8, x_2 = 1, x_3 = 7, x_4 = 3, x_5 = 6$. Give a measure of the spread of these data, for which $\bar{x} = 25/5 = 5$.

Solution The sample variance and the sample standard deviation give the spread of a set of measurements. You can use the given mean (which gives a measure of the center of the data) to find the sample variance by formula (3.1):

$$s^2 = \frac{\sum (x - \bar{x})^2}{n - 1}$$

$$= \frac{(8-5)^2 + (1-5)^2 + (7-5)^2 + (3-5)^2 + (6-5)^2}{5 - 1}$$

$$= \frac{9 + 16 + 4 + 4 + 1}{4} = \frac{34}{4} = 8.5$$

Then you can find the sample standard deviation from the sample variance by formula (3.2):

$$s = \sqrt{s^2} = \sqrt{8.5} = 2.92$$

PROBLEM 3-3 A sample of 6 mathematically minded students who applied for admission to a college had the following ACT verbal scores: 24, 26, 17, 20, 23, 22. (**a**) Find the sample mean of these ACT scores; (**b**) find the sample variance of these ACT scores by two different methods.

Solution The easiest—and most organized—way to solve this problem is to set up a table of all the terms from the formulas for the sample mean and sample variance:

$$\bar{x} = \frac{\sum x}{n} \quad \text{and} \quad s^2 = \frac{\sum (x - \bar{x})^2}{n - 1} = \frac{n \sum x^2 - (\sum x)^2}{n(n-1)}$$

x	x^2	$(x - \bar{x})$	$(x - \bar{x})^2$
24	576	2	4
26	676	4	16
17	289	−5	25
20	400	−2	4
23	529	1	1
22	484	0	0
$\sum x = 132$	$\sum x^2 = 2954$	$\sum (x - \bar{x}) = 0$	$\sum (x - \bar{x})^2 = 50$

Now all you have to do is substitute:

(**a**) $\bar{x} = \frac{132}{6} = 22$

(**b**) Using formula (3.1),

$$s^2 = \frac{50}{6 - 1} = 10$$

Using the working formula (3.3),

$$s^2 = \frac{6(2954) - (132)^2}{6(6-1)}$$

$$= \frac{17{,}724 - 17{,}424}{30} = \frac{300}{30} = 10$$

PROBLEM 3-4 A sample of 8 deposits made at a community bank yielded the following observations, which have been ordered:

$$41.17 < 137.93 < 156.68 < 195.23 < 248.61 < 305.23 < 525.16 < 2366.79$$

Find (**a**) the sample mean, (**b**) the median, (**c**) the sample variance (by two methods), (**d**) the sample standard deviation, and (**e**) the range.

Solution Notice first that these data aren't grouped. Then set up your table, including only necessary terms:

x	x^2	$(x - \bar{x})^2$
41.17	1,694.97	207,872.16
137.93	19,024.68	129,003.09
156.68	24,548.62	115,885.78
195.23	38,114.75	91,125.50
248.61	61,806.93	61,747.28
305.23	93,165.35	36,814.10
525.16	275,793.03	787.36
2366.79	5,601,694.90	3,495,740.70
$\sum x = 3976.80$	$\sum x^2 = 6,115,843.23$	$\sum(x - \bar{x})^2 = 4,138,975.97$

(a) $\bar{x} = \dfrac{3976.80}{8} = 497.10$

(b) $m = \dfrac{195.23 + 248.61}{2} = 221.92$

(c) Using formula (3.1),

$$s^2 = \frac{4,138,975.97}{7} = 591,282.28$$

Using the working formula (3.3),

$$s^2 = \frac{8(6,115,843.23) - (3976.80)^2}{8(8 - 1)}$$

$$= \frac{33,111,807.60}{56} = 591,282.28$$

(d) $s = \sqrt{591,282.28} = 768.95$

(e) The range is the difference between the largest and the smallest observations, $2366.79 - 41.17 = 2325.62$.

PROBLEM 3-5 The penalty minutes accrued by a sample of 7 forward-line hockey players who played more than 70 games were 66, 43, 49, 83, 47, 57, 159. Find **(a)** the sample mean, **(b)** the sample variance (choose your own method), **(c)** the sample standard deviation, and **(d)** the range.

Solution

x	x^2	$(x - \bar{x})$	$(x - \bar{x})^2$
66	4,356	−6	36
43	1,849	−29	841
49	2,401	−23	529
83	6,889	11	121
47	2,209	−25	625
57	3,249	−15	225
159	25,281	87	7569
504	46,234	0	9946

(a) $\bar{x} = \dfrac{504}{7} = 72$

(b) Using formula (3.1),

$$s^2 = \frac{9946}{6} = 1657.67$$

Using the working formula (3.3),

$$s^2 = \frac{7(46,234) - (504)^2}{7(7 - 1)} = \frac{69,622}{42} = 1657.67$$

note: If you'd used this method in the first place, you could have avoided calculating the $(x - \bar{x})$ and $(x - \bar{x})^2$ columns in your table.

(c) $s = \sqrt{1657.67} = 40.71$

(d) range $= 159 - 43 = 116$

PROBLEM 3-6 In a survey of a sample of 40 households, it was found that 2 households had 0 cars, 9 households had 1 car, 20 households had 2 cars, 7 households had 3 cars, and 2 households had 4 cars. Find **(a)** the sample variance and **(b)** the sample standard deviation for the number of cars per household.

Solution Notice that you're dealing with grouped discrete data, which means that you can set up a table whose columns represent the terms in formula (3.4) or (3.5). Naturally, you choose (3.5), so you don't have to do any extra calculations.

x	f	fx	fx^2
0	2	0	0
1	9	9	9
2	20	40	80
3	7	21	63
4	2	8	32
	$n = 40$	$\sum fx = 78$	$\sum fx^2 = 184$

(a) $$s^2 = \frac{n\sum fx^2 - (\sum fx)^2}{n(n-1)} = \frac{40(184) - (78)^2}{40(40-1)}$$

$$= \frac{7360 - 6084}{1560} = \frac{1276}{1560} = .818$$

(b) $s = \sqrt{s^2} = \sqrt{.818} = .904$

PROBLEM 3-7 In a sample of 100 rolls of a pair of fair 4-sided dice, 7 rolls yielded 2, 13 rolls yielded 3, 18 yielded 4, 25 yielded 5, 15 yielded 6, 15 yielded 7, and 7 yielded 8. Find **(a)** the sample variance and **(b)** the sample standard deviation of these data.

Solution

x	f	fx	fx^2
2	7	14	28
3	13	39	117
4	18	72	288
5	25	125	625
6	15	90	540
7	15	105	735
8	7	56	448
	100	501	2781

(a) $s^2 = \dfrac{100(2781) - (501)^2}{100(100-1)} = \dfrac{27{,}099}{9900} = 2.737$ **(b)** $s = \sqrt{2.737} = 1.654$

PROBLEM 3-8 A statistically minded farmer weighed the milk taken from a sample of 50 cows. She recorded her data (in lb) and drew up a frequency distribution table, grouping the data into 7 classes having a class width of 9.

42	53	38	40	48	44	44	47	20	38
50	41	32	37	67	14	43	38	36	42
29	21	51	45	73	15	50	71	49	29
25	37	39	28	17	64	76	42	52	50
40	38	48	57	21	49	46	32	28	24

Boundaries	Limits	Frequency	Class mark
13.5–22.5	14–22	6	18
22.5–31.5	23–31	6	27
31.5–40.5	32–40	12	36
40.5–49.5	41–49	14	45
49.5–58.5	50–58	7	54
58.5–67.5	59–67	2	63
67.5–76.5	68–76	3	72
		50	

Find **(a)** the sample variance for the grouped data, **(b)** the sample standard deviation for the grouped data, **(c)** the proportion of the weights within 1 standard deviation of the sample mean, **(d)** the proportion of the weights within 2 standard deviations of the sample mean, and **(e)** the range of these weights.

Solution You're dealing with grouped continuous data, so you can use either formula (3.6) or (3.7) for the approximate sample variance s_u^2 to set up your table. Since you have many data here, (3.7) is preferable.

(a) $s_u^2 = \dfrac{n\sum fu^2 - (\sum fu)^2}{n(n-1)}$

$= \dfrac{50(94{,}122) - (2052)^2}{50(49)}$

$= \dfrac{495{,}396}{2450} = 202.20$

(b) $s_u = \sqrt{s_u^2} = \sqrt{202.20} = 14.22$

u	f	fu	fu^2
18	6	108	1,944
27	6	162	4,374
36	12	432	15,552
45	14	630	28,350
54	7	378	20,412
63	2	126	7,938
72	3	216	15,552
	$n = 50$	$\sum fu = 2052$	$\sum fu^2 = 94{,}122$

(c) To find the proportion of weights that fall within 1 standard deviation, you find the mean first:

$$\bar{u} = \frac{\sum fu}{n} = \frac{2052}{50} = 41.04$$

Then use the mean to find the interval from $\bar{u} - ks$ to $\bar{u} + ks$:

$$(\bar{u} - 1s, \bar{u} + 1s) = (41.04 - 14.22, 41.04 + 14.22) = (26.82, 55.26)$$

and count the number of observations (i.e., raw data) >26.8 and <55.3. There are 36 such observations, so $36/50 = 72\%$ of the weights fall within 1 standard deviation.

(d) The interval $(41.04 - 2(14.22),\ 41.04 + 2(14.22)) = (12.60,\ 69.48)$ contains 47 observations, so $47/50 = 94\%$ of the weights fall within 2 standard deviations.

(e) Range $= 76 - 14 = 62$

PROBLEM 3-9 Brand X liquid soap is packaged in 1000-g bottles. A sample of 25 bottles was taken from the production line and weighed. The following data give the excess weight *over* 1000 g for each of the 25 bottles:

60 31 19 43 47 48 47 13 28 35 51 24 24
27 33 57 38 24 39 40 63 51 25 43 26

Find **(a)** the sample variance for the ungrouped data, **(b)** the sample variance for the grouped data (use a frequency distribution table having 6 classes with a class width of 10, beginning with weights 10–19), and **(c)** the proportion of weights within 2 standard deviations of the sample mean.

Solution

(a) For the ungrouped data, $\sum x = 936$ and $\sum x^2 = 39{,}352$. Thus by (3.3),

$$s_x^2 = \frac{n\sum x^2 - (\sum x)^2}{n(n-1)} = \frac{25(39{,}352) - (936)^2}{25(25-1)} = \frac{107{,}704}{600} = 179.51$$

(b)

Class limits	u	f	fu	fu^2
10–19	14.5	2	29.0	420.50
20–29	24.5	7	171.5	4,201.75
30–39	34.5	5	172.5	5,951.25
40–49	44.5	6	267.0	11,881.50
50–59	54.5	3	163.5	8,910.75
60–69	64.5	2	129.0	8,320.50
		25	932.5	39,686.25

$$s_u^2 = \frac{n\sum fu^2 - (\sum fu)^2}{n(n-1)} = \frac{25(39{,}686.25) - (932.5)^2}{25(25-1)} = \frac{122{,}600}{600} = 204.33$$

(c) With $\bar{x} = 936/25 = 37.44$ and $s_x = \sqrt{s_x^2} = \sqrt{179.51} = 13.40$, you consider the interval

$$(\bar{x} - 2s_x, \bar{x} + 2s_x) = (37.44 - 2(13.40), 37.44 + 2(13.40)) = (10.64, 64.24)$$

This interval contains all 25 observations, or 100%. Notice that Chebyshev's inequality guarantees at least 75%, and the empirical rule claims approximately 95% of the observations will lie within this interval.

Supplementary Exercises

PROBLEM 3-10 A fair 12-sided die was rolled 6 times, yielding the following sample observations: 4, 1, 8, 11, 3, 12. Find **(a)** $\sum x$ and $\sum x^2$, **(b)** the sample variance and sample standard deviation, and **(c)** the range.

Answer **(a)** $\sum x = 39, \sum x^2 = 355$ **(b)** $s^2 = 20.3, s = 4.51$ **(c)** range = 11

PROBLEM 3-11 An obstetrician used ultrasound examinations on a sample of 10 patients between the 16th and 25th weeks of pregnancy to measure the length (in mm) of the femur of each fetus. She obtained the following measurements: 45, 49, 35, 44, 45, 40, 52, 57, 62, 39. Find **(a)** $\sum x$ and $\sum x^2$, **(b)** the sample mean, **(c)** the median, **(d)** the sample variance and the sample standard deviation, and **(e)** the range.

Answer **(a)** $\sum x = 468, \sum x^2 = 22{,}530$ **(b)** $\bar{x} = 46.8$ **(c)** $m = 45$ **(d)** $s^2 = 69.73, s = 8.35$
(e) range = 27

PROBLEM 3-12 Given the following sample of moorhen clutch sizes,

Clutch size:	5	6	7	8	9	10	11	12	13
Frequency:	3	4	20	23	30	13	3	1	3

find **(a)** the sample variance, **(b)** the sample standard deviation, and **(c)** the range.

Answer **(a)** $s^2 = 2.43$ **(b)** $s = 1.56$ **(c)** range = 8

PROBLEM 3-13 The mg of tar in a sample of 7 brands of cigarettes advertised in a weekly magazine are 9, 12, 14, 11, 2, 10, 1. Find **(a)** the sample mean, **(b)** the sample variance, **(c)** the sample standard deviation, and **(d)** the range.

Answer **(a)** $\bar{x} = 8.43$ **(b)** $s^2 = 24.95$ **(c)** $s = 4.995$ **(d)** range = 13

PROBLEM 3-14 The maximum lung capacity (in liters) for a sample of 25 females and 25 males was measured and recorded:

Females					Males				
4.2	5.3	3.5	4.3	3.7	6.8	5.7	5.6	4.8	3.7
3.2	3.5	2.8	3.5	3.7	5.0	3.8	5.6	5.4	5.2
3.5	3.7	2.7	2.8	3.3	4.9	4.7	7.6	4.0	5.6
2.7	3.0	3.1	3.0	3.7	4.2	4.0	5.5	5.1	4.9
3.3	3.4	2.3	2.6	3.2	5.4	5.9	3.4	6.1	4.1

Find **(a)** the exact sample variance, **(b)** the exact sample standard deviation, and **(c)** the range for the 25 females. Find **(d)** the exact sample variance, **(e)** the exact sample standard deviation, and **(f)** the range for the 25 males.

Answer **(a)** $s^2 = .395$ **(b)** $s = .628$ **(c)** range = 3 **(d)** $s^2 = .973$ **(e)** $s = .986$ **(f)** range = 4.2

PROBLEM 3-15 The following weights (in kg) of a sample of 50 male freshmen are grouped in a frequency distribution table:

76.0	64.6	80.6	84.8	78.9	84.3	72.0
68.3	72.2	63.7	92.6	66.4	72.6	69.4
62.6	74.7	60.4	72.0	77.2	59.0	
75.0	54.1	90.8	66.1	68.1	69.1	
79.7	65.3	81.3	56.9	71.6	74.9	
73.0	88.0	59.5	62.8	75.0	71.3	
74.4	70.8	77.8	71.9	63.8	70.3	
66.0	74.0	70.7	69.6	75.3	63.4	

$\bar{x} = 71.66$

Class mark (u)	Frequency (f)	fu
55.75	3	167.25
62.75	11	690.25
69.75	17	1185.75
76.75	12	921.00
83.75	4	335.00
90.75	3	272.25
	50	3571.50

$\bar{u} = 71.43$

Find **(a)** $\sum fu$ and $\sum fu^2$ for the grouped data, **(b)** the sample variance and standard deviation for the grouped data, **(c)** $\sum x$ and $\sum x^2$ for the ungrouped data, and **(d)** the sample variance and standard deviation for the ungrouped data.

Answer **(a)** $\sum fu = 3571.50, \sum fu^2 = 258{,}793.13$ **(b)** $s_u^2 = 75.12, s_u = 8.667$
(c) $\sum x = 3582.8, \sum x^2 = 260{,}121.56$ **(d)** $s^2 = 69.2335, s = 8.321$

PROBLEM 3-16 The amounts of resistance for the following sample of 100 resistors were grouped in a frequency distribution table:

2.17	2.18	2.17	2.19	2.23	2.18	2.16	2.23	2.29	2.20
2.24	2.13	2.18	2.22	2.21	2.22	2.23	2.18	2.23	2.25
2.24	2.20	2.21	2.20	2.25	2.15	2.20	2.25	2.16	2.19
2.20	2.18	2.18	2.19	2.22	2.19	2.20	2.19	2.22	2.21
2.22	2.18	2.18	2.28	2.19	2.23	2.21	2.19	2.21	2.21
2.19	2.18	2.21	2.17	2.20	2.18	2.18	2.21	2.22	2.18
2.22	2.23	2.19	2.23	2.18	2.19	2.18	2.21	2.18	2.15
2.20	2.23	2.20	2.20	2.23	2.20	2.19	2.22	2.17	2.20
2.20	2.17	2.19	2.19	2.25	2.19	2.19	2.20	2.20	2.19
2.21	2.14	2.24	2.21	2.19	2.23	2.18	2.22	2.23	2.19

Class mark (u)	Frequency (f)	fu
2.135	2	4.270
2.155	4	8.620
2.175	21	45.675
2.195	33	72.435
2.215	20	44.300
2.235	14	31.290
2.255	4	9.020
2.275	1	2.275
2.295	1	2.295
	100	220.180

$\bar{u} = 2.2018$

Find (**a**) the sample standard deviation for the grouped data, (**b**) the percentage of the measurements that lie within 1 standard deviation of the sample mean, and (**c**) the percentage of the measurements that lie within 2 standard deviations of the sample mean.

Answer (**a**) $s_u = .028$ (**b**) 69% (**c**) 96%

PROBLEM 3-17 The lung capacity (in l) for a sample of 20 volleyball team members is given below, along with an appropriate frequency distribution table:

3.4	3.7	3.6	3.4	3.9
3.4	3.3	3.7	3.4	3.6
4.1	3.5	4.1	3.7	3.6
4.3	3.7	4.2	3.8	4.7

$\bar{x} = 3.755$

Class mark (u)	Frequency (f)	fu
3.4	6	20.4
3.7	8	29.6
4.0	3	12.0
4.3	2	8.6
4.6	1	4.6
	20	75.2

$\bar{u} = 3.76$

Find (**a**) the sample variance and (**b**) the sample standard deviation for the grouped data. Find (**c**) the sample variance and (**d**) the sample standard deviation for the ungrouped data.

Answer (**a**) $s_u^2 = .1194$ (**b**) $s_u = .3455$ (**c**) $s^2 = .1321$ (**d**) $s = .3634$

PROBLEM 3-18 The following populations for a sample of 6 Asian countries were given (in thousands): 16,786; 676,218; 151,720; 118,449; 50,740; 2472. The sample mean of these populations is 169,397.5. Find the sample standard deviation.

Answer $s = 254{,}954.61$

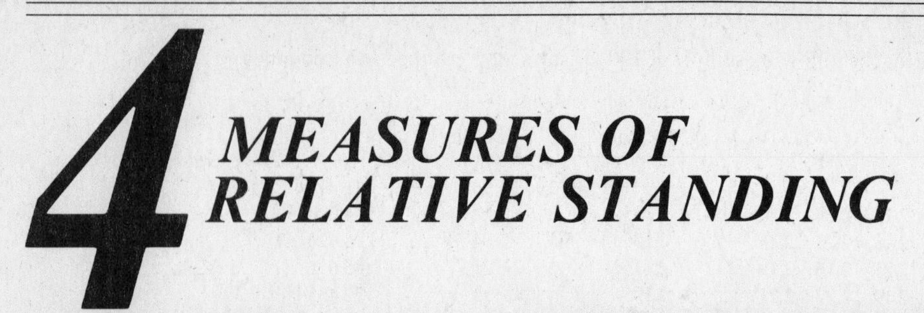

4 MEASURES OF RELATIVE STANDING

THIS CHAPTER IS ABOUT

☑ Quartiles, Percentiles, and Deciles
☑ z-Score
☑ Box-and-Whisker Diagrams

Measurements of relative standing give an indication of where an observation lies relative to the other observations in a data set.

4-1. Quartiles, Percentiles, and Deciles

Quartiles, percentiles, and deciles are points in an ordered set of data that divide the data into parts.

A. Quartiles

Given a set of n measurements, we can find three points, or **quartiles**, q_1, q_2, q_3, that divide the set into four quarters.

The **first quartile** q_1 is a point such that at least 25% of the measurements are at or below q_1 and approximately 75% of the measurements are above q_1.

The **second quartile** q_2 is a point corresponding to the median m, which divides the measurements into two equal parts.

The **third quartile** q_3 is a point such that at least 75% of the measurements are at or below q_3 and approximately 25% of the measurements are above q_3.

In order to calculate quartiles, we (1) order the data; (2) find the *location* of the quartile within the ordered set of data; (3) find the *value* of the quartile in terms of data values.

(1) Order the n observations from small to large, so that

$$x_1 \le x_2 \le \cdots \le x_j \le \cdots \le x_n$$

(i.e., let x_j be the jth observation in the ordered arrangement of x observations).

(2) Find the location of the quartile q_k by calculating $k(n+1)/4$; i.e., the location of the 1st quartile q_1 is $1(n+1)/4$, the location of q_2 is $2(n+1)/4$, and the location of q_3 is $3(n+1)/4$.

(3) Use the location of q_k to find the value of q_k:

(a) If $k(n+1)/4 = j$, where j is an integer, the value of q_k is equal to the value of x_j.

(b) If $k(n+1)/4 = j + f$, where j is an integer and $f = \frac{1}{4}, \frac{1}{2}, \frac{3}{4}$, the value of q_k is equal to $(x_j + x_{j+1})/2$.

This is easier done than said.

EXAMPLE 4-1 Seven rolls of a 12-sided die yielded the following outcomes: 3, 5, 8, 1, 5, 9, 11. What are the values of **(a)** the 1st quartile, **(b)** the 2nd quartile, and **(c)** the 3rd quartile?

Solution First, order the measurements:

$$x_1 = 1 \qquad x_2 = 3 \qquad x_3 = 5 \qquad x_4 = 5 \qquad x_5 = 8 \qquad x_6 = 9 \qquad x_7 = 11$$

(a) For the 1st quartile, $k = 1$, so $q_k = q_1$. Thus the location of the 1st quartile q_1 is $k(n + 1)/4 = 1(7 + 1)/4 = 2$. Since 2 is an integer, you know that q_1 must correspond to the 2nd outcome in the ordered list. Thus the value of the 1st quartile is $q_1 = x_2 = 3$.

(b) The location of the 2nd quartile q_2 is $2(7 + 1)/4 = 4$. Since 4 is an integer, the value of the 2nd quartile must correspond to the 4th measurement, which is the median: $q_2 = m = x_4 = 5$.

(c) The location of the 3rd quartile q_3 is $3(7 + 1)/4 = 24/4 = 6$. Thus the value of the 3rd quartile is $q_3 = x_6 = 9$.

---

note: The 2nd quartile q_2 always corresponds to the median m, so you can always find q_2 simply by finding m.

EXAMPLE 4-2 Nine ordered observations of the number of bets placed before losing \$5 in roulette were

$$x_1 = 7 \quad x_2 = 13 \quad x_3 = 15 \quad x_4 = 17 \quad x_5 = 29 \quad x_6 = 43 \quad x_7 = 95 \quad x_8 = 97 \quad x_9 = 701$$

Find the values of **(a)** q_1, **(b)** q_2, and **(c)** q_3.

Solution

(a) The location of q_1 is $1(9 + 1)/4 = 10/4 = 2.5 = 2 + 1/2$, which is between the 2nd and 3rd observations. Thus the value of the 1st quartile is

$$q_1 = \frac{x_2 + x_3}{2} = \frac{13 + 15}{2} = \frac{28}{2} = 14$$

(b) The location of $m = q_2$ is $2(9 + 1)/4 = 20/4 = 5$. Thus the median or 2nd quartile is $m = q_2 = x_5 = 29$.

(c) The location of q_3 is $3(9 + 1)/4 = 30/4 = 7.5 = 7 + 1/2$. Thus the value of the 3rd quartile is

$$q_3 = \frac{x_7 + x_8}{2} = \frac{95 + 97}{2} = \frac{192}{2} = 96$$

EXAMPLE 4-3 A pair of 6-sided dice was rolled 8 times, yielding the following (ordered) sums:

$$x_1 = 3 \quad x_2 = 5 \quad x_3 = 6 \quad x_4 = 6 \quad x_5 = 7 \quad x_6 = 7 \quad x_7 = 8 \quad x_8 = 10$$

Find the values of **(a)** q_1, **(b)** q_2, **(c)** q_3.

Solution

(a) The location of q_1 is $1(8 + 1)/4 = 2.25 = 2 + 1/4$. Thus the value of the 1st quartile is

$$q_1 = \frac{x_2 + x_3}{2} = \frac{5 + 6}{2} = 5.5$$

(b) The location of the median is $2(8 + 1)/4 = 18/4 = 4.5 = 4 + 1/2$. Thus the value of the median is

$$m = q_2 = \frac{x_4 + x_5}{2} = \frac{6 + 7}{2} = 6.5$$

(c) The location of q_3 is $3(8 + 1)/4 = 27/4 = 6.75 = 6 + 3/4$. Thus the value of the 3rd quartile is

$$q_3 = \frac{x_6 + x_7}{2} = \frac{7 + 8}{2} = 7.5$$

For large sets of data, the worst problem is ordering the data, so a stem-and-leaf diagram can be helpful in locating the quartiles.

EXAMPLE 4-4 The weights of 24 "three-pound" bags of apples were as follows:

3.26	3.62	3.39	3.12	3.53	3.30	3.10	3.26	3.19	3.22	3.14	3.39
3.31	3.21	3.49	3.41	3.02	3.17	3.20	3.12	3.42	3.36	3.21	3.26

Find the values of **(a)** the 1st quartile and **(b)** the 3rd quartile.

Solution Construct a stem-and-leaf diagram for the weights of the apples, ordering the stems and leaves as shown.

	Stems	Leaves
Row 1	3.0	2
Row 2	3.1	0 2 2 4 7 9
Row 3	3.2	0 1 1 2 6 6 6
Row 4	3.3	0 1 6 9 9
Row 5	3.4	1 2 9
Row 6	3.5	3
Row 7	3.6	2

(a) The location of q_1 is $1(24 + 1)/4 = 6.25$, so to find the value of the 1st quartile, you need the values of x_6 and x_7. Counting the leaves in the stem-and-leaf diagram, you see that the 6th and 7th leaves are 7 and 9. Thus $x_6 = 3.17$ and $x_7 = 3.19$. It follows that the value of the 1st quartile is

$$q_1 = \frac{x_6 + x_7}{2} = \frac{3.17 + 3.19}{2} = 3.18$$

(b) The location of q_3 is $3(24 + 1)/4 = 18.75$, so you need the values of x_{18} and x_{19}. Counting the leaves to find the 18th and 19th leaves, you see that $x_{18} = 9$ and $x_{19} = 9$. Since $x_{18} = x_{19} = 3.39$, it follows that the value of the 3rd quartile is

$$q_3 = \frac{3.39 + 3.39}{2} = 3.39$$

B. Percentiles and deciles

Percentiles: Given a set of n measurements, if k is an integer between 1 and 99 inclusive, the ***k*th percentile** is a point p_k such that at least $k\%$ of the measurements fall at or below this point and about $(100 - k)\%$ of the measurements fall above this point.

Deciles: When k is divisible by 10 (i.e., $k = 10, 20, 30, 40, 50, 60, 70, 80, 90$), the percentile p_k is called a **decile**. (For example, the 30th percentile p_{30} is the 3rd decile d_3.)

To find percentiles (and deciles) for small sets of data, we use a procedure similar to that for finding quartiles. First order the n observations:

$$x_1 \le x_2 \le \cdots \le x_j \le \cdots \le x_n$$

Then find the location of the kth percentile p_k by calculating $k(n + 1)/100$, where $k = 1, 2, \ldots, 99$.

- If $k(n + 1)/100 = j$, where j is an integer, then the value of the kth percentile is $p_k = x_j$.
- If $k(n + 1)/100 = j + f$, where j is an integer and f is a fraction between 0 and 1, then the value of the kth percentile is $p_k = (x_j + x_{j+1})/2$.

EXAMPLE 4-5 In a survey of textbook costs in the natural sciences, the cost per credit hour was calculated for 50 texts, yielding the following ordered data:

3.89	3.98	4.32	4.63	5.98	6.32	3.32	6.32	6.78	7.38
7.49	7.49	7.65	7.65	7.83	7.98	8.32	8.65	8.65	8.98
8.98	9.32	9.32	9.48	9.65	10.10	10.23	10.48	10.57	10.65
10.78	10.87	11.17	11.32	11.32	11.73	11.98	12.03	12.12	12.14
12.30	12.32	12.97	13.32	13.48	14.22	15.48	16.63	17.48	32.95

What are the values of **(a)** the 58th percentile, **(b)** the 32nd percentile, **(c)** the 80th percentile, and **(d)** the 8th decile?

Solution

(a) The location of the 58th percentile is $k(n + 1)/100 = 58(50 + 1)/100 = 29.58$. Thus the value of the 58th percentile is

$$p_{58} = \frac{x_{29} + x_{30}}{2} = \frac{10.57 + 10.65}{2} = \frac{21.22}{2} = 10.61$$

(b) The location of the 32nd percentile is $32(50 + 1)/100 = 16.32$. Thus the 32nd percentile is

$$p_{32} = \frac{x_{16} + x_{17}}{2} = \frac{7.98 + 8.32}{2} = \frac{16.30}{2} = 8.15$$

(c) The location of the 80th percentile is $80(50 + 1)/100 = 40.8$. Thus the 80th percentile is

$$p_{80} = \frac{x_{40} + x_{41}}{2} = \frac{12.14 + 12.30}{2} = \frac{24.44}{2} = 12.22$$

(d) The 8th decile is equal to the 80th percentile. That is, $d_8 = p_{80} = 12.22$.

C. Calculating percentiles for large sets of continuous data

Percentiles, deciles, and quartiles are most meaningful for very large data sets; for example, SAT and ACT national exam scores are divided into percentiles, so that the relative standing of each score is clear.

To calculate percentiles for large sets of continuous data, we may use the cumulative relative frequency ogive curve (Section 1-4B). To do this, we have to realize that the cumulative relative frequency is in fact a percentage, which gives us the location of the percentile. So, first we locate the desired percentile on the vertical (cumulative relative frequency) axis, and draw a horizontal line until it intersects the ogive curve. Then, we can find the value of the percentile by drawing a vertical line from that point of intersection down to the horizontal (class boundaries) axis. The point at which the vertical line intersects the horizontal axis is the desired point, i.e., the number at or below which the desired percentage of measurements will fall.

note: This procedure works for quartiles and deciles too. Simply change quartiles and deciles to percentiles, and go on from there.

EXAMPLE 4-6 The ogive curve for the heights (in cm) of 50 women is given in Fig. 4-1a. Find **(a)** the 20th percentile, **(b)** the 2nd decile, **(c)** the 62nd percentile, **(d)** the 75th percentile, and **(e)** the 3rd quartile.

Solution (see Fig. 4-1b)

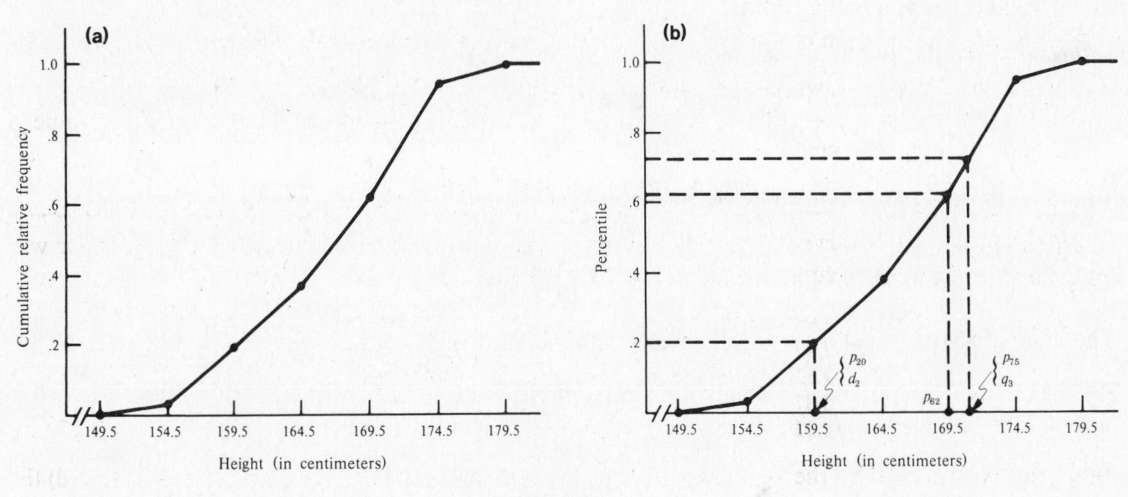

Figure 4-1

(a) Draw a horizontal line on the graph, beginning at .20 on the vertical axis and ending at the intersection with the ogive curve. Then draw a vertical line from that intersection to the horizontal axis. This line crosses the horizontal axis at 159.5, so $p_{20} = 159.5$.

(b) The 2nd decile is equal to the 20th percentile, so $d_2 = 159.5$.

(c) Draw a horizontal line from .62 on the vertical axis to its intersection with the ogive curve. Draw a vertical line from this intersection down to the horizontal axis at 169.5. Thus $p_{62} = 169.5$.

(d) Draw a horizontal line from .75 to its intersection with the ogive curve, and a vertical line from this intersection to the horizontal axis at approximately 171. Thus $p_{75} = 171$.

(e) The 3rd quartile equals the 75th percentile, so $q_3 = p_{75} = 171$.

4-2. z-Score

The **z-score** gives the relative position of an observation x with respect to the sample mean $\bar{x}$ in terms of the number of standard deviations s. Thus, if $\bar{x}$ is the sample mean and s is the sample standard deviation of a set of observations $x_1, x_2, \ldots, x_n$, the z-score of any observation x is

z-SCORE

$$z = \frac{x - \bar{x}}{s} \qquad \textbf{(4.1)}$$

A positive z-score indicates that an observation x is above the sample mean, and a negative z-score indicates that an observation x is below the sample mean.

EXAMPLE 4-7 For a statistics test, the sample mean was 79.9 and the sample standard deviation was 12.8. What would your z-score be if you made **(a)** a 94 or **(b)** a 64 on the test?

Solution

(a) The z-score associated with $x = 94$ is

$$z = \frac{x - \bar{x}}{s} = \frac{94 - 79.9}{12.8} = \frac{14.1}{12.8} = 1.10$$

Thus $x = 94$ is 1.10 standard deviations above the mean. (Good job!)

(b) The z-score associated with $x = 64$ is

$$z = \frac{64 - 79.9}{12.8} = -\frac{15.9}{12.8} = -1.24$$

Thus $x = 64$ is 1.24 standard deviations below the mean. (*Not* so good: Read this Outline *twice*.)

EXAMPLE 4-8 Suppose you've just taken two statistics tests. On Test I, $\bar{x} = 75$ and $s = 15$, while on Test II, $\bar{x} = 65$ and $s = 10$. On Test I you score a 90 and on Test II you score an 85. Which test score would you keep if you had a choice?

Solution Find the z-score associated with $x = 90$ on Test I:

$$z = \frac{90 - 75}{15} = 1$$

Then find the z-score associated with $x = 85$ on Test II:

$$z = \frac{85 - 65}{10} = 2$$

The score of 85 on Test II is better, relatively, than the score of 90 on Test I, so you should keep the 85.

EXAMPLE 4-9 For the apple weight data in Example 4-4, $\bar{x} = 3.28$ and $s = .14$. Find the z-scores for **(a)** $x = 3.02$ and **(b)** $x = 3.62$.

Solution

(a) The z-score for $x = 3.02$ is

$$z = \frac{3.02 - 3.28}{.14} = \frac{-.26}{.14} = -1.86$$

(b) The z-score for $x = 3.62$ is

$$z = \frac{3.62 - 3.28}{.14} = \frac{.34}{.14} = 2.43$$

If you know the z-score, $\bar{x}$, and s, you can find the corresponding value of x by rearranging formula (4.1):

$$x = \bar{x} + zs \qquad \textbf{(4.2)}$$

EXAMPLE 4-10 If $\bar{x} = 72$, $s = 14$, and your z-score is $z = 1.5$ on a particular test, what test score did you receive?

Solution $x = \bar{x} + zs$

$$= 72 + (1.5)(14) = 72 + 21 = 93$$

4-3. Box-and-Whisker Diagrams

A **box-and-whisker diagram** is a visual representation of a set of data that depicts the maximum, the minimum, and the three quartiles of the data set. To draw a box-and-whisker diagram, first order a set of data $x_1 \leq x_2 \leq \cdots \leq x_n$, so that x_1 is the minimum and x_n is the maximum. Then calculate the values of the 1st quartile q_1, the median $m = q_2$, and the 3rd quartile q_3. Now draw a horizontal axis scaled to the data. Above this axis draw a horizontal line—the *left whisker*—from x_1 to q_1, and another horizontal line—the *right whisker*—from q_3 to x_n. Finally, draw a rectangle—a *box*—with ends through quartiles q_1 and q_3 and a vertical line (through the box) at the median m.

EXAMPLE 4-11 In Example 4-1, you ordered the 7 rolls of a 12-sided die as

$$x_1 = 1 \quad x_2 = 3 \quad x_3 = 5 \quad x_4 = 5 \quad x_5 = 8 \quad x_6 = 9 \quad x_7 = 11$$

and you found that $q_1 = 3$, $m = 5$, $q_3 = 9$. Construct a box-and-whisker diagram for these data.

Solution It's clear that the minimum is $x_1 = 1$ and the maximum is $x_7 = 11$, so you choose a horizontal axis that slightly exceeds these values at both ends. Draw the left whisker above the axis from $x_1 = 1$ to $q_1 = 3$, and the right whisker from $q_3 = 9$ to $x_7 = 11$. Then draw a box with ends at $q_1 = 3$ and $q_3 = 9$ and drop a vertical line through the box at $m = 5$. And so you have the box-and-whisker diagram shown in Fig. 4-2.

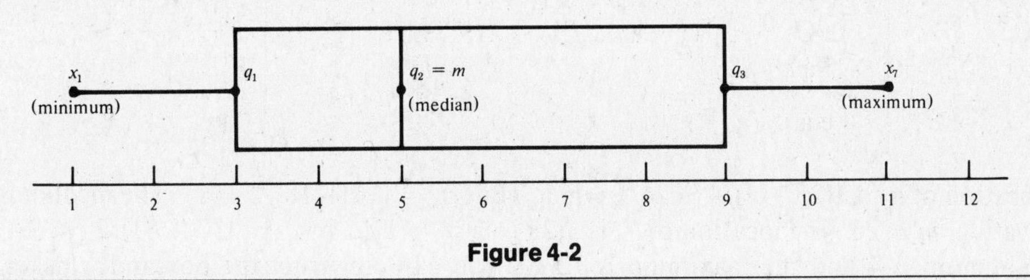

Figure 4-2

EXAMPLE 4-12 In Example 4-2 there were 9 observations of the number of bets placed before losing $5 in roulette:

$$7 \quad 13 \quad 15 \quad 17 \quad 29 \quad 43 \quad 95 \quad 97 \quad 701$$

and $q_1 = 14$, $m = 29$, and $q_3 = 96$. Construct a box-and-whisker diagram for these data.

Solution The minimum is $x_1 = 7$ and the maximum is $x_9 = 701$, so you obtain the box-and-whisker diagram shown in Fig. 4-3.

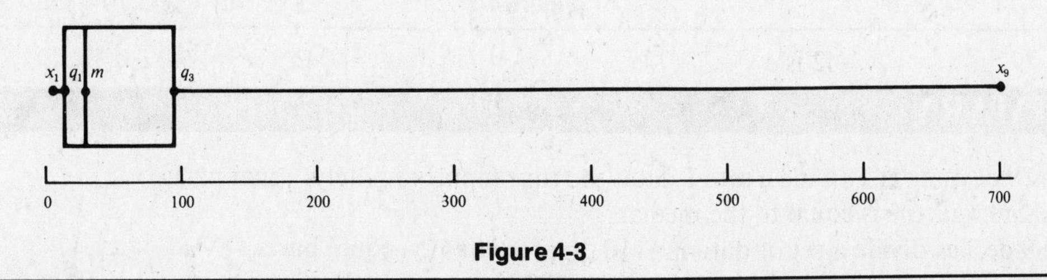

Figure 4-3

note: By comparing the box-and-whisker diagrams in Figs. 4-2 and 4-3, you can see that the data in Example 4-11 are relatively symmetric, while the data in Example 4-12 are asymmetric. The asymmetric or *skewed* data are indicated by the long right whisker and the placement of the vertical line in the box in Fig. 4-3.

EXAMPLE 4-13 Construct a box-and-whisker diagram for the data in Example 4-3.

Solution The minimum is 3, the maximum is 10, and $q_1 = 5.5, m = 6.5, q_3 = 7.5$. So you can construct the box-and-whisker diagram shown in Fig. 4-4.

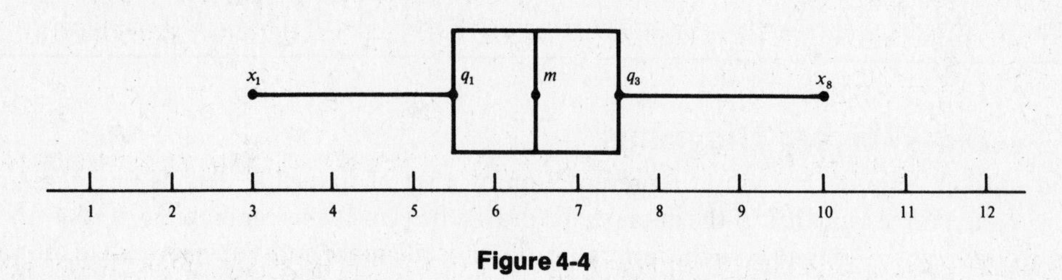

Figure 4-4

EXAMPLE 4-14 The weights (in g) of 25 fish of a certain species are

23	15	44	37	16	18	26	31	49	11	14	23	
29	61	36	26	28	41	72	4	6	20	30	21	52

(a) Use a stem-and-leaf diagram to order these data. **(b)** Find q_1, m, and q_3. **(c)** Draw a box-and-whisker diagram for the weights.

Solution

(a) Using integer stems, you have the following:

Stems	Leaves
0	4 6
1	5 6 8 1 4
2	3 6 3 9 6 8 0 1
3	7 1 6 0
4	4 9 1
5	2
6	1
7	2

Ordering the leaves, you have the following:

Stems	Leaves
0	4 6
1	1 4 5 6 8
2	0 1 3 3 6 6 8 9
3	0 1 6 7
4	1 4 9
5	2
6	1
7	2

(b) The location of q_1 is $1(25 + 1)/4 = 26/4 = 6.5$. Thus $q_1 = (16 + 18)/2 = 17$. The median is the 13th observation, $m = 26$. The location of q_3 is $3(25 + 1)/4 = 19.5$, so $q_3 = (37 + 41)/2 = 39$.

(c) The minimum is 4 and the maximum is 72. So you can construct the box-and-whisker diagram shown in Fig. 4-5.

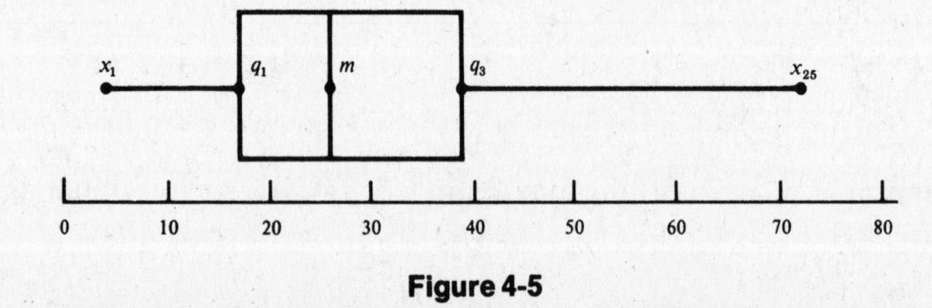

Figure 4-5

SUMMARY

1. The three quartiles divide a set of data into four (approximately) equal parts.
2. The 2nd quartile is equal to the median.
3. The 9 deciles divide a set of data into 10 (approximately) equal parts.
4. The 99 percentiles divide a set of data into 100 (approximately) equal parts.
5. The z-score gives the position of an observation of x relative to the sample mean $\bar{x}$ in terms of the number of standard deviations s: $z = (x - \bar{x})/s$.
6. A box-and-whisker diagram depicts data in terms of the three quartiles, the minimum, and the maximum.

RAISE YOUR GRADES

Can you . . . ?

☑ define and calculate values of quartiles
☑ define and calculate values of deciles
☑ define and calculate values of percentiles
☑ define and calculate a z-score
☑ draw a box-and-whisker diagram

RAPID REVIEW

1. The number of quartiles is ______________.
2. At least ______________% of a set of measurements falls at or below the 1st quartile.
3. Approximately 25% of a set of measurements falls at or above the ______________ quartile.
4. If $x_1 = 2$, $x_2 = 5$, $x_3 = 7$, $x_4 = 10$, $x_5 = 13$, $x_6 = 18$, $x_7 = 19$, then $q_1 =$ **(a)** ______________, $m =$ **(b)** ______________, $q_3 =$ **(c)** ______________.
5. The 60th percentile is equal to the ______________ decile.
6. The 4th decile is equal to the ______________ percentile.
7. The 3rd quartile is equal to the ______________ percentile.
8. The median is equal to the **(a)** ______________ quartile, **(b)** ______________ decile, and **(c)** ______________ percentile.
9. If $\bar{x} = 50$ and $s = 10$, then the z-score associated with $x = 63$ is $z =$ ______________.
10. If $\bar{x} = 50$, $s = 10$, and the z-score of x is 2, then $x =$ ______________.
11. The box-and-whisker diagram uses the three quartiles, the **(a)** ______________, and the **(b)** ______________.

Answers (1) 3 (2) 25% (3) 3rd (4) (a) 5 (b) 10 (c) 18 (5) 6th (6) 40th (7) 75th (8) (a) 2nd (b) 5th (c) 50th (9) 1.3 (10) 70 (11) (a) minimum (b) maximum

SOLVED PROBLEMS

Math Review

PROBLEM 4-1 If $k = 3$, $n = 79$, and $j = k(n + 1)/4$, solve for j.

Solution Substitute in the values for k and n and evaluate:

$$j = \frac{k(n+1)}{4}$$

$$= \frac{3(79+1)}{4} = \frac{3(80)}{4} = \frac{240}{4} = 60$$

PROBLEM 4-2 If $k = 86$, $n = 249$, and $j = k(n + 1)/100$, solve for j.

Solution

$$j = \frac{k(n+1)}{100}$$

$$= \frac{86(249+1)}{100} = \frac{86(250)}{100} = \frac{21{,}500}{100} = 215$$

PROBLEM 4-3 If $z = (x - \bar{x})/s$, $\bar{x} = 80$, $s = 13.6$, and $z = 1.25$, solve for x.

Solution

$$z = \frac{x - \bar{x}}{s}$$

$$1.25 = \frac{(x - 80)}{13.6}$$

$$1.25(13.6) = x - 80$$

$$17 + 80 = x$$

$$x = 97$$

PROBLEM 4-4 If $z = (x - \bar{x})/s$, solve for x in terms of z, $\bar{x}$, and s.

Solution Rearrange:

$$z = \frac{x - \bar{x}}{s}$$

$$zs = x - \bar{x}$$

$$zs + \bar{x} = x$$

that is,

$$x = \bar{x} + zs$$

Quartiles, Percentiles, and Deciles

PROBLEM 4-5 A farmer weighed the milk produced in a day by each of 7 cows and obtained the following weights in pounds:

61 27 58 28 48 30 78

(a) Order these weights from low to high. Find **(b)** the 1st quartile, **(c)** the median, and **(d)** the 3rd quartile.

Solution

(a) The ordered weights are

$$x_1 = 27 \quad x_2 = 28 \quad x_3 = 30 \quad x_4 = 48 \quad x_5 = 58 \quad x_6 = 61 \quad x_7 = 78$$

(b) The position of the 1st quartile $q_k = q_1$ is $k(n + 1)/4 = 1(7 + 1)/4 = 2$. The value of the 1st quartile is $q_1 = x_2 = 28$.

(c) The position of the median $m = q_2$ is $2(7 + 1)/4 = 16/4 = 4$. The value of the median is $m = x_4 = 48$.

(d) The position of the 3rd quartile q_3 is $3(7 + 1)/4 = 24/4 = 6$. The value of the 3rd quartile is $q_3 = x_6 = 61$.

PROBLEM 4-6 Eleven students in an English class were weighed, and the following weights (in lb) were recorded: 138, 131, 112, 101, 166, 161, 147, 143, 194, 186, 176. Find **(a)** the 1st quartile, **(b)** the median, and **(c)** the 3rd quartile.

Solution The ordered weights were

101 112 131 138 143 147 161 166 176 186 194

(a) The position of the 1st quartile is $1(11 + 1)/4 = 3$. The value of the 1st quartile is $q_1 = x_3 = 131$, where x_3 is the 3rd weight among the ordered weights.

(b) The position of the median is $2(11 + 1)/4 = 6$. The value of the median is $m = x_6 = 147$.

(c) The position of the 3rd quartile is $3(11 + 1)/4 = 9$. The value of the 3rd quartile is $q_3 = x_9 = 176$.

PROBLEM 4-7 Flexibility of 15 inches is equivalent to touching your toes without bending your knees. As an instructor of the Jacques Le Limb Health, Fitness, and First Aid Society, you're asked to measure the flexibility in inches of 23 freshman males. You record the following observations:

12.0	18.5	21.5	15.5	19.2	22.0	15.5	20.0	22.0	16.0	20.0	22.5
16.5	20.2	23.0	17.0	21.0	23.2	18.0	21.2	23.5	18.5	21.3	

Find **(a)** the 1st quartile, **(b)** the median, and **(c)** the 3rd quartile.

Solution To order the data you might set up a stem-and-leaf diagram, as shown.

Stems	Leaves
12	0
13	
14	
15	5 5
16	0 5
17	0
18	0 5 5
19	2
20	0 0 2
21	0 2 3 5
22	0 0 5
23	0 2 5

(a) The position of q_1 is $1(23 + 1)/4 = 6$, so $q_1 = x_6 = 17.0$, where x_6 is the 6th observation in the ordered measurements.
(b) The position of $m = q_2$ is $2(23 + 1)/4 = 12$, so $m = q_2 = x_{12} = 20.0$.
(c) The position of q_3 is $3(23 + 1)/4 = 18$, so $q_3 = x_{18} = 22.0$.

PROBLEM 4-8 The following ordered data represent the number of inches traveled by the back wheels of four bicycles for one revolution of the pedals in different gear combinations:

33.4	34.4	35.4	36.6	37.8	39.1	40.5	42.0	42.5
43.6	43.9	44.5	45.3	46.8	47.2	48.4	49.3	50.1
51.5	52.0	54.0	56.2	56.7	58.5	59.6	61.0	63.0
63.8	66.7	66.9	70.2	70.8	73.9	75.6	78.0	81.0
82.6	87.2	87.8	93.6	100.3	108.0			

Find **(a)** the 1st quartile, **(b)** the 3rd quartile, **(c)** the 45th percentile, and **(d)** the 6th decile.

Solution

(a) The position of q_1 is $1(42 + 1)/4 = 10.75$, which is between the 10th and the 11th measurements. Thus the 1st quartile is

$$q_1 = \frac{x_{10} + x_{11}}{2} = \frac{43.6 + 43.9}{2} = \frac{87.5}{2} = 43.75$$

(b) The position of q_3 is $3(42 + 1)/4 = 32.25$. Thus the 3rd quartile is

$$q_3 = \frac{x_{32} + x_{33}}{2} = \frac{70.8 + 73.9}{2} = \frac{144.7}{2} = 72.35$$

(c) The position of the 45th percentile p_{45} is $45(42 + 1)/100 = 19.35$. Thus the 45th percentile is

$$p_{45} = \frac{x_{19} + x_{20}}{2} = \frac{51.5 + 52.0}{2} = \frac{103.5}{2} = 51.75$$

(d) The position of the 6th decile d_6 is $6(42 + 1)/10 = 25.8$. Thus the 6th decile is

$$d_6 = \frac{x_{25} + x_{26}}{2} = \frac{59.6 + 61.0}{2} = \frac{120.6}{2} = 60.3$$

PROBLEM 4-9 You are the manager of Benny's Savings and Loan, and you want to know what age groups your depositors fall into. So you take a survey. The 200 participants in your survey have the following age distribution:

Class boundaries	Age	Frequency	Cumulative frequency	Cumulative relative frequency
9.5–19.5	10–19	10	10	.05
19.5–29.5	20–29	30	40	.20
29.5–39.5	30–39	54	94	.47
39.5–49.5	40–49	50	144	.72
49.5–59.5	50–59	24	168	.84
59.5–69.5	60–69	18	186	.93
69.5–79.5	70–79	12	198	.99
79.5–89.5	80–89	2	200	1.00
		200		

Find **(a)** the median, **(b)** the 2nd decile, **(c)** the 78th percentile, **(d)** the 84th percentile, and **(e)** the 96th percentile.

Solution Draw the relative frequency ogive curve, as in Fig. 4-6. Locate on the vertical axis the points .50, .20, .78, .84, and .96. From these points draw horizontal lines until they intersect the ogive curve. From these points of intersection draw vertical lines until they intersect the horizontal axis. The answers, which you read from the horizontal axis, are **(a)** $m = 40.7$, **(b)** $d_2 = 29.5$, **(c)** $p_{78} = 54.5$, **(d)** $p_{84} = 59.5$, and **(e)** $p_{96} = 74.5$.

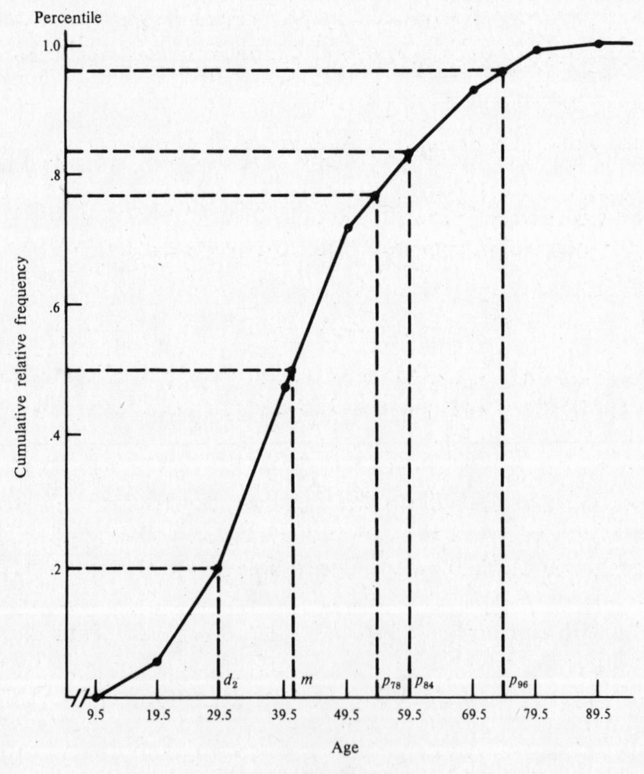

Figure 4-6

z-Score

PROBLEM 4-10 The heights of 50 women were measured in centimeters. If $\bar{x} = 166.12$ and $s = 6.26$, what are the z-scores for the heights **(a)** 175 and **(b)** 157?

Solution

(a) The z-score for $x = 175$ is

$$z = \frac{x - \bar{x}}{s} = \frac{175 - 166.12}{6.26} = 1.42$$

(b) The z-score for $x = 157$ is

$$z = \frac{157 - 166.12}{6.26} = -1.46$$

PROBLEM 4-11 In a random sample of 50 textbooks, the sample mean of the costs was 25.93 and the sample standard deviation was 10.33. Find the z-scores for costs of **(a)** 36.85 and **(b)** 11.95.

Solution

(a) For a cost of 36.85,

$$z = \frac{x - \bar{x}}{s} = \frac{36.85 - 25.93}{10.33} = 1.06$$

(b) For a cost of 11.95,

$$z = \frac{x - \bar{x}}{s} = \frac{11.95 - 25.93}{10.33} = -1.35$$

PROBLEM 4-12 The percentage of body fat for 219 freshman males who joined the Jacques Le Limb Society was measured, yielding a sample mean of 12.87 and a sample standard deviation of 4.64. Find the z-scores for **(a)** $x = 16.2$ and **(b)** $x = 5.2$.

Solution

(a) The z-score for $x = 16.2$ is

$$z = \frac{16.2 - 12.87}{4.64} = .72$$

(b) The z-score for $x = 5.2$ is

$$z = \frac{5.2 - 12.87}{4.64} = -1.65$$

PROBLEM 4-13 You're told that the average cost of a textbook is $\bar{x} = 25.93$ with a standard deviation of $s = 10.33$. If the z-score of the cost of your book is $z = -1.45$, what's the cost of your book?

Solution The cost x of your book is

$$x = \bar{x} + zs = 25.93 + (-1.45)(10.33) = 25.93 - 14.98 = 10.95$$

PROBLEM 4-14 A fast-food restaurant with an interest in the "stay slim" market listed 10 choices from its menu, giving the following numbers of calories per serving in hundreds (e.g., 33 is 330 calories, 56 is 560 calories, etc.):

33 37 42 51 54 56 57 61 67 72

For these data, **(a)** calculate $\bar{x}$ and s, **(b)** find the z-scores for $x = 61$ and $x = 42$, and **(c)** find the number of calories for a menu item if the z-score is $z = -1.27$ and if $z = 1.51$.

Solution

(a) To find the mean and standard deviation, you first construct a table. (Of course, you have to find the sample mean before you can calculate the second and third columns.) The sample mean is

$$\bar{x} = \frac{530}{10} = 53$$

The sample variance is $s^2 = 1428/9 = 158.67$, so the standard deviation is

$$s = \sqrt{158.67} = 12.60$$

x	$x - \bar{x}$	$(x - \bar{x})^2$
33	−20	400
37	−16	256
42	−11	121
51	−2	4
54	1	1
56	3	9
57	4	16
61	8	64
67	14	196
72	19	361
530	0	1428

(b) The z-score for $x = 61$ is

$$z = \frac{61 - 53}{12.60} = .63$$

and the z-score for $x = 42$ is

$$z = \frac{42 - 53}{12.60} = -.87$$

(c) If the z-score equals -1.27, then $x = 53 + (-1.27)(12.60) = 37.00$ or 370 calories. If the z-score equals 1.51, then $x = 53 + (1.51)(12.60) = 72.03$ or 720 calories. (You may decide which item to select!)

Box-and-Whisker Diagrams

PROBLEM 4-15 Draw a box-and-whisker diagram for the flexibility measurements that are given in Problem 4-7.

Solution You need 5 characteristics to construct a box-and-whisker diagram: q_1, m, q_3, the maximum, and the minimum. You calculated $q_1 = 17.0$, $m = 20.0$, and $q_3 = 22.0$ in Problem 4-7. You find the maximum and minimum from the ordered data: The minimum is $x_1 = 12.0$, and the maximum is $x_{23} = 23.5$. The box-and-whisker diagram resulting from these characteristics is shown in Fig. 4-7.

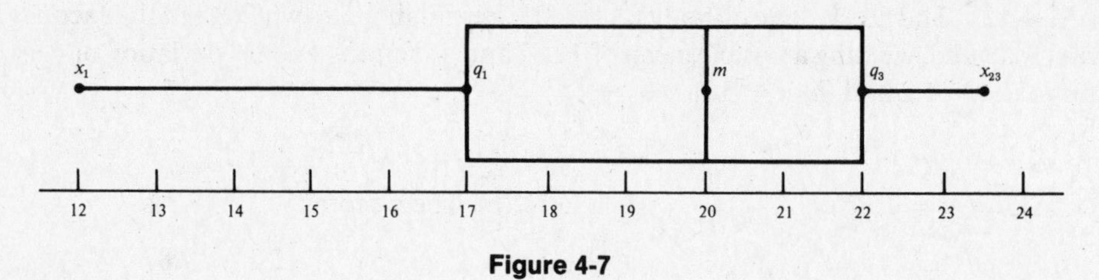

Figure 4-7

PROBLEM 4-16 Draw a box-and-whisker diagram using the bicycle measurements in Problem 4-8.

Solution The minimum is $x_1 = 33.4$, $q_1 = 43.75$, $m = 55.1$, $q_3 = 72.35$, and the maximum is $x_{42} = 108.0$. The box-and-whisker diagram is shown in Fig. 4-8.

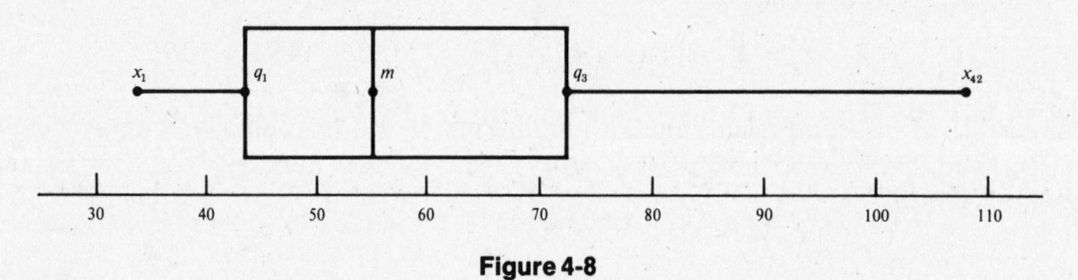

Figure 4-8

PROBLEM 4-17 The following data give the infant mortality (death within the first year) per 1000 infants for 50 countries from Africa, North America, and Europe:

125.3—Algeria	126.6—Burundi	154.3—Chad
134.5—Congo	150.0—Ethiopia	107.3—Ghana
92.0—Kenya	120.9—Mali	114.4—Morocco
140.5—Nigeria	127.0—Rwanda	152.6—Senegal
100.6—South Africa	131.1—Sudan	114.7—Togo
106.5—Tunisia	100.5—Uganda	116.6—Zaire
140.0—Zanzibar	78.8—Zimbabwe	22.4—Bahamas
9.6—Canada	18.0—Costa Rica	17.3—Cuba
73.1—Dominican Republic	44.0—El Salvador	32.4—Greenland
15.4—Grenada	16.2—Jamaica	59.8—Mexico
96.5—Nicaragua	36.2—Panama	18.6—Puerto Rico
11.2—United States	22.5—Virgin Islands	12.8—Austria
11.7—Belgium	9.3—France	14.3—Greece
10.6—Ireland	12.7—Italy	8.1—Netherlands
7.5—Norway	26.0—Portugal	10.3—Spain
6.8—Sweden	7.6—Switzerland	11.4—Scotland
13.5—Northern Ireland	33.5—Yugoslavia	

Draw a box-and-whisker diagram for these data.

Solution First, order the data for the 50 countries.

6.8	7.5	7.6	8.1	9.3	9.6	10.3	10.6	11.2	11.4
11.7	12.7	12.8	13.5	14.3	15.4	16.2	17.3	18.0	18.6
22.4	22.5	26.0	32.4	33.5	36.2	44.0	59.8	73.1	78.8
92.0	96.5	100.5	100.6	106.5	107.3	114.4	114.7	116.6	120.9
125.3	126.6	127.0	131.1	134.5	140.0	140.5	150.0	152.6	154.3

The ordered data give you $x_1 = 6.8$ and $x_{50} = 154.3$ for the minimum and maximum. But you have to calculate the three quartiles. The position of q_1 is $1(50 + 1)/4 = 12.75$, so

$$q_1 = \frac{x_{12} + x_{13}}{2} = \frac{12.7 + 12.8}{2} = 12.75$$

The 2nd quartile q_2 is the median m, so

$$q_2 = m = \frac{x_{25} + x_{26}}{2} = \frac{33.5 + 36.2}{2} = 34.85$$

The position of q_3 is $3(50 + 1)/4 = 38.25$, so

$$q_3 = \frac{x_{38} + x_{39}}{2} = \frac{114.7 + 116.6}{2} = 115.65$$

The box-and-whisker diagram is shown in Fig. 4-9.

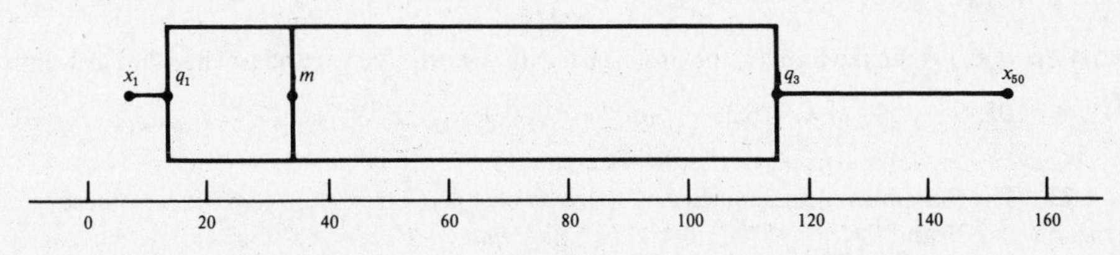

Figure 4-9

Supplementary Exercises

PROBLEM 4-18 The weights of soap in fifteen "1000-g" bottles were

1019 1013 1024 1024 1025 1043 1028 1027
1039 1043 1047 1035 1033 1040 1026

Find **(a)** the 1st quartile, **(b)** the median, and **(c)** the 3rd quartile.

Answer **(a)** $q_1 = 1024$ **(b)** $m = 1028$ **(c)** $q_3 = 1040$

PROBLEM 4-19 A random sample of 20 grocery orders at a food store revealed the following costs per order:

23.71 32.39 40.17 75.58 34.65 5.13 8.22 13.77 15.36 11.81
4.67 32.15 8.02 52.50 52.03 58.72 37.26 53.23 11.07 18.54

Find **(a)** the 1st quartile, **(b)** the median, and **(c)** the 3rd quartile.

Answer **(a)** $q_1 = 11.44$ **(b)** $m = 28.05$ **(c)** $q_3 = 46.10$

PROBLEM 4-20 An obstetrician used ultrasound examinations on 10 patients to measure the widest diameter (in mm) of the fetal head and recorded the following measurements:

44 68 57 48 68 66 47 53 60 51

Find **(a)** the 1st quartile, **(b)** the median, and **(c)** the 3rd quartile.

Answer **(a)** $q_1 = 47.5$ **(b)** $m = 55$ **(c)** $q_3 = 67$

PROBLEM 4-21 The lengths of time (in minutes) required for each of 73 women to complete a 25-km race were

92.30	96.65	100.36	105.44	107.71	108.02	110.07
112.03	112.15	117.19	118.07	118.08	118.32	119.36
119.91	121.64	123.12	123.58	126.16	126.30	128.23
128.45	128.95	129.77	129.96	130.47	130.69	131.57
132.26	134.69	136.22	136.87	137.40	138.03	138.32
139.18	140.88	141.15	141.41	142.32	143.39	144.08
144.43	145.10	145.96	146.12	146.24	146.25	146.27
146.67	148.32	149.52	150.65	151.45	153.89	156.57
157.07	157.93	157.99	158.89	158.91	159.57	161.05
161.41	162.72	166.57	177.59	180.70	182.44	186.37
193.51	193.52	204.10				

Find (**a**) the 1st quartile, (**b**) the median, (**c**) the 3rd quartile, (**d**) the 7th decile, and (**e**) the 32nd percentile.

Answer (**a**) $q_1 = 124.87$ (**b**) $m = 140.88$ (**c**) $q_3 = 155.23$ (**d**) $d_7 = 148.92$ (**e**) $p_{32} = 129.36$

PROBLEM 4-22 The lengths of time (in minutes) for each of 81 men to run a 25-km race were

80.67	83.96	84.66	86.22	88.22	88.83	91.12	93.55	95.29	96.35	97.25
97.85	99.25	99.72	100.72	101.35	101.67	104.14	104.98	105.76	106.22	107.42
107.81	108.73	109.47	109.59	110.01	110.65	111.28	111.47	112.12	112.20	112.52
113.52	114.11	114.31	114.86	115.10	115.32	116.03	116.26	116.82	117.15	117.65
118.62	118.97	119.64	120.66	121.10	122.07	123.18	123.55	124.29	124.81	125.67
127.11	127.49	128.31	128.87	129.58	130.45	131.51	131.61	132.02	133.32	133.78
136.03	136.38	137.76	138.77	139.73	140.73	142.65	143.23	144.41	149.88	153.87
154.73	165.68	172.93	201.27							

Find (**a**) the 1st quartile, (**b**) the median, (**c**) the 3rd quartile, (**d**) the 7th decile, and (**e**) the 32nd percentile.

Answer (**a**) $q_1 = 105.99$ (**b**) $m = 116.26$ (**c**) $q_3 = 130.98$ (**d**) $d_7 = 127.90$ (**e**) $p_{32} = 109.80$

PROBLEM 4-23 The maximum lung capacity of a group of college freshman females was measured (in liters). The sample mean was 3.36 and the sample standard deviation was .63. Find the z-scores for (**a**) a capacity of 2.3 and (**b**) a capacity of 4.3.

Answer (**a**) $z = -1.68$ (**b**) $z = 1.49$

PROBLEM 4-24 The maximum lung capacity of a group of college freshman males was measured (in liters), yielding a sample mean of 5.08 and a sample standard deviation of .99. Find the z-scores for (**a**) a capacity of 3.7 and (**b**) a capacity of 6.8.

Answer (**a**) $z = -1.39$ (**b**) $z = 1.74$

PROBLEM 4-25 For the lung capacities in Problem 4-23 ($\bar{x} = 3.36$ and $s = .63$), if a woman's z-score is $z = -.89$, what is her maximum lung capacity?

Answer $x = 2.8$

PROBLEM 4-26 For the lung capacities in Problem 4-24 ($\bar{x} = 5.08$ and $s = .99$), if a man's z-score is $z = -.99$, what is his maximum lung capacity?

Answer $x = 4.1$

PROBLEM 4-27 Draw a box-and-whisker diagram for the soap weights in Problem 4-18.

Answer See Fig. 4-10.

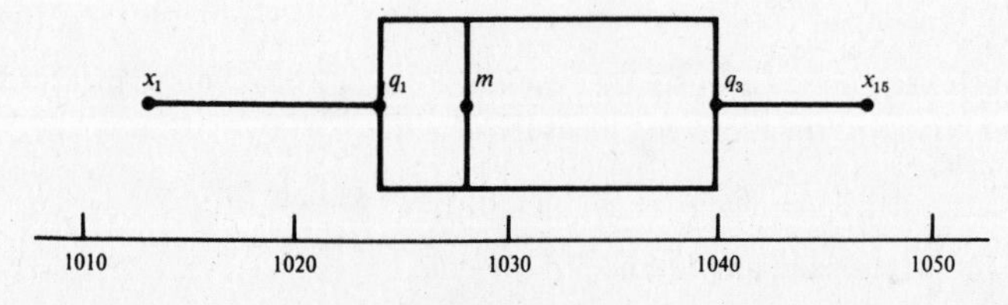

Figure 4-10

PROBLEM 4-28 Draw a box-and-whisker diagram for the grocery orders in Problem 4-19.

Answer See Fig. 4-11.

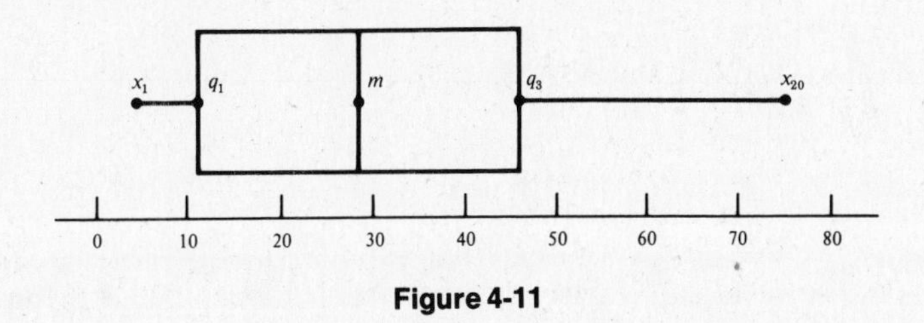

Figure 4-11

PROBLEM 4-29 Draw a box-and-whisker diagram for the head sizes given in Problem 4-20.

Answer See Fig. 4-12.

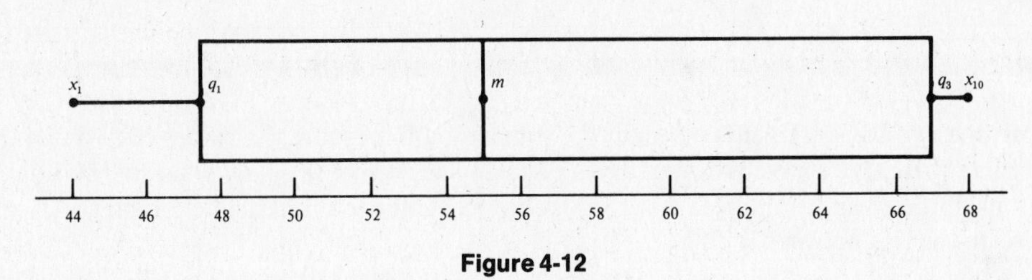

Figure 4-12

PROBLEM 4-30 Draw a box-and-whisker diagram for the lengths of time for the women to complete the 25-km race as given in Problem 4-21.

Answer See Fig. 4-13.

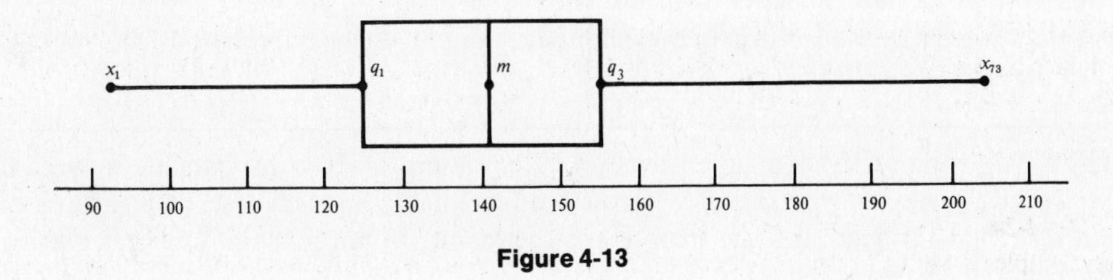

Figure 4-13

PROBLEM 4-31 Draw a box-and-whisker diagram for the lengths of time for the men to complete the 25-km race as given in Problem 4-22.

Answer See Fig. 4-14.

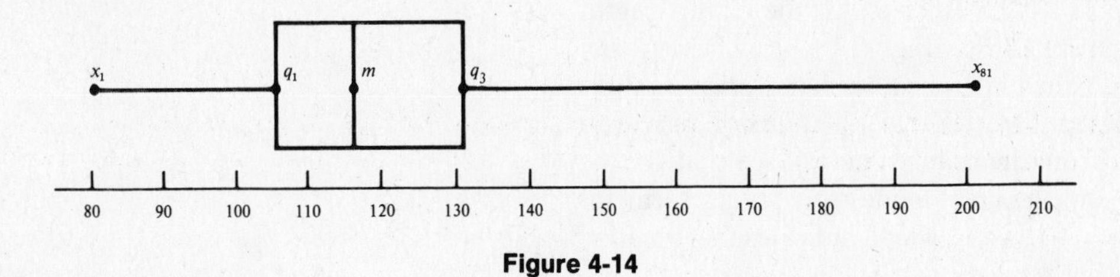

Figure 4-14

5 BIVARIATE DATA

THIS CHAPTER IS ABOUT

☑ Scatter Diagrams
☑ Correlation Coefficient
☑ Linear Regression

When you take two measurements on each of n subjects, you measure the values of two variables, say x and y; in other words, you have **bivariate data**. You can use graphs to discover useful information about the relationship between x and y. It is also possible to compute numerical characteristics to describe possible relationships between x and y.

5-1. Scatter Diagrams

We often want to determine whether there *is* a relationship between x and y—i.e., whether the value of y depends in some way on the value of x. One way of determining whether the two variables are related is to think of the paired measures $(x_1, y_1), (x_2, y_2), \ldots, (x_n, y_n)$ as (x, y)-coordinates and plot them as points on a graph, so that we have a **scatter diagram**. The scatter diagram gives us a graphical illustration of whether or not there is a relationship between x and y—and if there is a relationship, whether it is, for example, a linear relationship.

EXAMPLE 5-1 An obstetrician did ultrasound examinations on 10 of his patients between their 16th and 25th weeks of pregnancy to check the growth of each fetus. In each case, he measured the widest diameter x of the fetal head and the length y of the femur (in mm). Make a scatter diagram of the following paired measurements.

$$(56,45)\quad(65,49)\quad(47,35)\quad(57,44)\quad(62,45)$$
$$(48,40)\quad(68,52)\quad(75,57)\quad(79,62)\quad(49,39)$$

Solution Scale the horizontal axis from 45 to 80 for the head-diameter measurements, and the vertical axis from 30 to 65 for the femur-length measurements. (See Fig. 5-1.) Then, using the ordered pairs as (x, y)-coordinates, place a point on the graph for each of the 10 pairs of measurements. From the graph, you can see that the x-values and the corresponding y-values seem to increase along a line, which indicates that x and y are *linearly* related.

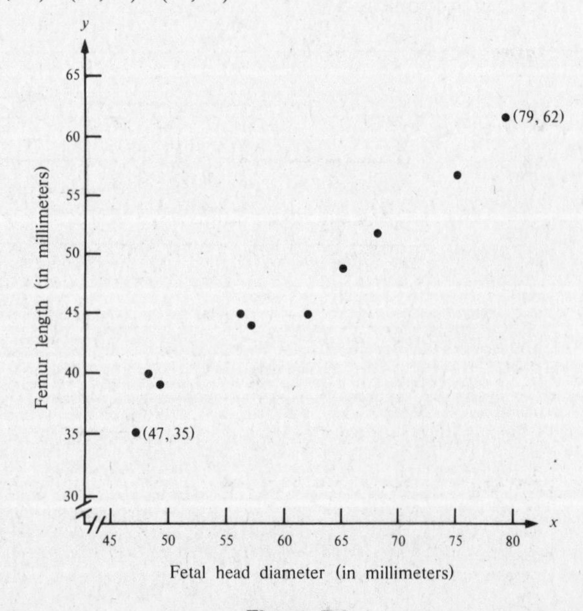

Figure 5-1

We determine the *strength* of a linear relationship between x and y by how well the data fit a line. The slope of the line indicates whether the relationship is *negative* or *positive*—if the y-values increase as the x-values increase, the relationship is positive; if the y-values decrease as the x-values increase, the relationship is negative. Notice that on the scatter diagram in Fig. 5-1 the x-values and the corresponding y-values seem to increase along a line, and the data fit the line quite well. This indicates that there is a strong positive linear relationship between x and y.

EXAMPLE 5-2 The American College Testing (ACT) score x in English and the score y in mathematics were recorded for each of 10 young men applying for admission to a college. The following paired data were noted:

$$(23,31) \quad (21,15) \quad (20,23) \quad (17,25) \quad (21,29) \quad (27,27) \quad (17,19) \quad (24,20) \quad (28,32) \quad (19,23)$$

Make a scatter diagram for these data. What can you tell about the relationship between x and y from the graph?

Solution Scale both the x- and y-axis from 14 to 32, plotting the points on the graph as indicated by each of the paired scores, as shown in Fig. 5-2. Now you can see that there seems to be a positive relationship between x and y, since larger y-values appear to be associated with larger x-values. But if you compare Fig. 5-2 with Fig. 5-1, you see that the positive relationship evident in Fig. 5-2 is not as strong as that in Fig. 5-1; that is, the plotted points are scattered and do not fit closely to a line.

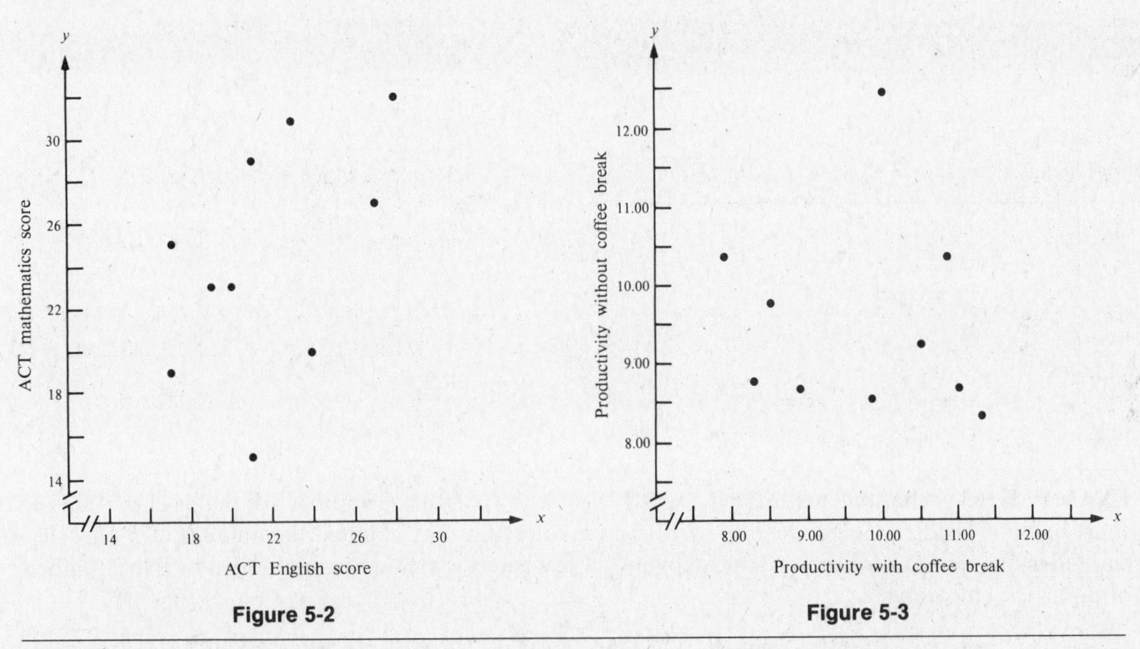

Figure 5-2 **Figure 5-3**

EXAMPLE 5-3 An ambitious restaurant manager wanted to find out whether or not her waiters would be more productive after a short coffee break, so she conducted an experiment. Choosing 10 waiters, she measured their productivity by averaging the amount of money shown on the checks each waiter turned in on two Friday evenings. On the first Friday evening, each waiter had a 15-minute coffee break and turned in an average of x amount of money; on the second Friday evening, each waiter went without a coffee break and turned in an average of y amount of money, as follows:

$$(9.78, 8.52) \quad (10.47, 9.23) \quad (8.56, 9.79) \quad (10.86, 10.36) \quad (11.33, 8.34)$$
$$(10.99, 8.70) \quad (7.91, 10.35) \quad (8.90, 8.63) \quad (10.04, 12.44) \quad (8.28, 8.76)$$

What conclusions can you draw about productivity and coffee breaks?

Solution Make a scatter diagram of the data, scaling both axes from 7.50 to 12.50, as in Fig. 5-3. Notice that there doesn't seem to be much relation between x and y—the value of y doesn't increase or decrease as the value of x increases. You can assume that there is no relationship between x and y—that is, it appears likely from this small study that productivity is based on something other than whether or not a waiter takes a coffee break.

EXAMPLE 5-4 Fifteen male students who participated in a health/fitness program were quite naturally curious about their results in the program. They compared their weights when the program began with their weights when it ended to see if they had lost or gained. The following ordered pairs (in kg) give the pre-weight x and the weight change y for each of the students. (A negative weight change corresponds to a loss of weight.)

(88.3, 1.0) (79.0, −.8) (71.5, .9) (80.0, −2.0) (78.3, −1.0) (59.3, 3.2) (71.7, −1.1) (57.2, −.2)
(63.0, 2.0) (71.2, −1.2) (72.3, 2.1) (76.2, .2) (62.6, 2.8) (95.6, −1.3) (64.2, 2.6)

Make a scatter diagram for these data and indicate what relationship pre-weight has to weight change.

Solution (See Fig. 5-4.) Notice that for these data the value of y decreases as the value of x increases, so there seems to be a negative relationship between pre-weight and weight change. That is, the lighter students tended to gain weight, while the heavier students tended to lose weight.

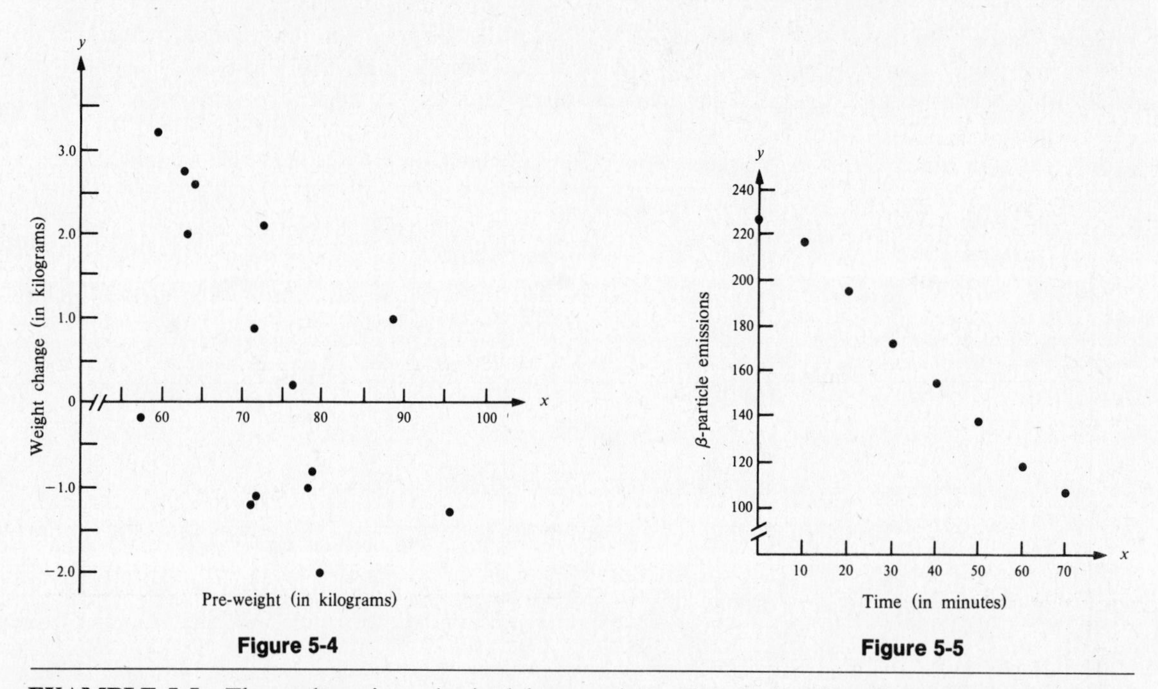

Figure 5-4

Figure 5-5

EXAMPLE 5-5 The students in a physics lab were given a lump of radioactive material that had a short half-life. Using a Geiger counter at 10-minute intervals, they counted the number of β particles emitted in one second and recorded the following data, where x is the time in minutes and y is the number of β-particle emissions.

(0, 226) (10, 216) (20, 195) (30, 172) (40, 154) (50, 138) (60, 119) (70, 106)

Make a scatter diagram for these data, and indicate what relationship there is between time and the number of β-particle emissions.

Solution You see from the scatter diagram in Fig. 5-5 that as the value of x increases, the corresponding value of y decreases in a linear pattern, which indicates there is a very strong negative linear relationship between x and y. Compare this diagram to Fig. 5-1, in which the diagram indicates a very strong positive linear relationship.

EXAMPLE 5-6 In 1976 the age x (in years) and death rate y per 1000 people in the Soviet Union were

(.5, 31.1) (2, 8.7) (7, .7) (12, .5) (17, 1.0) (22, 1.7) (27, 2.1) (32, 3.0)
(37, 3.8) (42, 5.3) (47, 6.9) (52, 9.3) (57, 13.4) (62, 18.9) (67, 28.0)

Make a scatter diagram for these data. What can you say about the relationship between age and death rate?

Solution (See Fig. 5-6.) Notice that although there appears to be a corresponding change in the values of x and y, the relationship is nonlinear. The death rate—not surprisingly—is highest for infants and the very old.

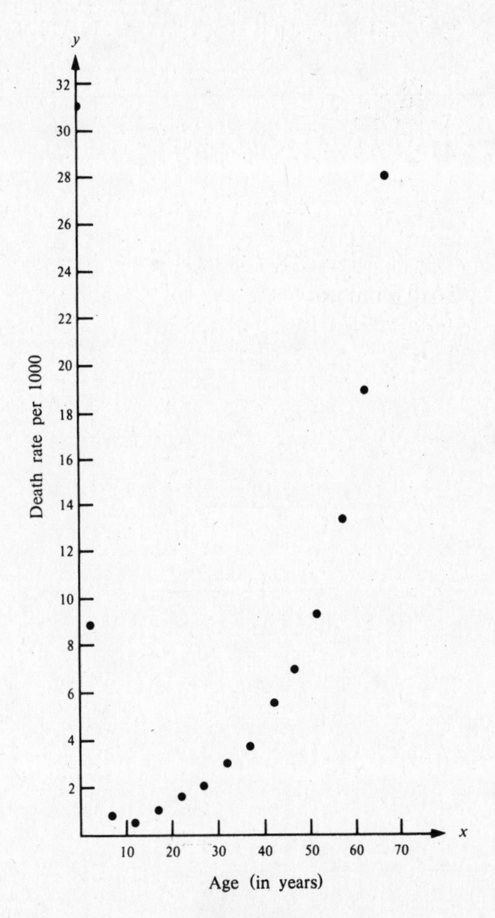

Figure 5-6

5-2. Correlation Coefficient

When we have paired data that are linearly related, we use the linear correlation coefficient to give a quantitative measure of how strong the linear relationship is between x and y—i.e., to what degree x and y are related.

For a set of paired data $(x_1, y_1), (x_2, y_2), \ldots, (x_n, y_n)$ the **linear correlation coefficient** r is defined by

LINEAR CORRELATION COEFFICIENT

$$r = \frac{\sum(x - \bar{x})(y - \bar{y})/(n - 1)}{s_x s_y} \tag{5.1}$$

where s_x and s_y are the respective sample standard deviations of the observations of x and y.

To simplify the calculation of r, we can use the following formula, which eliminates finding s_x and s_y and doing the repeated $x - \bar{x}$ and $y - \bar{y}$ computations.

LINEAR CORRELATION COEFFICIENT (calculating form)

$$r = \frac{n(\sum xy) - (\sum x)(\sum y)}{\sqrt{n(\sum x^2) - (\sum x)^2}\sqrt{n(\sum y^2) - (\sum y)^2}} \tag{5.2}$$

note:
- It's always true that $-1 \le r \le 1$.
- If $r = 1$, then all of the points lie on a line whose slope is positive—a perfect positive correlation.
- If $r = -1$, then all of the points lie on a line whose slope is negative—a perfect negative correlation.

EXAMPLE 5-7 Find the linear correlation coefficient for the following four ($n = 4$) pairs of numbers: (2, 3), (4, 5), (1, 1), (5, 3).

Solution Construct the following table to help in calculating r:

x	y	xy	x^2	y^2	$x - \bar{x}$	$(x - \bar{x})^2$	$y - \bar{y}$	$(y - \bar{y})^2$	$(x - \bar{x})(y - \bar{y})$
2	3	6	4	9	−1	1	0	0	0
4	5	20	16	25	1	1	2	4	2
1	1	1	1	1	−2	4	−2	4	4
5	3	15	25	9	2	4	0	0	0
12	12	42	46	44	0	10	0	8	6

To use formula (5.1), you must first calculate

$$\bar{x} = \frac{12}{4} = 3, \qquad \bar{y} = \frac{12}{4} = 3, \qquad s_x^2 = \frac{\sum(x - \bar{x})^2}{n - 1} = \frac{10}{3}, \qquad s_y^2 = \frac{\sum(y - \bar{y})^2}{n - 1} = \frac{8}{3}$$

and the table tells you that $\sum(x - \bar{x})(y - \bar{y}) = 6$. Then from formula (5.1):

$$r = \frac{\sum(x - \bar{x})(y - \bar{y})/(n - 1)}{s_x s_y}$$

$$= \frac{6/3}{\sqrt{10/3}\sqrt{8/3}} = \frac{2}{\dfrac{(\sqrt{10})(2\sqrt{2})}{3}}$$

$$= \frac{3}{\sqrt{10}\sqrt{2}} = .67$$

Alternatively, formula (5.2) gives you

$$r = \frac{n(\sum xy) - (\sum x)(\sum y)}{\sqrt{n(\sum x^2) - (\sum x)^2}\sqrt{n(\sum y^2) - (\sum y)^2}}$$

$$= \frac{4(42) - 12(12)}{\sqrt{4(46) - (12)^2}\sqrt{4(44) - (12)^2}} = \frac{24}{(2\sqrt{10})(4\sqrt{2})}$$

$$= \frac{3}{\sqrt{10}\sqrt{2}} = .67$$

EXAMPLE 5-8 Find the correlation coefficient for the 10 observations of the fetal head diameter x and the femur length y, as given in Example 5-1.

Solution First construct the table as shown. Then use formula (5.2) to avoid computing $\bar{x}$, $\bar{y}$, s_x, and s_y.

x	y	xy	x^2	y^2
56	45	2,520	3,136	2,025
65	49	3,185	4,225	2,401
47	35	1,645	2,209	1,225
57	44	2,508	3,249	1,936
62	45	2,790	3,844	2,025
48	40	1,920	2,304	1,600
68	52	3,536	4,624	2,704
75	57	4,275	5,625	3,249
79	62	4,898	6,241	3,844
49	39	1,911	2,401	1,521
606	468	29,188	37,858	22,530

$$r = \frac{10(29{,}188) - 606(468)}{\sqrt{10(37{,}858) - (606)^2}\sqrt{10(22{,}530) - (468)^2}}$$

$$= \frac{8272}{\sqrt{11{,}344}\sqrt{6276}}$$

$$= \frac{8272}{8437.7} = .98$$

This value of r is very close to 1.0, which indicates that there is a very strong, although not perfect, positive correlation between x and y. You can see from the graph in Fig. 5-1 that the data do indeed fall very near to a line of positive slope.

EXAMPLE 5-9 Find the correlation coefficient for the ACT English score x and the ACT mathematics score y for each of the 10 students in Example 5-2.

x	y	xy	x^2	y^2
23	31	713	529	961
21	15	315	441	225
20	23	460	400	529
17	25	425	289	625
21	29	609	441	841
27	27	729	729	729
17	19	323	289	361
24	20	480	576	400
28	32	896	784	1024
19	23	437	361	529
217	244	5387	4839	6224

Solution Construct the table as shown. Then, by formula (5.2),

$$r = \frac{10(5387) - 217(244)}{\sqrt{10(4839) - (217)^2}\sqrt{10(6224) - (244)^2}}$$

$$= \frac{922}{\sqrt{1301}\sqrt{2704}}$$

$$= \frac{922}{1875.61} = .492$$

There is a somewhat positive correlation, which is illustrated in Fig. 5-2.

EXAMPLE 5-10 Find the correlation coefficient for the productivity of the 10 waiters as given in Example 5-3.

Solution Construct the table as shown. Then

x	y	xy	x^2	y^2
9.78	8.52	83.33	95.65	72.59
10.47	9.23	96.64	109.62	85.19
8.56	9.79	83.80	73.27	95.84
10.86	10.36	112.51	117.94	107.33
11.33	8.34	94.49	128.37	69.56
10.99	8.70	95.61	120.78	75.69
7.91	10.35	81.87	62.57	107.12
8.90	8.63	76.81	79.21	74.48
10.04	12.44	124.90	100.80	154.75
8.28	8.76	72.53	68.56	76.74
97.12	95.12	922.49	956.77	919.29

$$r = \frac{10(922.49) - 97.12(95.12)}{\sqrt{10(956.77) - (97.12)^2}\sqrt{10(919.29) - (95.12)^2}}$$

$$= \frac{-13.15}{\sqrt{135.41}\sqrt{145.09}}$$

$$= \frac{-13.15}{140.17} = -.094$$

Here r is close to 0, which indicates that x and y are unrelated. The scatter diagram for these data (Fig. 5-3) also indicates that the pairs are unrelated.

EXAMPLE 5-11 Find the correlation coefficient between the pre-weight x and the weight change y for the 15 students as given in Example 5-4.

Solution Construct the table as shown.

x	y	xy	x^2	y^2
88.3	1.0	88.30	7,796.89	1.00
79.0	−.8	−63.20	6,241.00	.64
71.5	.9	64.35	5,112.25	.81
80.0	−2.0	−160.00	6,400.00	4.00
78.3	−1.0	−78.30	6,130.89	1.00
59.3	3.2	189.76	3,516.49	10.24
71.7	−1.1	−78.87	5,140.89	1.21
57.2	−.2	−11.44	3,271.84	.04
63.0	2.0	126.00	3,969.00	4.00
71.2	−1.2	−85.44	5,069.44	1.44
72.3	2.1	151.83	5,227.29	4.41
76.2	.2	15.24	5,806.44	.04
62.6	2.8	175.28	3,918.76	7.84
95.6	−1.3	−124.28	9,139.36	1.69
64.2	2.6	166.92	4,121.64	6.76
1090.4	7.2	376.15	80,862.18	45.12

$$r = \frac{15(376.15) - 1090.4(7.2)}{\sqrt{15(80{,}862.18) - (1090.4)^2}\sqrt{15(45.12) - (7.2)^2}}$$

$$= \frac{-2208.63}{\sqrt{23{,}960.54}\sqrt{624.96}}$$

$$= \frac{-2208.63}{3869.67} = -.571$$

Here r is negative, indicating a negative correlation. Notice that in the scatter diagram (Fig. 5-4), as x increases y tends to decrease.

EXAMPLE 5-12 Using the Geiger counter data in Example 5-5, find the correlation coefficient between the time x and the number of β-particle emissions y.

Solution Construct the table as shown.

x	y	xy	x^2	y^2
0	226	0	0	51,076
10	216	2,160	100	46,656
20	195	3,900	400	38,025
30	172	5,160	900	29,584
40	154	6,160	1,600	23,716
50	138	6,900	2,500	19,044
60	119	7,140	3,600	14,161
70	106	7,420	4,900	11,236
280	1326	38,840	14,000	233,498

$$r = \frac{8(38{,}840) - 280(1326)}{\sqrt{8(14{,}000) - (280)^2}\sqrt{8(233{,}498) - (1326)^2}}$$

$$= \frac{-60{,}560}{\sqrt{33{,}600}\sqrt{109{,}708}}$$

$$= \frac{-60{,}560}{60{,}713.99} = -.997$$

This is a nearly perfect negative correlation, as you can see in Fig. 5-5.

5-3. Linear Regression

A. Least-squares regression line: Slope-intercept form

When the data in a set of paired data display a linear relationship, we often want to find the straight line that best fits the data. We can do this by using the *slope-intercept form* of the equation of a straight line, which is

$$y = b_0 + b_1 x$$

where b_0 is the y-intercept (the point at which the line crosses the y-axis) and b_1 is the slope of the line. But we don't know b_0 and b_1, so we have to *estimate* them. We can use the **method of least squares** to approximate b_0 and b_1, so that we get a line in which the sum of the squares of the vertical distances from the points to the line is a minimum. This line is the **least-squares regression line.**

If a set of paired data $(x_1, y_1), (x_2, y_2), \ldots, (x_n, y_n)$ is given, the estimates of b_0 and b_1 given by the method of least squares are

METHOD OF LEAST SQUARES

$$\text{Slope:} \quad \hat{b}_1 = \frac{n(\sum xy) - (\sum x)(\sum y)}{n(\sum x^2) - (\sum x)^2} \tag{5.3}$$

$$y\text{-Intercept:} \quad \hat{b}_0 = \bar{y} - \hat{b}_1 \bar{x} \tag{5.4}$$

The resulting equation for the least-squares regression line is

LEAST-SQUARES REGRESSION LINE (slope-intercept form)

$$\hat{y} = \hat{b}_0 + \hat{b}_1 x \tag{5.5}$$

note: $\hat{b}_1$ and $\hat{b}_0$ are read "b_1 hat" and "b_0 hat." The hat (^) indicates that we're estimating the values of b_0 and b_1.

EXAMPLE 5-13 Given the set of ordered pairs (2, 3), (4, 5), (1, 1), (5, 3), **(a)** find the least-squares regression line for the data; **(b)** draw the scatter diagram and the line on the same graph (see Example 5-7).

Solution

(a) You find the estimated slope of the line $\hat{b}_1$ and the estimated y-intercept $\hat{b}_0$ by the method of least squares [formulas (5.3) and (5.4)]. You've already found the values of $\bar{x}$ and $\bar{y}$ in Example 5-7.

$$\hat{b}_1 = \frac{n(\sum xy) - (\sum x)(\sum y)}{n(\sum x^2) - (\sum x)^2} = \frac{4(42) - 12(12)}{4(46) - (12)^2} = \frac{24}{40} = .6$$

$$\hat{b}_0 = \bar{y} - \hat{b}_1 \bar{x} = \frac{12}{4} - .6\left(\frac{12}{4}\right) = 3 - 1.8 = 1.2$$

Substitute these values into the equation for the least-squares regression line (5.5):

$$\hat{y} = \hat{b}_0 + \hat{b}_1 x = 1.2 + .6x$$

(b) Draw the scatter diagram by plotting the given ordered pairs as points on a graph. Then use the equation for the regression line to find two points that will define the line and allow you to draw it. You know that the y-intercept $(0, y)$ has $\hat{y} = \hat{b}_0 = 1.2$, so $(0, 1.2)$ is one point. Then you can pick any value for x, say 5, and substitute it into the regression-line equation to get another value for y:

$$\hat{y} = \hat{b}_0 + \hat{b}_1 x = 1.2 + .6x = 1.2 + .6(5) = 4.2$$

This gives you your other point $(5, 4.2)$. Now draw the regression line as shown in Fig. 5-7.

note: It's always true that $(\bar{x}, \bar{y})$ lies on the regression line, as in Fig. 5-7, where $(\bar{x}, \bar{y}) = (3, 3)$.

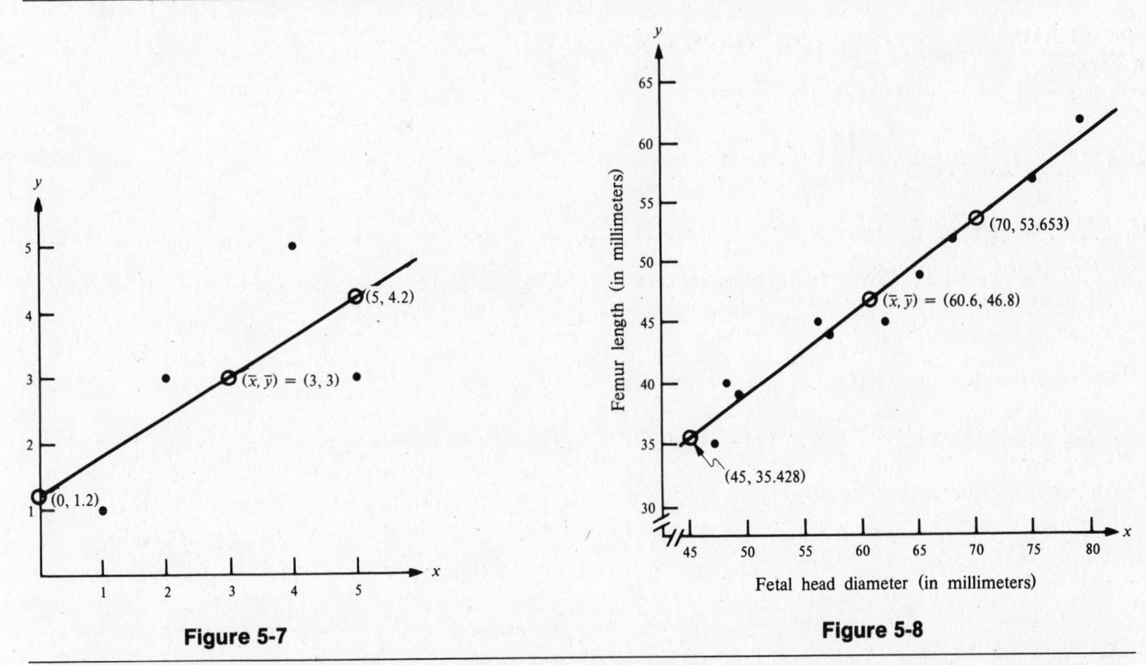

Figure 5-7 **Figure 5-8**

EXAMPLE 5-14 **(a)** Find the least-squares regression line for the fetal-measurement data in Examples 5-1 and 5-8. **(b)** Graph this line on the scatter diagram given in Fig. 5-1.

Solution

(a) Using the tabulated data in Example 5-8, calculate

$$\hat{b}_1 = \frac{10(29{,}188) - 606(468)}{10(37{,}858) - (606)^2} = \frac{8272}{11{,}344} = .729$$

$$\hat{b}_0 = \frac{468}{10} - .729\left(\frac{606}{10}\right)$$

$$= 46.800 - .729(60.6) = 46.800 - 44.177$$

$$= 2.623$$

So the least-squares regression line is

$$y = 2.623 + 0.729x$$

(b) The y-intercept $(0, 2.623)$ doesn't fall on the graph because the scale is too large (see Fig. 5-1). But that's not a problem. You know that $(\bar{x}, \bar{y})$ must fall on the least-squares regression line, so point $(60.6, 46.8)$ is one point on the line. To find another point, pick a value of x near the left or right side of the graph and find the corresponding value of y. Thus, if you pick $x = 45$,

$$y = 2.623 + .729(45) = 35.428$$

or if $x = 70$,

$$y = 2.623 + .729(70) = 53.653$$

Now you can draw the regression line using any two of the three points you've found, as in Fig. 5-8.

EXAMPLE 5-15 **(a)** Find the least-squares regression line for the ACT scores given in Examples 5-2 and 5-9. **(b)** Graph this least-squares regression line on the scatter diagram given in Fig. 5-2.

Solution

(a) Using the tabulated data in Example 5-9, calculate

$$\hat{b}_1 = \frac{10(5387) - 217(244)}{10(4839) - (217)^2} = \frac{922}{1301} = .709$$

$$\hat{b}_0 = \frac{244}{10} - .709\left(\frac{217}{10}\right) = 24.4 - 15.385$$

$$= 9.015$$

So the least-squares regression line is

$$\hat{y} = 9.015 + .709x$$

(b) To draw the least-squares regression line shown in Fig. 5-9, you can use any two of the following points:

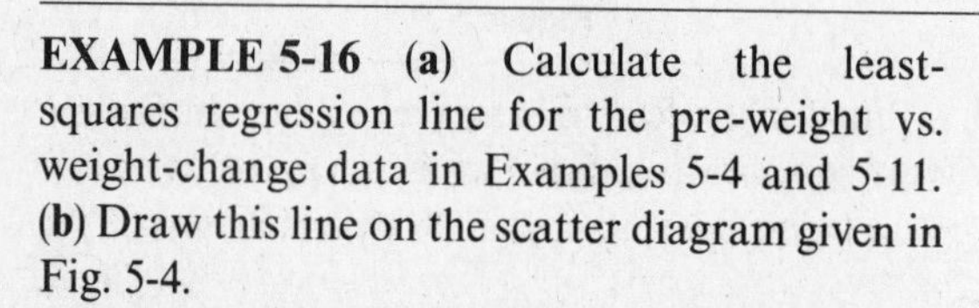

x	$\hat{y}$
16	20.359
28	28.867
21.7	24.400

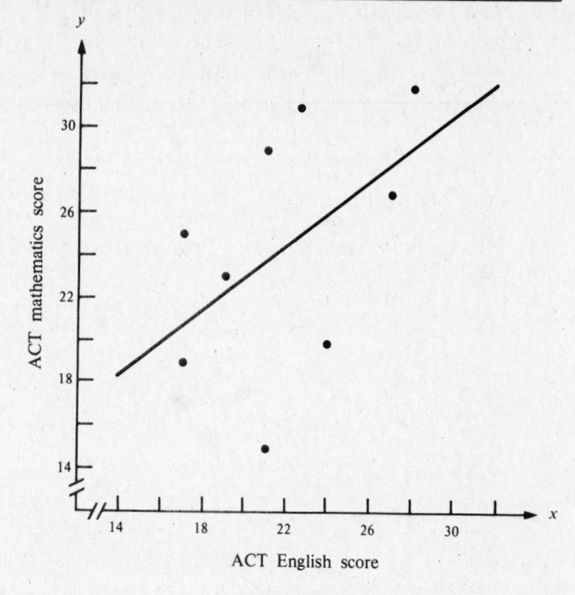

Figure 5-9

EXAMPLE 5-16 **(a)** Calculate the least-squares regression line for the pre-weight vs. weight-change data in Examples 5-4 and 5-11. **(b)** Draw this line on the scatter diagram given in Fig. 5-4.

Solution

(a) From the table in Example 5-11 you have

$$\hat{b}_1 = \frac{15(376.15) - 1090.4(7.2)}{15(80,862.18) - (1090.4)^2} = \frac{-2208.63}{23,960.5}$$

$$= -.092$$

$$\hat{b}_0 = \frac{7.2}{15} - (-.092)\left(\frac{1090.4}{15}\right) = .48 + 6.688$$

$$= 7.168$$

Thus $\quad \hat{y} = 7.168 - .092x$

(b) Picking any two points:

x	$\hat{y}$
60.00	1.65
100.00	−2.03
72.69	.48

you can draw the regression line shown in Fig. 5-10.

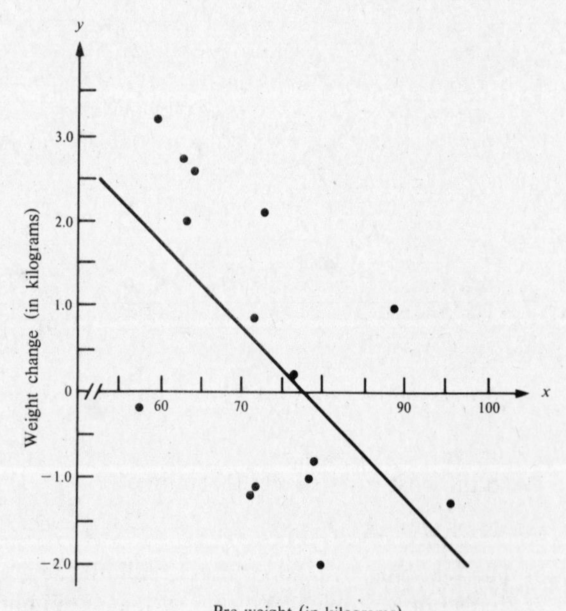

Figure 5-10

EXAMPLE 5-17 **(a)** Calculate the least-squares regression line for the Geiger counter data in Examples 5-5 and 5-12. **(b)** Draw the line on the scatter diagram given in Fig. 5-5.

Solution

(a) $\hat{b}_1 = \dfrac{8(38{,}840) - 280(1326)}{8(14{,}000) - (280)^2} = \dfrac{-60{,}560}{33{,}600}$

$= -1.802$

$\hat{b}_0 = \dfrac{1326}{8} - (-1.802)\left(\dfrac{280}{8}\right)$

$= 165.75 + 63.07 = 228.82$

Thus $\hat{y} = 228.82 - 1.802x$.

(b) The y-intercept is $y = 228.82$. The point $(\bar{x}, \bar{y}) = (35, 165.75)$, which you calculate from the data in Example 5-12, also lies on the line. Use these two points to draw the line on the scatter diagram, as in Fig. 5-11.

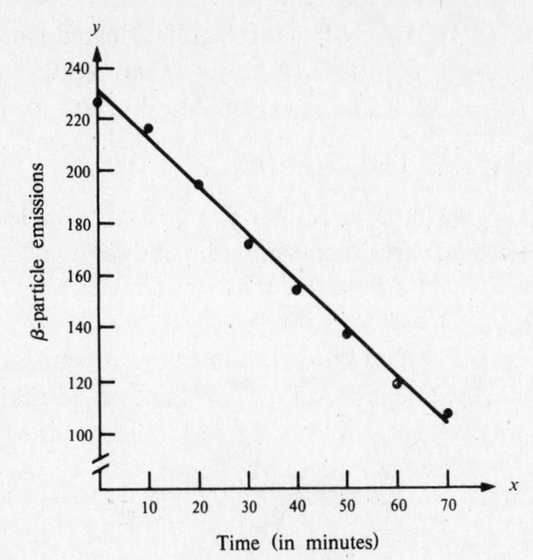

Figure 5-11

B. Least-squares regression line: Point-slope form

When we've already calculated (or been given) the sample means and standard deviations for the paired data in a set, we can find the least-squares regression line by using the *point-slope form* of the linear equation:

LEAST-SQUARES REGRESSION LINE (point-slope form)

$$\hat{y} = \bar{y} + r\left(\frac{s_y}{s_x}\right)(x - \bar{x}) \tag{5.6}$$

where the slope of the line—the coefficient of x—is expressed as $r(s_y/s_x)$. This form shows us how the linear correlation coefficient r affects the slope of the regression line. And we can also see that when r is positive, the slope must be positive; and when r is negative, the slope must be negative.

note: The quotient s_y/s_x is always positive because standard deviations are always positive (see Chapter 3).

EXAMPLE 5-18 Find the equation for the least-squares regression line from the data given in Example 5-7.

Solution From Example 5-7 you know that

$$\bar{x} = 3, \qquad \bar{y} = 3, \qquad s_x = \sqrt{\frac{10}{3}}, \qquad s_y = \sqrt{\frac{8}{3}}, \qquad r = \frac{3}{\sqrt{20}}$$

Thus from formula (5.6):

$$\hat{y} = \bar{y} + r\left(\frac{s_y}{s_x}\right)(x - \bar{x}) = 3 + \left(\frac{3}{\sqrt{20}}\right)\left(\frac{\sqrt{8/3}}{\sqrt{10/3}}\right)(x - 3) = 3 + \left(\frac{3\sqrt{8}}{\sqrt{200}}\right)(x - 3)$$

$$= 3 + \left(\frac{3}{\sqrt{25}}\right)(x - 3) = 3 + .6(x - 3) = 3 + .6x - 1.8$$

$$= 1.2 + .6x$$

Notice that $\hat{y} = 1.2 + .6x$ brings you back to the slope-intercept form ($\hat{y} = \hat{b}_0 + \hat{b}_1 x$), which you found in Example 5-13.

EXAMPLE 5-19 Eight golfers tested 2 brands of golf balls by hitting the balls from the tee in random order: 3 golf balls of brand A and 3 golf balls of brand B. The average number of yards (x, y) for the 3 drives for brand A (x) and the 3 drives for brand B (y) were

(235, 236) (198, 217) (196, 203) (210, 218) (229, 236) (234, 251) (248, 254) (230, 227)

Given that $\bar{x} = 222.50$, $\bar{y} = 230.25$, $s_x = 18.883$, $s_y = 17.483$, and $r = .921$, **(a)** find the equation for the least-squares regression line and **(b)** draw the least-squares regression line on the scatter diagram.

Solution

(a) Since you know the standard deviations and the linear correlation coefficient, you can find the least-squares regression line by the point-slope formula (5.6):

$$\hat{y} = \bar{y} + r\left(\frac{s_y}{s_x}\right)(x - \bar{x})$$

$$= 230.25 + .921\left(\frac{17.483}{18.883}\right)(x - 222.50)$$

$$= 40.521 + .853x$$

(b) To graph the least-squares regression line, draw the scatter diagram and then solve the slope-intercept equation for $\hat{y}$, using values of x taken from the graph.

x	$\hat{y}$
200.00	211.12
250.00	253.77
222.50	230.31

The least-squares regression line shown in Fig. 5-12 can be drawn from any two of these points.

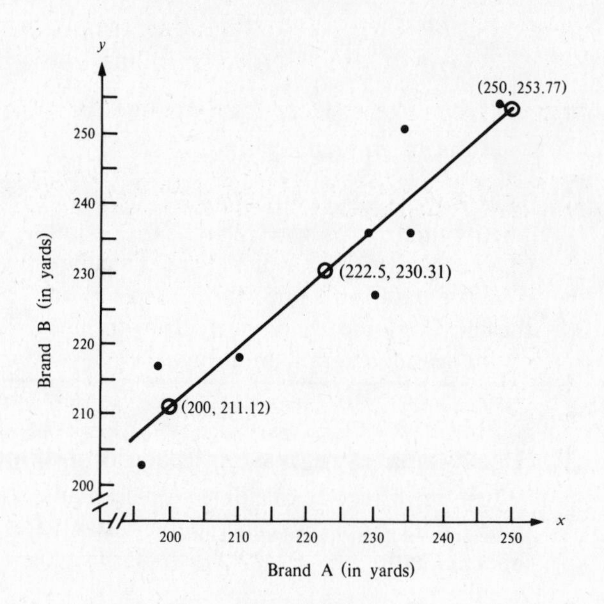

Figure 5-12

SUMMARY

1. A set of paired data is represented in coordinate (x, y) form by $(x_1, y_1), (x_2, y_2), \ldots, (x_n, y_n)$.
2. A *scatter diagram* gives a graphical presentation of paired data, which are plotted as points (x, y) on a graph.
3. The *correlation coefficient* r measures the linear relationship between paired measurements x and y in a set of paired data:

$$r = \frac{n(\sum xy) - (\sum x)(\sum y)}{\sqrt{n(\sum x^2) - (\sum x)^2}\sqrt{n(\sum y^2) - (\sum y)^2}} = \frac{\sum(x - \bar{x})(y - \bar{y})/(n - 1)}{s_x s_y}$$

4. It's always true that $-1 \leq r \leq 1$: When r is positive, the slope of the regression line is positive, when r is negative, the slope is negative.
5. The *method of least squares* yields the equation of the "best-fitting" line for a set of paired data.
6. The best-fitting line for a set of paired data is called the *least-squares regression line*, whose equation can be written in two forms:

Slope-intercept: $\hat{y} = \hat{b}_0 + \hat{b}_1 x$ (where $\hat{b}_1$ = estimated slope; $\hat{b}_0$ = estimated y-intercept)

Point-slope: $\hat{y} = \bar{y} + r\left(\frac{s_y}{s_x}\right)(x - \bar{x})$ (where $r(s_y/s_x)$ = slope)

RAISE YOUR GRADES

Can you . . . ?

☑ plot a set of points to make a scatter diagram
☑ define and calculate the correlation coefficient
☑ find the values of $\hat{b}_0$ and $\hat{b}_1$ in the slope-intercept form of the equation for the least-squares regression line
☑ find the value of $\hat{y}$ in the point-slope form of the equation for the least-square regression line, given the standard deviations, means, and the correlation coefficient
☑ graph the least-squares regression line

RAPID REVIEW

1. Paired data can be presented graphically using a ____________.
2. It's always true that (a) ____________ $\leq r \leq$ (b) ____________, where r is the linear correlation coefficient.
3. If r is close to 0, there is no ____________ relationship between the values of x and y.
4. If $r = 1$, all of the points lie on a line with ____________ slope.
5. If $r = -1$, all of the points lie on a line with ____________ slope.
6. In the equation $y = b_0 + b_1 x$, the term b_0 is the ____________.
7. In the equation $y = b_0 + b_1 x$, the term b_1 is the ____________ of the line.
8. The point-slope form of the least-squares regression line is
   $\hat{y} =$ (a) ____________ $+ r\left(\frac{s_y}{s_x}\right)(x -$ (b) ____________).
9. The sign of ____________ determines whether the slope of the regression line is positive or negative.
10. If $\bar{x} = 80, \bar{y} = 70, s_x = 15, s_y = 30$, and $r = .8$, then the equation of the least-squares regression line is ____________.

Answers **(1)** scatter diagram **(2)** **(a)** -1 **(b)** 1 **(3)** linear **(4)** positive **(5)** negative **(6)** y-intercept **(7)** slope **(8)** **(a)** $\bar{y}$ **(b)** $\bar{x}$ **(9)** r **(10)** $\hat{y} = 1.6x - 58$

SOLVED PROBLEMS

Math Review

PROBLEM 5-1 If $y = .5x + 2$, **(a)** give the y-intercept, **(b)** give the slope, and **(c)** graph the line.

Solution

(a) The y-intercept is the constant term 2, which is found by setting $x = 0$ in the equation and solving for y.

(b) The slope is .5, the coefficient of x.

(c) (See Fig. 5-13.) The y-intercept $(0, 2)$ is one point on the line. Find a second point on the line using an arbitrary value of x, say $x = 6$, and draw a line through them.

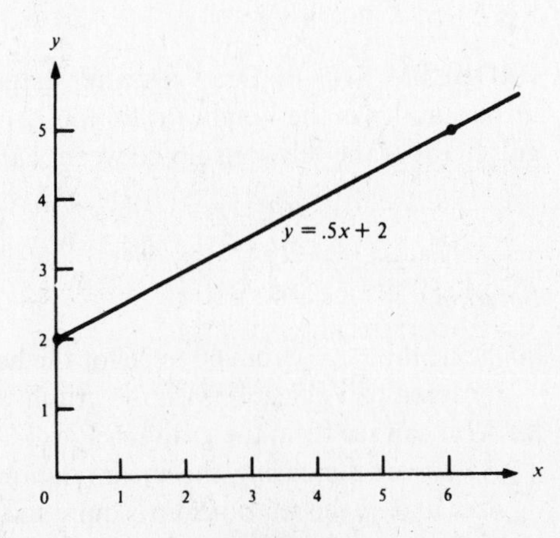

Figure 5-13

PROBLEM 5-2 If $y = 3 - .5x$, (**a**) give the y-intercept, (**b**) give the slope, and (**c**) graph the line.

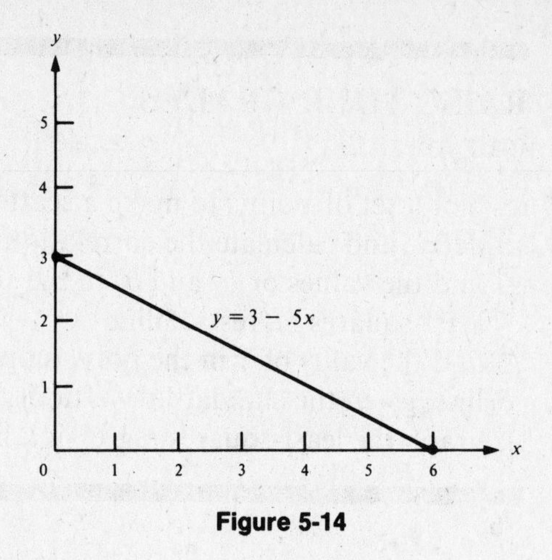

Figure 5-14

Solution

(**a**) The y-intercept is 3.
(**b**) The slope is $-.5$.
(**c**) The point $(0, 3)$ is on the line. If $x = 6$, then $y = 3 - .5(6) = 0$, so $(6, 0)$ is on the line. Plot these two points and draw a line through them, as in Fig. 5-14.

Scatter Diagrams

PROBLEM 5-3 A pair of 8-sided dice was rolled 5 times, so that each roll had 2 outcomes—a small one and a large one (or possibly equal outcomes). Let x be the smaller outcome and y be the larger outcome of each roll. Draw a scatter diagram for 5 pairs of observations:

$$(1,2) \quad (2,5) \quad (3,4) \quad (4,8) \quad (5,6)$$

Solution Scale both the x- and the y-axes from 0 to 8. Place a dot on the graph to represent each one of the five outcomes, using their (x, y) values as coordinate points, as in Fig. 5-15.

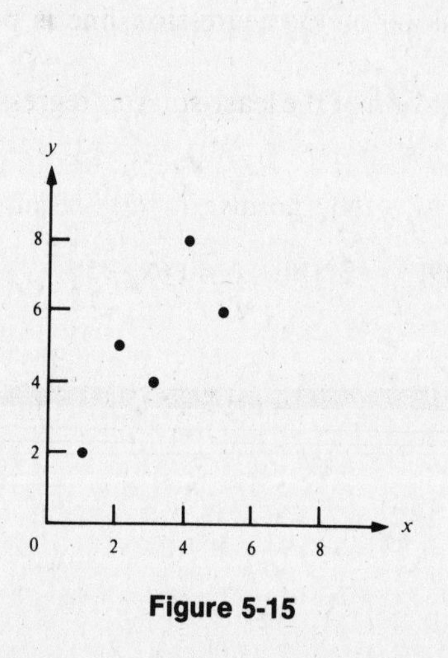

Figure 5-15

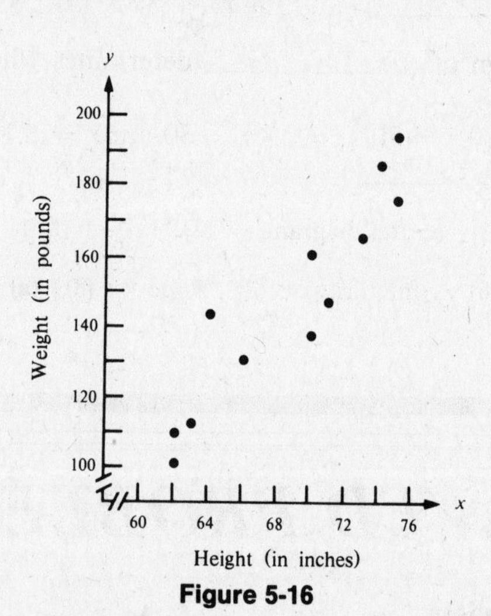

Figure 5-16

PROBLEM 5-4 (**a**) Draw a scatter diagram for the following pairs of data, where x is the height (in inches) and y is the weight (in lb) for 12 students in an English class. (**b**) What can you tell from the graph about the relationship between x and y?

$$\begin{array}{cccccc} (63,112) & (70,161) & (75,176) & (74,186) & (66,131) & (70,138) \\ (62,101) & (64,143) & (73,166) & (75,194) & (71,147) & (62,110) \end{array}$$

Solution

(**a**) Scale the x-axis from 60 to 76 for the heights, and the y-axis from 100 to 200 for the weights. Plot the ordered pairs as points on the graph, as shown in the scatter diagram (Fig. 5-16).
(**b**) You can see from the graph in Fig. 5-16 that the x and y values tend to increase in a line. Since the values are increasing, the linear relationship between x and y is positive. Also, the points fit fairly well to a line, so the relationship is quite strong. You can say that there seems to be a strong positive linear relationship between x and y.

PROBLEM 5-5 The following table gives the year x and the motor vehicle death rate y per 100 million vehicle miles for the years 1966 to 1981.

Year (x):	1966	1967	1968	1969	1970	1971	1972	1973
Death rate (y):	5.70	5.50	5.40	5.21	4.88	4.57	4.43	4.24

Year (x):	1974	1975	1976	1977	1978	1979	1980	1981
Death rate (y):	3.59	3.45	3.33	3.35	3.39	3.45	3.48	3.29

(**a**) Make a scatter diagram for these data. (**b**) Based on the scatter diagram, do x and y appear to be linearly related?

Solution

(**a**) Place the years along the x-axis and the death rates along the y-axis, as in Fig. 5-17.

(**b**) The data for the years 1966–1976 do appear to be linearly related. The data for the years 1976–1980 also appear to be linearly related, but in a line different from that of the preceding years.

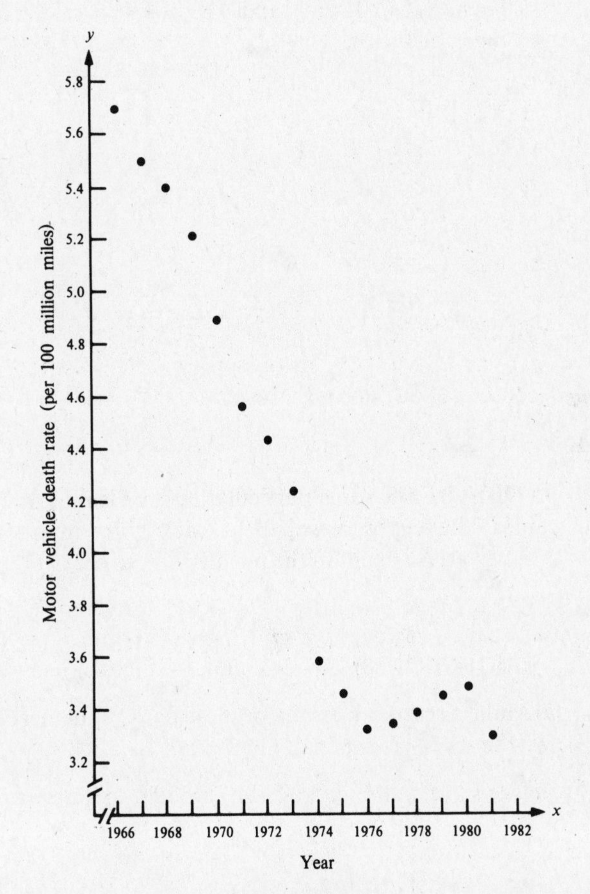

Figure 5-17

PROBLEM 5-6 Each of 10 women applied for admission to a college with the following American College Testing (ACT) score x in English and the score y in mathematics:

(23, 25) (20, 19) (26, 25) (22, 19) (26, 20)
(23, 19) (20, 27) (19, 21) (29, 27) (22, 28)

Make a scatter diagram for these data.

Solution See Fig. 5-18.

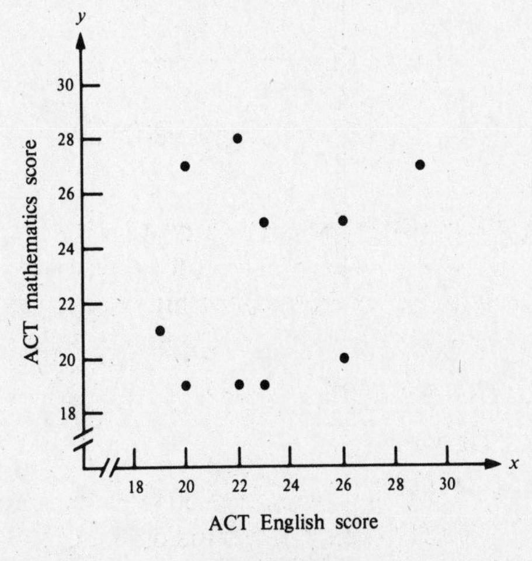

Figure 5-18

PROBLEM 5-7 Fourteen corporate executives participated in an exercise program in which one goal was to lose weight. The following pairs of data give each executive's weight x at the beginning of the program and the resulting weight change y (in lb). Positive y denotes a gain in weight and negative y denotes a loss.

(207, 2) (236, −17) (183, 3) (170, −5) (215, −3) (178, −12) (179, −11)
(215, −2) (207, −7) (190, −7) (215, −9) (158, −4) (165, −6) (163, −1)

Make a scatter diagram for these data.

Solution See Fig. 5-19.

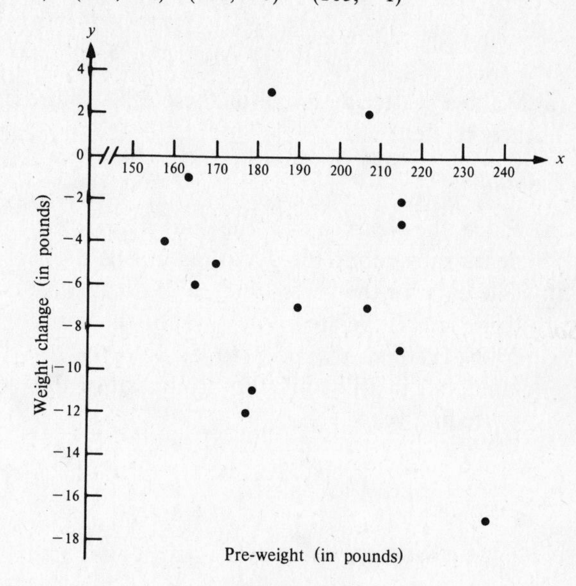

Figure 5-19

PROBLEM 5-8 In a particular hockey league, there were 21 forwards who each played more than 70 games. The record keepers kept track of the number of points x each player scored and the number of minutes y each spent in the penalty box to see if there was a correlation between the two. The following ordered pairs were recorded:

(89, 43) (93, 29) (53, 57) (46, 8) (66, 64) (47, 10) (61, 121) (61, 32) (102, 73) (72, 55) (93, 114)
(126, 104) (51, 88) (107, 43) (67, 66) (60, 42) (59, 49) (95, 83) (76, 47) (40, 56) (20, 159)

(a) Make a scatter diagram for these data. **(b)** Is there a relationship between the number of points scored and the number of penalty minutes?

Solution **(a)** See Fig. 5-20. **(b)** There doesn't appear to be a linear relationship between x and y.

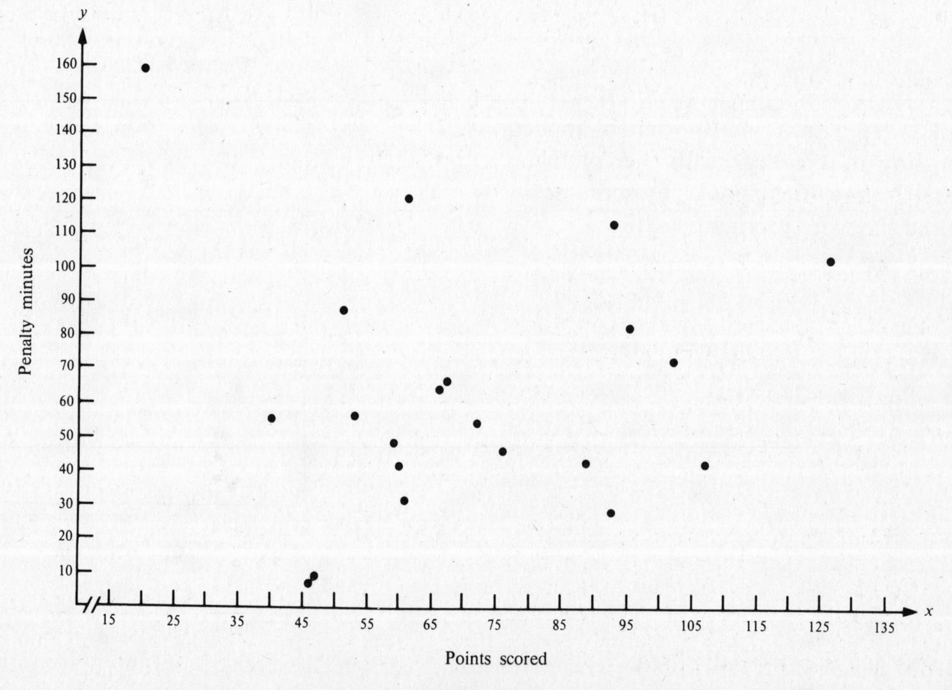

Figure 5-20

PROBLEM 5-9 The United Nations Population and Vital Statistics Report lists the estimated birth rate x and death rate y for 22 African countries as follows:

Algeria	(47.4, 14.2)	Benin	(48.8, 19.1)
Burundi	(42.0, 20.4)	Chad	(44.1, 24.1)
Congo	(44.6, 19.0)	Ethiopia	(49.8, 25.2)
Gambia	(47.5, 22.9)	Guinea	(46.1, 20.7)
Ivory Coast	(47.5, 18.2)	Kenya	(53.8, 14.4)
Liberia	(49.8, 20.9)	Morocco	(45.4, 13.6)
Namibia	(42.0, 13.5)	Nigeria	(49.8, 17.8)
Sierra Leone	(45.5, 19.2)	South Africa	(37.9, 10.3)
Sudan	(45.8, 19.2)	Uganda	(44.7, 14.4)
Tanzania	(46.3, 15.8)	Zanzibar	(47.0, 21.0)
Zaire	(46.2, 18.7)	Zimbabwe	(47.3, 13.6)

Make a scatter diagram for these data.

Solution See Fig. 5-21.

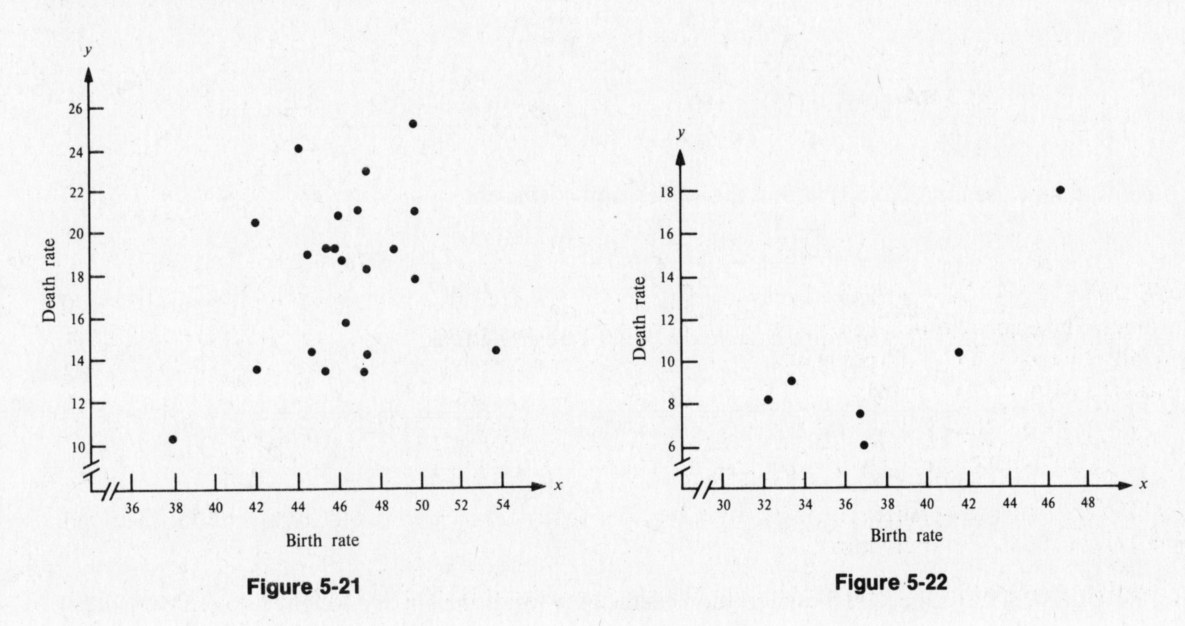

Figure 5-21

Figure 5-22

PROBLEM 5-10 The estimated birth rate x and death rate y for 6 South American countries are as follows:

Bolivia	(46.6, 18.0)	Brazil	(33.3, 9.1)
Colombia	(32.1, 8.2)	Ecuador	(41.6, 10.4)
Paraguay	(36.7, 7.6)	Venezuela	(36.9, 6.1)

Make a scatter diagram for these data.

Solution See Fig. 5-22.

PROBLEM 5-11 The birth rate x and death rate y for 18 European countries are as follows:

Austria	(12.5, 12.0)	Belgium	(12.2, 11.4)
Bulgaria	(14.0, 10.7)	Denmark	(10.3, 10.8)
Finland	(13.7, 9.0)	France	(14.7, 10.1)
Greece	(14.3, 8.7)	Hungary	(12.5, 13.5)
Ireland	(20.3, 9.4)	Italy	(10.9, 9.4)
Luxembourg	(12.0, 11.6)	Netherlands	(12.0, 8.2)
Poland	(19.4, 9.2)	Romania	(17.1, 10.0)
Spain	(11.5, 9.3)	Switzerland	(11.5, 9.3)
England	(12.8, 11.7)	Scotland	(12.8, 12.6)

Make a scatter diagram for these data.

Solution See Fig. 5-23.

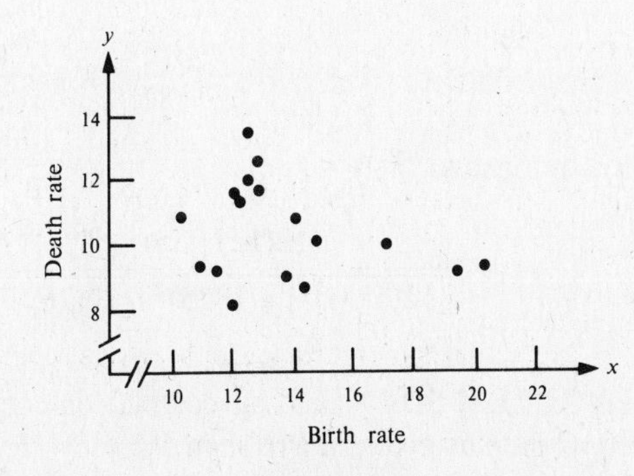

Figure 5-23

Correlation Coefficient

PROBLEM 5-12 Find the correlation coefficient r for the data in Problem 5-3.

Solution First construct the following table:

x	y	xy	x^2	y^2	$x-\bar{x}$	$y-\bar{y}$	$(x-\bar{x})(y-\bar{y})$
1	2	2	1	4	−2	−3	6
2	5	10	4	25	−1	0	0
3	4	12	9	16	0	−1	0
4	8	32	16	64	1	3	3
5	6	30	25	36	2	1	2
15	25	86	55	145	0	0	11

Also calculate the sample means and the sample standard deviations:

$$\bar{x} = \frac{15}{5} = 3, \qquad \bar{y} = \frac{25}{5} = 5$$

$$s_x^2 = \frac{5(55) - (15)^2}{5(4)} = \frac{50}{20} = 2.5 \qquad s_y^2 = \frac{5(145) - (25)^2}{5(4)} = \frac{100}{20} = 5$$

You can now use formula (5.1) to find the correlation coefficient:

$$r = \frac{\sum(x-\bar{x})(y-\bar{y})/(n-1)}{s_x s_y} = \frac{11/4}{\sqrt{2.5}\sqrt{5}} = .778$$

If you use formula (5.2), you don't have to do the extra calculations:

$$r = \frac{n(\sum xy) - (\sum x)(\sum y)}{\sqrt{n(\sum x^2) - (\sum x)^2}\sqrt{n(\sum y^2) - (\sum y)^2}} = \frac{5(86) - 15(25)}{\sqrt{5(55) - (15)^2}\sqrt{5(145) - (25)^2}}$$

$$= \frac{430 - 375}{\sqrt{50}\sqrt{100}} = \frac{55}{\sqrt{5000}} = .778$$

PROBLEM 5-13 Calculate the correlation coefficient r for the height and weight data in Problem 5-4.

Solution

x	y	xy	x^2	y^2
63	112	7,056	3,969	12,544
70	161	11,270	4,900	25,921
75	176	13,200	5,625	30,976
74	186	13,764	5,476	34,596
66	131	8,646	4,356	17,161
70	138	9,660	4,900	19,044
62	101	6,262	3,844	10,201
64	143	9,152	4,096	20,449
73	166	12,118	5,329	27,556
75	194	14,550	5,625	37,636
71	147	10,437	5,041	21,609
62	110	6,820	3,844	12,100
825	1765	122,935	57,005	269,793

From formula (5.2):

$$r = \frac{12(122{,}935) - 825(1765)}{\sqrt{12(57{,}005) - (825)^2}\sqrt{12(269{,}793) - (1765)^2}} = \frac{19{,}095}{\sqrt{3435}\sqrt{122{,}291}} = \frac{19{,}095.0}{20{,}495.6} = .93$$

PROBLEM 5-14 Find the correlation coefficient r for the ACT English and mathematics scores of the 10 women as given in Problem 5-6.

Solution

x	y	xy	x^2	y^2
23	25	575	529	625
20	19	380	400	361
26	25	650	676	625
22	19	418	484	361
26	20	520	676	400
23	19	437	529	361
20	27	540	400	729
19	21	399	361	441
29	27	783	841	729
22	28	616	484	784
230	230	5318	5380	5416

So, $$r = \frac{10(5318) - 230(230)}{\sqrt{10(5380) - (230)^2}\sqrt{10(5416) - (230)^2}} = \frac{280}{\sqrt{900}\sqrt{1260}} = \frac{280}{1064.89} = .26$$

PROBLEM 5-15 Find r for the weights and weight changes of the executives as given in Problem 5-7.

Solution

x	y	xy	x^2	y^2
207	2	414	42,849	4
236	−17	−4,012	55,696	289
183	3	549	33,489	9
170	−5	−850	28,900	25
215	−3	−645	46,225	9
178	−12	−2,136	31,684	144
179	−11	−1,969	32,041	121
215	−2	−430	46,225	4
207	−7	−1,449	42,849	49
190	−7	−1,330	36,100	49
215	−9	−1,935	46,225	81
158	−4	−632	24,964	16
165	−6	−990	27,225	36
163	−1	−163	26,569	1
2681	−79	−15,578	521,041	837

So, $$r = \frac{14(-15{,}578) - (2681)(-79)}{\sqrt{14(521{,}041) - (2681)^2}\sqrt{14(837) - (-79)^2}} = \frac{-6293}{\sqrt{106{,}813}\sqrt{5477}} = -.26$$

PROBLEM 5-16 Find r for the hockey-player data given in Problem 5-8.

Solution

x	y	xy	x^2	y^2
89	43	3,827	7,921	1,849
93	29	2,697	8,649	841
53	57	3,021	2,809	3,249
46	8	368	2,116	64
66	64	4,224	4,356	4,096
47	10	470	2,209	100
61	121	7,381	3,721	14,641
61	32	1,952	3,721	1,024
102	73	7,446	10,404	5,329
72	55	3,960	5,184	3,025
93	114	10,602	8,649	12,996
126	104	13,104	15,876	10,816
51	88	4,488	2,601	7,744
107	43	4,601	11,449	1,849
67	66	4,422	4,489	4,356
60	42	2,520	3,600	1,764
59	49	2,891	3,481	2,401
95	83	7,885	9,025	6,889
76	47	3,572	5,776	2,209
40	56	2,240	1,600	3,136
20	159	3,180	400	25,281
1484	1343	94,851	118,036	113,659

So,

$$r = \frac{21(94{,}851) - 1484(1343)}{\sqrt{21(118{,}036) - (1484)^2}\sqrt{21(113{,}659) - (1343)^2}} = \frac{-1141}{\sqrt{276{,}500}\sqrt{583{,}190}} = -.003$$

Linear Regression

PROBLEM 5-17 **(a)** Find the least-squares regression line for the dice data in Problems 5-3 and 5-12. **(b)** Draw this regression line on the scatter diagram.

Solution

(a) First find the estimated slope of the line $\hat{b}_1$ using formula (5.3):

$$\hat{b}_1 = \frac{n(\sum xy) - (\sum x)(\sum y)}{n(\sum x^2) - (\sum x)^2} = \frac{5(86) - 15(25)}{5(55) - (15)^2} = \frac{430 - 375}{275 - 225} = \frac{55}{50} = 1.1$$

Then, using formula (5.4), figure the estimated y-intercept $\hat{b}_0$ by

$$\hat{b}_0 = \bar{y} - \hat{b}_1\bar{x} = \frac{25}{5} - 1.1\left(\frac{15}{5}\right) = 5 - 3.3 = 1.7$$

Finally, use formula (5.5) to find the least-squares regression line:

$$\hat{y} = \hat{b}_0 + \hat{b}_1 x = 1.7 + 1.1x$$

If you use the alternate formula (5.6), then

$$\hat{y} = \bar{y} + r\left(\frac{s_y}{s_x}\right)(x - \bar{x}) = 5 + .778\sqrt{\frac{5}{2.5}}(x - 3)$$

$$= 5 + 1.1(x - 3) = 5 + 1.1x - 3.3 = 1.7 + 1.1x$$

(b) To draw the least-squares regression line on the scatter diagram, you need two points. You know that the estimated y-intercept $\hat{b}_0$ is 1.7 (when $x = 0$, $y = 1.7$) so you can use that as one point. You also know from Problem 5-12 that $(\bar{x}, \bar{y}) = (3, 5)$, and you can use this ordered pair as the second point. Now you can draw the least-squares regression line, as shown in Fig. 5-24.

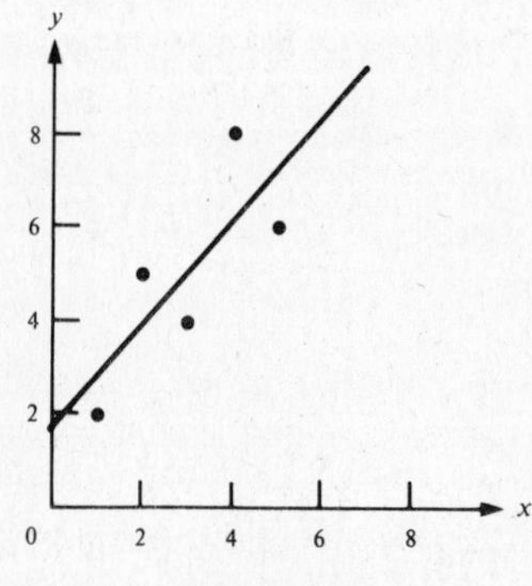

Figure 5-24

PROBLEM 5-18 **(a)** Find the least-squares regression line for the height and weight data in Problems 5-4 and 5-13. **(b)** Draw this line on the scatter diagram.

Solution

(a)

$$\hat{b}_1 = \frac{12(122{,}935) - 825(1765)}{12(57{,}005) - (825)^2} = \frac{19{,}095}{3435} = 5.56$$

$$\hat{b}_0 = \frac{1765}{12} - 5.56\left(\frac{825}{12}\right) = -235.17$$

So, from formula (5.5) the least-squares regression line is $\hat{y} = -235.17 + 5.56x$.

(b) To graph the regression line, substitute a value for x into formula (5.5) to get the corresponding value for $\hat{y}$. Perhaps you should choose $x = 60$ since it's the lowest x value on the graph:

$$\hat{y} = -235.17 + 5.56(60) = 98.43$$

Then substitute, say, $x = 75$ into formula (5.5) to get the corresponding $\hat{y}$-value:

$$\hat{y} = -235.17 + 5.56(75) = 181.83$$

Now plot the two points (60, 98.43) and (75, 181.83) on the graph, as shown in Fig. 5-25.

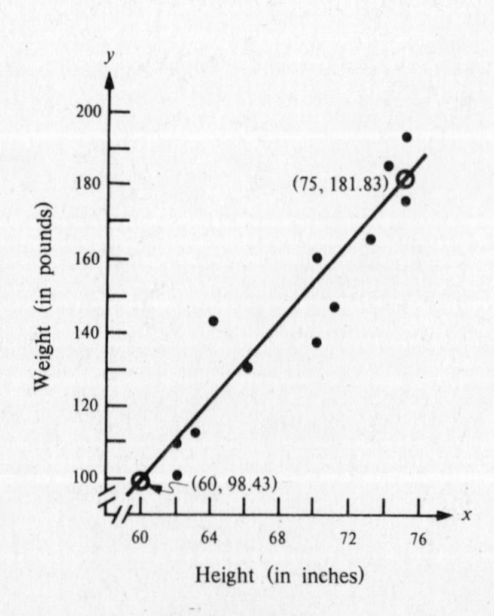

Figure 5-25

PROBLEM 5-19 (**a**) Find the least-squares regression line for the executives' weight data in Problems 5-7 and 5-15. (**b**) Draw this line on the scatter diagram.

Solution

(**a**) From formulas (5.3) and (5.4):

$$\hat{b}_1 = \frac{14(-15{,}578) - (2681)(-79)}{14(521{,}041) - (2681)^2} = \frac{-6293}{106{,}813} = -.059$$

$$\hat{b}_0 = \frac{-79}{14} - (-.059)\left(\frac{2681}{14}\right) = 5.656$$

And from (5.5):

$$\hat{y} = 5.656 - .059x$$

(**b**) Substitute two values for x into (5.5) to get the corresponding values for $\hat{y}$—you might use $x = 155$ and $x = 240$, so that

$$\hat{y} = 5.656 - .059(155) = -3.489$$

and

$$\hat{y} = 5.656 - .059(240) = -8.504$$

Plot these points on the graph and connect them (see Fig. 5-26).

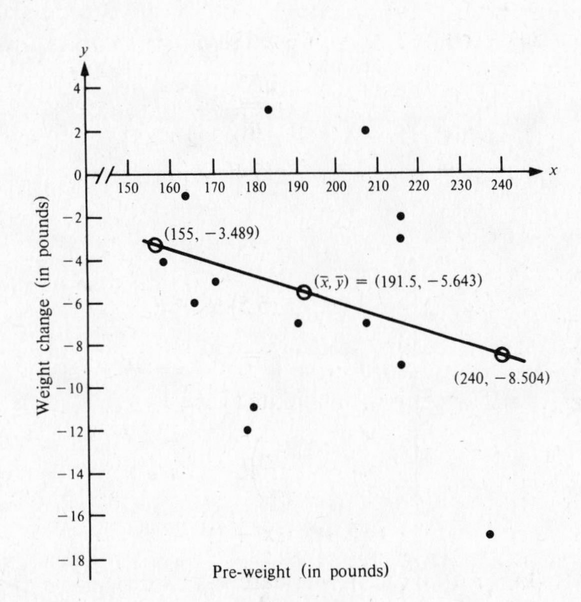

Figure 5-26

PROBLEM 5-20 Fifteen women in their first year of college were measured in a health/fitness program. The weight x (in kg) and the percentage of body fat y for each woman were recorded:

(63.3, 23.2) (62.0, 28.1) (80.9, 30.4) (57.0, 27.4) (55.7, 23.0)
(56.3, 16.9) (59.6, 20.3) (80.3, 35.4) (57.2, 22.4) (69.2, 28.8)
(59.2, 23.5) (51.5, 24.1) (73.0, 30.3) (60.6, 14.6) (61.9, 17.5)

For these data find (**a**) the correlation coefficient r and (**b**) the least-squares regression line. (**c**) Make a scatter diagram and draw the least-squares regression line on the graph.

Solution Construct a table as shown.

x	y	xy	x^2	y^2
63.3	23.2	1,468.56	4,006.89	538.24
62.0	28.1	1,742.20	3,844.00	789.61
80.9	30.4	2,459.36	6,544.81	924.16
57.0	27.4	1,561.80	3,249.00	750.76
55.7	23.0	1,281.10	3,102.49	529.00
56.3	16.9	951.47	3,169.69	285.61
59.6	20.3	1,209.88	3,552.16	412.09
80.3	35.4	2,842.62	6,448.09	1253.16
57.2	22.4	1,281.28	3,271.84	501.76
69.2	28.8	1,992.96	4,788.64	829.44
59.2	23.5	1,391.20	3,504.64	552.25
51.5	24.1	1,241.15	2,652.25	580.81
73.0	30.3	2,211.90	5,329.00	918.09
60.6	14.6	884.76	3,672.36	213.16
61.9	17.5	1,083.25	3,831.61	306.25
947.7	365.9	23,603.49	60,967.47	9384.39

(a) Use (5.2):

$$r = \frac{15(23{,}603.49) - (947.7)(365.9)}{\sqrt{15(60{,}967.47) - (947.7)^2}\sqrt{15(9384.39) - (365.9)^2}} = \frac{7288.92}{\sqrt{16{,}376.76}\sqrt{6883.04}}$$

$$= \frac{7288.92}{10{,}617.06} = .687$$

(b) Use (5.3) and (5.4):

$$\hat{b}_1 = \frac{15(23{,}603.49) - (947.7)(365.9)}{15(60{,}967.47) - (947.7)^2} = \frac{7288.92}{16{,}376.76} = .445$$

$$\hat{b}_0 = \frac{365.9}{15} - .445\left(\frac{947.7}{15}\right) = -3.722$$

Then by (5.5): $\hat{y} = -3.722 + .445x$

(c) See Fig. 5-27.

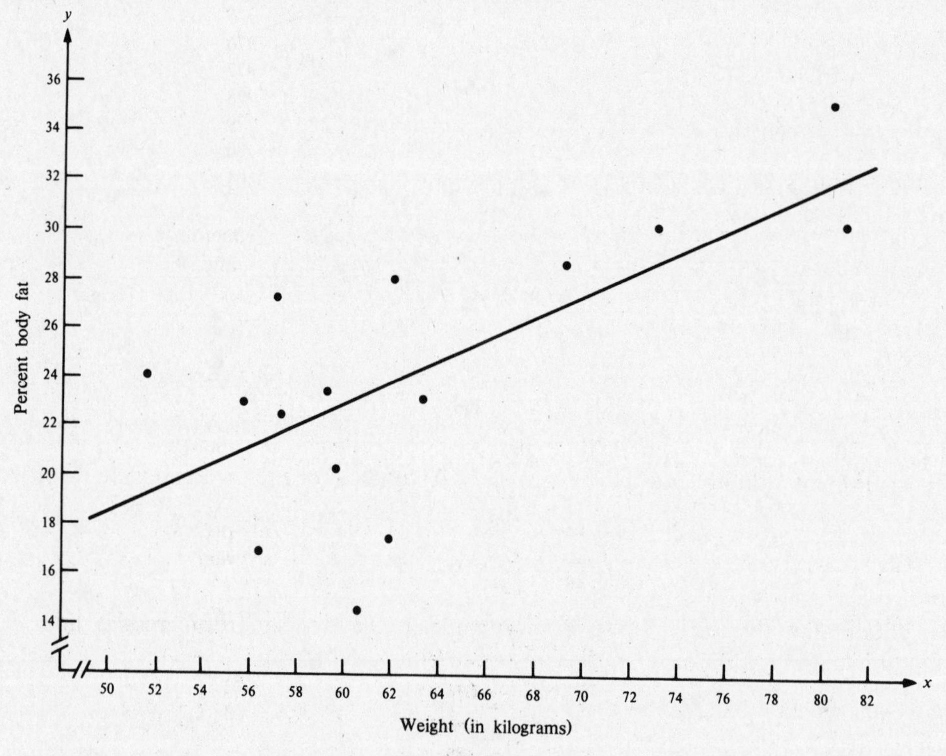

Figure 5-27

Supplementary Exercises

PROBLEM 5-21 The following paired data give the average milligrams x of tar and the average milligrams y of nicotine per cigarette calculated by the FTC method for 12 brands of cigarettes:

(11, .8) (17, 1.1) (10, .9) (2, .2) (14, 1.1) (2, .3)
(5, .5) (17, 1.3) (17, 1.4) (8, .6) (9, .8) (15, 1.0)

Use these data to find (a) $\sum x$, $\sum x^2$, $\sum y$, $\sum y^2$, $\sum xy$; (b) the correlation coefficient r; and (c) the least-squares regression line. (d) Plot the points and the least-squares regression line on the same graph.

Answer **(a)** $\sum x = 127, \sum x^2 = 1687, \sum y = 10, \sum y^2 = 9.9, \sum xy = 128.3$ **(b)** $r = .969$ **(c)** $\hat{y} = .140 + .006x$
(d) See Fig. 5-28.

PROBLEM 5-22 Each of 10 golfers hit 3 balls of brand A and 3 balls of brand B in random order. The following paired values (x, y) give the averages of the total yards covered in the air and on the ground by the brand A and the brand B balls.

(240, 257) (221, 229) (265, 279) (238, 248) (223, 226)
(256, 263) (262, 260) (272, 274) (276, 281) (256, 259)

For these data, find **(a)** the correlation coefficient r and **(b)** the least-squares regression line. **(c)** Plot these points and the regression line on the same graph.

Answer **(a)** $r = .955$ **(b)** $\hat{y} = 24.618 + .929x$ **(c)** See Fig. 5-29.

PROBLEM 5-23 These tabulated data for the years 1970–1981 give the percentage of unemployment in Sweden for men x and women y between the ages of 20 and 24.

(a) Find the correlation coefficient r between x and y.
(b) Find the equation of the least-squares regression line.
(c) Plot these points and the regression line on the same graph.

Answer **(a)** $r = .810$
(b) $\hat{y} = 1.792 + .617x$
(c) See Fig. 5-30.

Year	Men (x)	Women (y)
1970	2.5	2.4
1971	3.7	3.8
1972	4.2	4.9
1973	4.2	4.7
1974	2.7	3.8
1975	2.1	3.5
1976	2.2	3.4
1977	2.9	3.5
1978	4.3	4.3
1979	3.6	3.8
1980	3.5	3.9
1981	4.8	4.6

PROBLEM 5-24 The maximum lung capacity (in liters) was figured for 20 members of a women's college volleyball team before and after the season. Find **(a)** the correlation coefficient r, and **(b)** the least-squares regression line. **(c)** Plot the points and the line on the same graph.

Pre-season (x):	3.4	3.7	3.6	3.4	3.9	3.4	3.3	3.7	3.4	3.6	4.1	3.5	4.1	3.7	3.6	4.3	3.7	4.2	3.8	4.7
Post-season (y):	3.4	3.8	4.1	3.7	3.9	3.4	3.7	3.9	3.4	4.5	4.2	3.6	3.8	4.1	3.7	4.6	3.8	4.3	4.1	4.6

Answer **(a)** $r = .762$ **(b)** $\hat{y} = .972 + .788x$ **(c)** See Fig. 5-31.

PROBLEM 5-25 Researchers studied the effect of exercise on the systolic blood pressure of 8 males, 19–20 years of age. They measured the blood pressure of the youths before (x) and after (y) several minutes of running up and down stairs.
Find **(a)** the correlation coefficient r and **(b)** the least-squares regression line. **(c)** Plot the points and the regression line on the same graph.

Before (x):	120	119	122	120	121	119	115	122
After (y):	150	165	177	157	165	167	159	163

Answer **(a)** $r = .350$ **(b)** $\hat{y} = 15.296 + 1.232x$ **(c)** See Fig. 5-32.

PROBLEM 5-26 The weight x (in kg) and the percentage y of body fat of 15 male freshmen were measured at the beginning of a health/fitness program. Use these data to find **(a)** the correlation coefficient r, and **(b)** the least-squares regression line. **(c)** Plot the points and the regression line on the same graph.

Weight (x):	66.1	68.7	69.1	79.7	65.3	81.3	56.9	71.6	74.9	73.0	88.0	59.5	62.8	75.0	71.3
% Fat (y):	19.7	11.9	9.7	17.3	10.8	12.5	18.0	16.9	10.7	10.8	17.2	22.8	11.6	13.5	10.6

Answer **(a)** $r = -.185$ **(b)** $\hat{y} = 20.626 - .090x$ **(c)** See Fig. 5-33.

PROBLEM 5-27 The percentage x of body fat of 10 male freshmen was measured at the beginning of a health/fitness program. The change y in this percentage was calculated at the end of the program. (A negative y indicates a loss of body fat.) Use these data to find **(a)** the correlation coefficient r and **(b)** the least-squares regression line. **(c)** Plot these points and the regression line on the same graph.

x:	9.5	15.4	11.7	13.7	12.5	10.8	11.0	9.4	15.0	6.8
y:	−2.0	−2.7	−.3	1.0	−.9	.1	−.5	−.6	−4.0	−.4

Answer **(a)** $r = -.429$ **(b)** $\hat{y} = 1.689 - .235x$ **(c)** See Fig. 5-34.

PROBLEM 5-28 For the birth rate x and death rate y for the 22 African countries listed in Problem 5-9, find **(a)** the correlation coefficient r, and **(b)** the least-squares regression line.

Answer **(a)** $r = .288$ **(b)** $\hat{y} = 2.285 + .339x$

PROBLEM 5-29 For the birth rate x and the death rate y for the 6 South American countries listed in Problem 5-10, find **(a)** the correlation coefficient r, and **(b)** the least-squares regression line.

Answer **(a)** $r = .809$ **(b)** $y = -13.970 + .630x$

PROBLEM 5-30 For the birth rate x and the death rate y for the 18 European countries listed in Problem 5-11, find **(a)** the correlation coefficient and **(b)** the least-squares regression line.

Answer **(a)** $r = -.264$ **(b)** $y = 12.297 - .141x$

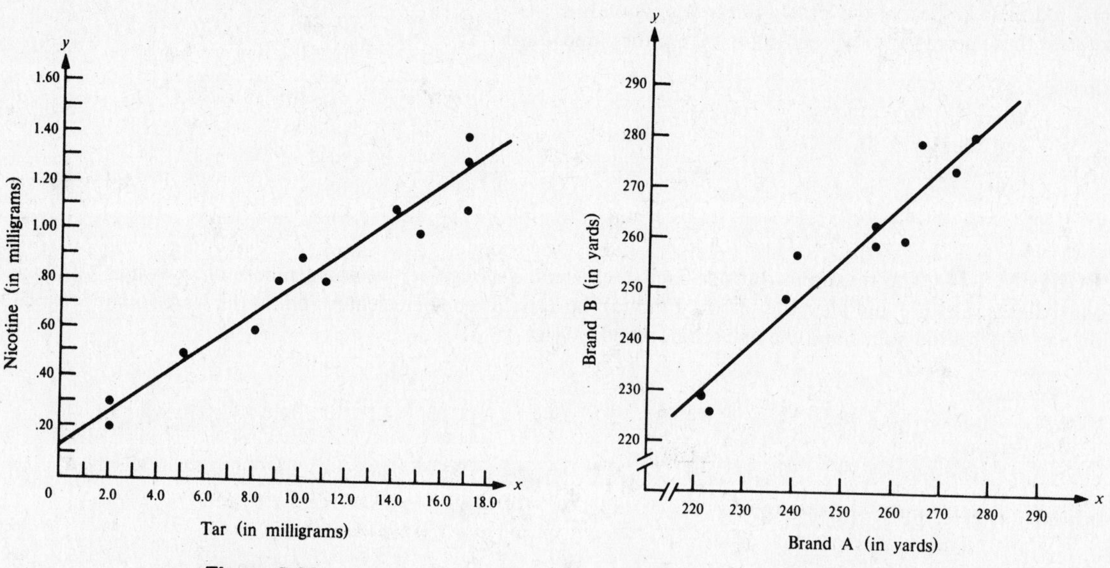

Figure 5-28

Figure 5-29

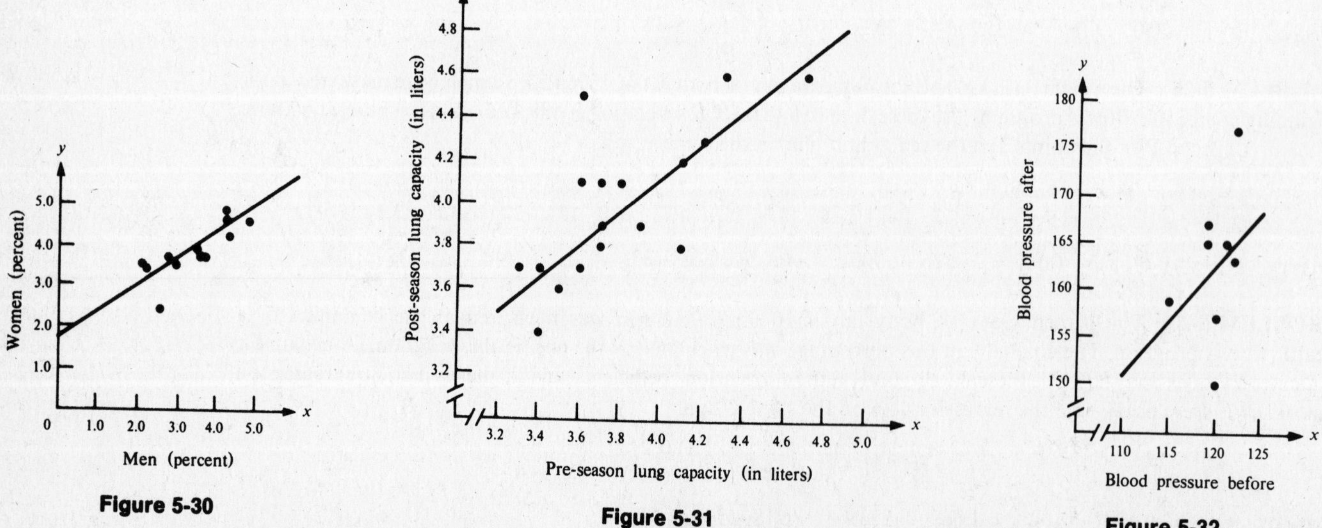

Figure 5-30

Figure 5-31

Figure 5-32

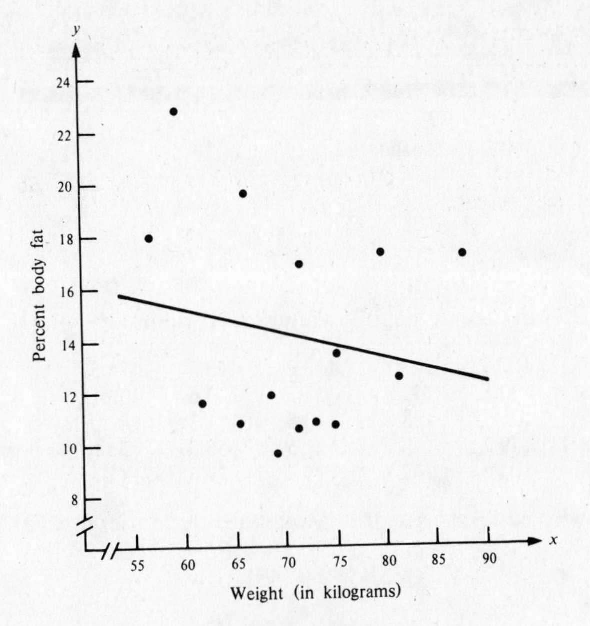

Figure 5-33

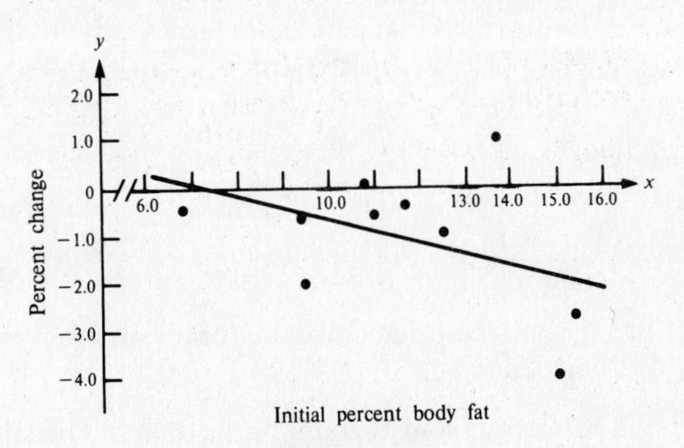

Figure 5-34

EXAM 1

1. A 4-sided die was rolled 24 times, yielding the following outcomes:

1	4	2	4	3	3	2	1	2	4	3	1
3	4	4	2	1	1	2	4	1	3	1	2

Group these data into a frequency distribution.

2. Fifty women aged 35 through 39 completed a 25-km race in the following times (in minutes):

109.1	111.0	111.5	113.6	114.1	122.6	123.1	123.2	123.5	124.5
125.3	127.4	128.1	128.2	130.8	132.5	132.8	133.3	133.6	135.1
136.8	137.0	137.7	138.0	138.3	138.7	138.8	140.5	141.3	142.0
142.5	144.7	145.2	145.4	147.8	149.7	150.3	151.5	153.5	155.6
156.9	158.5	160.4	162.9	163.8	165.6	167.8	175.5	179.6	187.7

Group these data into a frequency distribution with equal class widths. Make the boundaries for the first class 107.25–119.25.

3. Construct a histogram for the data in Question 1.

4. Construct a histogram for the data in Question 2.

5. The weights (in kilograms) of 25 female college freshmen in a health/fitness program were

58.0 51.0 57.0 57.4 49.8 42.7 53.8 47.0 61.0 65.5 54.5 64.5
64.3 67.0 60.5 68.8 71.0 51.2 65.6 63.5 63.4 56.3 45.7 72.3 70.5

The weights (in kilograms) of 25 male college freshmen in a health/ fitness program were

72.2 75.4 68.2 59.0 63.0 74.0 66.1 82.5 71.0 74.5 80.0 78.0
58.4 77.7 75.3 85.0 73.0 79.3 79.1 72.0 54.0 85.5 69.5 83.4 72.9

Make a two-sided stem-and-leaf diagram for these data, with the weights of the women on the left and the weights of the men on the right. Use as stems 4∗, 4•, 5∗, 5•,...,8∗, 8•.

6. Make a one-sided stem-and-leaf diagram for the 50 weights in Question 5.

7. Draw the empirical distribution function for the data in Question 1.

8. Draw the cumulative relative frequency ogive curve for the data in Question 2.

9. Five rolls of a fair 12-sided die were 8, 3, 12, 7, 5. Find **(a)** the sample mean and **(b)** the median.

10. Find the sample mean for the data in Question 1.

11. Find the sample mean for the grouped data in Question 2.

12. Find the mode for the data in Question 1.

13. Find **(a)** the sample variance and **(b)** the standard deviation for the data in Question 9.

14. Find **(a)** the sample variance and **(b)** the standard deviation for the data in Question 1.

15. Find the variance for the grouped data in the tabulation shown.

Class limits	Frequency
21–23	1
24–26	3
27–29	9
30–32	5
33–35	2

16. If $n = 360$ students took a test for which the sample mean is $\bar{x} = 74.5$ and the standard deviation is $s = 12$, at least how many students had scores between 50.5 and 98.5?

17. Find the range for all of the weights in Question 5.

18. For the 50 race times in Question 2, find **(a)** the first quartile q_1, **(b)** the median m, **(c)** the third quartile q_3, and **(d)** the 60th percentile.

19. Given $\bar{x} = 83.7$ and $s = 10$, find the z-score for $x = 91$.

20. If $\bar{x} = 83.7$, $s = 10$, and the z-score is $z = -1.5$, what is the value of x?

21. Draw a box-and-whisker diagram for the data in Question 5, making use of the stem-and-leaf diagram drawn in Question 6.

22. A 6-sided die was rolled twice. Let x equal the outcome on the first roll and y equal the sum of the rolls. This experiment was repeated five times with the following results: (4, 6), (2, 3), (6, 11), (1, 5), (3, 9). Make a scatter diagram for these data.

23. Find the correlation coefficient r for the data in Question 22.

24. **(a)** Find the least-squares regression line for the data in Question 22. **(b)** Graph this line on the scatter diagram.

25. If $\bar{x} = 50$, $\bar{y} = 70$, $s_x = 5$, $s_y = 9$, and $r = -.65$, find the equation of the least-squares regression line.

Answers to Exam 1

1. (Section 1-1)

Outcome	Tally	Frequency
1	𝍸 ||	7
2	𝍸 |	6
3	𝍸	5
4	𝍸 |	6
		24

2. (Section 1-1)

Class boundaries	Tally	Frequency
107.25–119.25	𝍸	5
119.25–131.25	𝍸 𝍸	10
131.25–143.25	𝍸 𝍸 𝍸 |	16
143.25–155.25	𝍸 |||	8
155.25–167.25	𝍸 ||	7
167.25–179.25	||	2
179.25–191.25	||	2
		50

3. (Section 1-2) See Fig. E1-1.

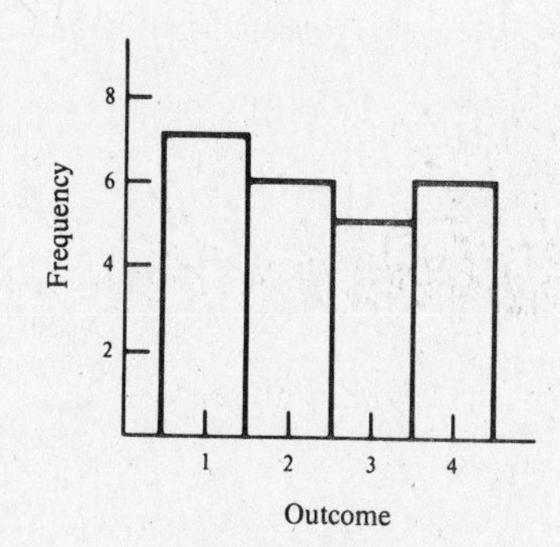

Figure E1-1

4. (Section 1-2) See Fig. E1-2.

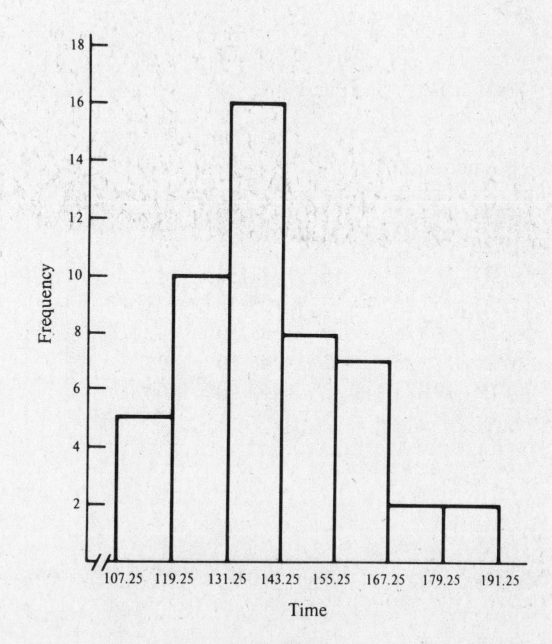

Figure E1-2

5. (Section 1-3)

Women	Stems	Men
27	4∗	
57 70 98	4·	
12 45 38 10	5∗	40
63 74 70 80	5·	90 84
34 35 05 43 45 10	6∗	30
56 88 70 55	6·	82 61 95
05 23 10	7∗	22 40 10 45 30 20 29
	7·	54 80 77 53 93 91
	8∗	25 00 34
	8·	50 55

6. (Section 1-3)

Stems	Leaves
4∗	27
4·	98 70 57
5∗	10 38 45 12 40
5·	80 70 74 63 90 84
6∗	10 45 43 05 35 34 30
6·	55 70 88 56 82 61 95
7∗	10 23 05 22 40 10 45 30 20 29
7·	54 80 77 53 93 91
8∗	25 00 34
8·	50 55

7. (Section 1-4) See Fig. E1-3.

Outcome (x)	Freq. (f)	Cum. rel. freq.
1	7	7/24 = .29
2	6	13/24 = .54
3	5	18/24 = .75
4	6	24/24 = 1.00

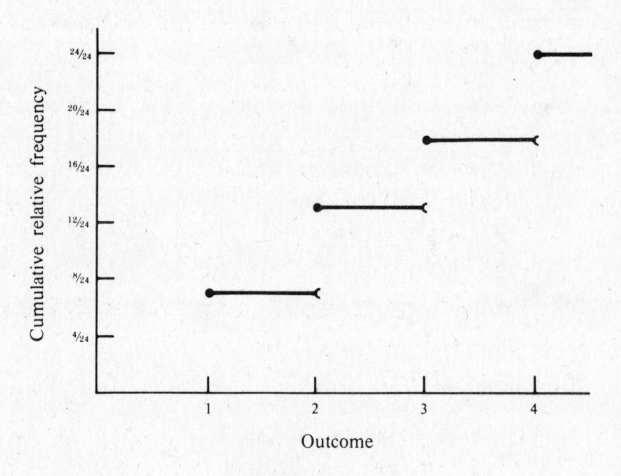

Figure E1-3

8. (Section 1-4) See Fig. E1-4.

Class boundaries	Freq. (f)	Cum. rel. freq.
107.25–119.25	5	5/50 = .10
119.25–131.25	10	15/50 = .30
131.25–143.25	16	31/50 = .62
143.25–155.25	8	39/50 = .78
155.25–167.25	7	46/50 = .92
167.25–179.25	2	48/50 = .96
179.25–191.25	2	50/50 = 1.00

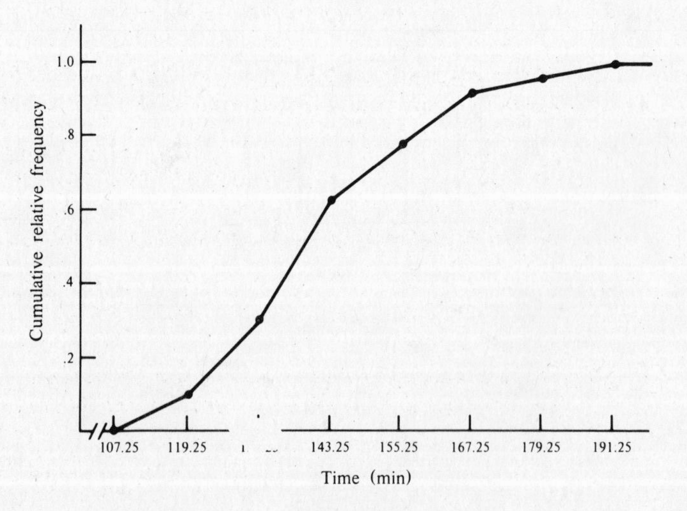

Figure E1-4

9. **(a)** (Section 2-1) $\bar{x} = \dfrac{\sum x}{n} = \dfrac{8 + 3 + 12 + 7 + 5}{5} = \dfrac{35}{5} = 7$

(b) (Section 2-3) Ordering the data yields $3 < 5 < 7 < 8 < 12$. When the number of observations is odd, the median is the middle observation, so the median is $m = 7$.

10. (Section 2-2) $\bar{x} = \dfrac{\sum fx}{n} = \dfrac{58}{24} = 2.42$

11. (Section 2-2) $\bar{u} = \dfrac{\sum fu}{n} = \dfrac{7054.50}{50} = 141.09$

12. (Section 2-4) The mode is the observation that occurs most frequently, so for these data the mode is equal to 1.

13. (Section 3-1)

(a) Using formula (3.1): $s^2 = \dfrac{\sum(x - \bar{x})^2}{n - 1} = \dfrac{46}{4} = 11.5$

Using formula (3.3): $s^2 = \dfrac{n\sum x^2 - (\sum x)^2}{n(n-1)} = \dfrac{5(291) - (35)^2}{5(4)} = \dfrac{230}{20} = 11.5$

(b) $s = \sqrt{s^2} = \sqrt{11.5} = 3.39$

14. (Section 3-2)

(a) Using formula (3.4): $s^2 = \dfrac{\sum f(x - \bar{x})^2}{n - 1} = \dfrac{31.83}{23} = 1.384$

Using formula (3.5): $s^2 = \dfrac{n\sum fx^2 - (\sum fx)^2}{n(n-1)} = \dfrac{24(172) - (58)^2}{24(23)} = \dfrac{764}{552} = 1.384$

(b) Using formula (3.2), the standard deviation is $s = \sqrt{1.384} = 1.176$.

15. (Section 3-2) $s_u^2 = \dfrac{n\sum fu^2 - (\sum fu)^2}{n(n-1)} = \dfrac{20(16{,}532) - (572)^2}{20(19)} = \dfrac{3456}{380} = 9.095$

16. (Section 3-3) The score 50.5 is two standard deviations below the mean and the score 98.5 is two standard deviations above the mean, so $k = 2$. Chebyshev's inequality guarantees that at least $1 - 1/(k^2) = 1 - 1/(2^2) = 1 - (1/4) = 3/4 = 75\%$ of the scores will fall within the interval (50.5, 98.5). So at least $(360)(75\%) = 270$ scores are between 50.5 and 98.5.

17. (Section 3-4) The range is the difference between the maximum and minimum values in a set of data. Here the range is $85.5 - 42.7 = 42.8$.

18. (Section 4-1)

(a) The *location* of q_1 is $k(n + 1)/4$, or $1(50 + 1)/4 = 51/4 = 12.75$.
The *value* of q_1 is therefore the average of the 12th and 13th times:

$$q_1 = \frac{127.4 + 128.1}{2} = 127.75$$

(b) The location of the median m (or q_2) is $2(50 + 1)/4 = 102/4 = 25.5$.
The value of the median is the average of the 25th and 26th times:

$$m = \frac{138.3 + 138.7}{2} = 138.5$$

(c) Because $3(51)/4 = 38.25$, the third quartile q_3 is the average of the 38th and 39th times:

$$q_3 = \frac{151.5 + 153.5}{2} = 152.5$$

(d) The location of the 60th percentile p_{60} (or 6th decile, d_6) $= k(n + 1)/100 = 60(51)/100 = 30.6$.
The value of p_{60} is therefore the average of the 30th and 31st times:

$$p_{60} = d_6 = \frac{142.0 + 142.5}{2} = 142.25$$

19. (Section 4-2) $z = \dfrac{x - \bar{x}}{s} = \dfrac{91 - 83.7}{10} = \dfrac{7.3}{10} = .73$

20. (Section 4-2) $x = \bar{x} + zs = 83.7 + (-1.5)(10) = 68.7$

21. (Section 4-3) To draw this diagram, you must find the minimum and maximum values, q_1, $m = q_2$, and q_3:

minimum = 42.7
maximum = 85.5
$q_1 = 57.7$
$m = q_2 = 66.55$
$q_3 = 74.25$

See Fig. E1-5.

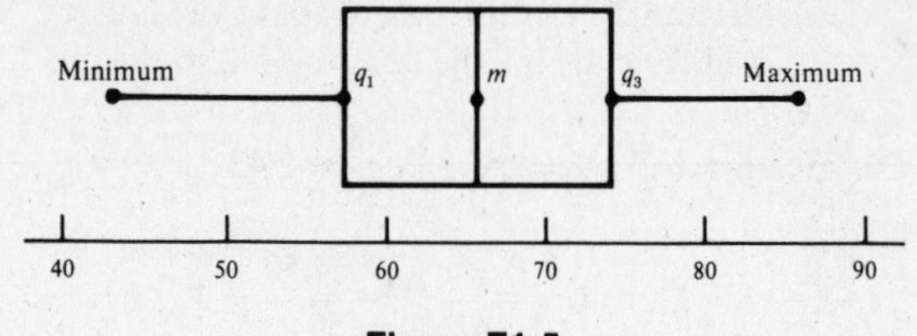

Figure E1-5

22. (Section 5-1) To make this diagram, simply plot each paired measurement as (x, y)-coordinates on a graph. See Fig. E1-6.

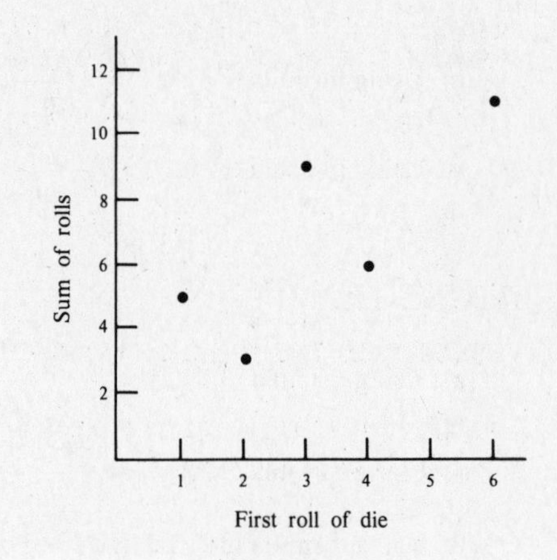

Figure E1-6

23. (Section 5-2)

$$r = \frac{n(\sum xy) - (\sum x)(\sum y)}{\sqrt{n(\sum x^2) - (\sum x)^2}\sqrt{n(\sum y^2) - (\sum y)^2}}$$

$$= \frac{5(128) - 16(34)}{\sqrt{5(66) - (16)^2}\sqrt{5(272) - (34)^2}}$$

$$= \frac{96}{\sqrt{74}\sqrt{204}} = .781$$

24. (Section 5-3)

(a) Using the data you derived in Question 23 and formula (5.3), find the slope of the least-squares regression line:

$$\hat{b}_1 = \frac{n(\sum xy) - (\sum x)(\sum y)}{n(\sum x^2) - (\sum x)^2}$$

$$= \frac{5(128) - 16(34)}{5(66) - (16)^2} = \frac{96}{74} = 1.30$$

By formula (5.4), you find the y-intercept of the line:

$$\hat{b}_0 = \bar{y} - \hat{b}_1\bar{x} = \frac{34}{5} - \left(\frac{96}{74}\right)\left(\frac{16}{5}\right) = 6.8 - 4.15 = 2.65$$

Thus by formula (5.5) the least-squares regression line is

$$\hat{y} = \hat{b}_0 + \hat{b}_1 x = 2.65 + 1.30x$$

(b) Graph this line by plotting any two pairs of points that satisfy the line's equation, then drawing a line through those two points. See Fig. E1-7.

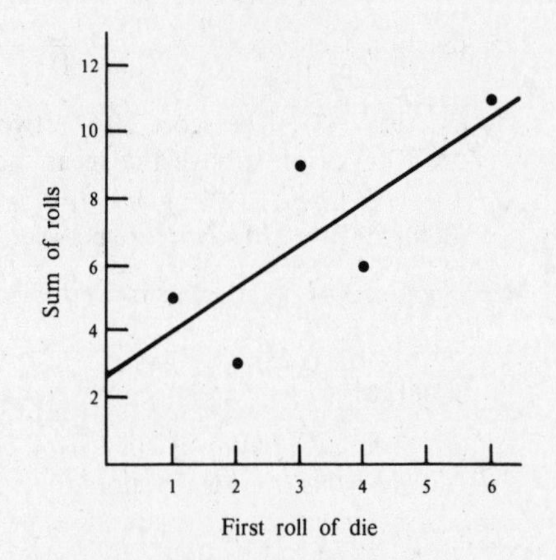

Figure E1-7

25. (Section 5-3) Using formula (5.6), you find

$$y = \bar{y} + r\frac{s_y}{s_x}(x - \bar{x}) = 70 + (-.65)\left(\frac{9}{5}\right)(x - 50) = 70 - 1.17(x - 50)$$

$$= 70 - 1.17x + 58.5 = 128.5 - 1.17x$$

6 PROBABILITY

THIS CHAPTER IS ABOUT

- ☑ The Concept of Probability
- ☑ Properties of Probability
- ☑ Methods of Enumeration
- ☑ Probabilities for Intersections of Events

In order to interpret statistical inferences correctly, you need an understanding of probability since it is probability that provides the vehicle for making inferences from data.

6-1. The Concept of Probability

A. Sample spaces

In statistics, almost anything that has consequences (outcomes) can be an *experiment*. An experiment for which we can describe *all* the possible outcomes—even though we cannot predict a *particular* outcome with certainty—is called a **random experiment**. The *set* of all possible outcomes of any random experiment is called the **sample space**. We denote a sample space with S, and often describe S by collecting the outcomes of the random experiment inside braces { }.

EXAMPLE 6-1 The cups of coffee on a buffet table have been poured from 2 coffee urns. One urn contains regular coffee and the other contains decaffeinated coffee. If you select one of these cups of coffee at random, what is the sample space?

Solution The sample space is the set of all the possible outcomes of the random experiment of selecting a cup of coffee:

$$S = \{\text{regular coffee, decaffeinated coffee}\}$$

EXAMPLE 6-2 You roll a fair 6-sided die once and note the outcome on the die. Describe the sample space.

Solution The sample space is $S = \{1, 2, 3, 4, 5, 6\}$.

EXAMPLE 6-3 You roll a 4-sided die twice. Describe the sample space.

Solution You can describe S by using the following list of 16 possible outcomes. The first number in each pair is the outcome on the first roll, and the second number is the outcome on the second roll.

1,1	1,2	1,3	1,4
2,1	2,2	2,3	2,4
3,1	3,2	3,3	3,4
4,1	4,2	4,3	4,4

or $S = \{(1,1), (1,2), (1,3), (1,4), (2,1), (2,2), (2,3), (2,4), (3,1), (3,2), (3,3), (3,4), (4,1), (4,2), (4,3), (4,4)\}$.

EXAMPLE 6-4 To promote the sale of Freakies Cereal, a plastic model of one of the 7 Freakies characters is put into each box. The customer will get one particular plastic Freakie with each box of Freakies purchased. List the sample space.

Solution Assuming that you know the names of the Freakies (probably a false assumption), you can list the sample space as

$$S = \{\text{Bossmoss, Gargle, Hamhose, Grumble, Goody-Goody, Snorkeldorf, Cowmumble}\}$$

EXAMPLE 6-5 In some states you can win a lottery by correctly identifying a 4-digit integer that's selected at random. List the sample space for this experiment.

Solution
$$S = \{0000, 0001, 0002, \ldots, 9999\}$$
where the 3 dots stand for all of the integers between 0002 and 9999.

EXAMPLE 6-6 Educators and programmers have written a computer program to help school children learn the 50 state capitals. The computer selects a state at random; then the child is asked to name the capital of that state. List the sample space.

Solution S is a listing of the 50 states.

EXAMPLE 6-7 What's the sample space if you select a card from a standard deck of playing cards?

Solution S is a listing of the 52 cards in a deck.

EXAMPLE 6-8 When a 6-sided die is rolled 6 times, you can say that a match occurs if side k is rolled on the kth roll. "Success" is the possibility that at least one match will occur on the 6 rolls, and "failure" is the possibility that no matches will occur on the 6 rolls. Describe the sample space.

Solution The outcomes here aren't the 6 rolls of the die, but the possibilities that at least one match or that no matches will occur on the 6 rolls. That is, there are two possible outcomes: $S = \{\text{Success, Failure}\}$.

EXAMPLE 6-9 The distribution of students by class standing in a statistics class is

Freshmen	5
Sophomores	37
Juniors	42
Seniors	16
	100

Select a student at random. List the sample space if you want to know only the class standing of the selected student.

Solution
$$S = \{\text{Freshman, Sophomore, Junior, Senior}\}$$

EXAMPLE 6-10 You roll a pair of 6-sided dice until a sum of 7 is observed. Describe the sample space if you want to know how many rolls it takes to get a 7.

Solution $S = \{1, 2, 3, 4, \ldots\}$, where the 3 dots denote the rest of the positive integers. [Later we will see that the probabilities associated with large integer values are very small.]

Notice that the likelihood of observing a specific event within a sample space depends on different things. In Examples 6-2 to 6-7, each of the outcomes listed has the same chance of being observed; but in Example 6-9, it is more likely that a student selected at random is a junior than a senior because of the uneven distribution. Then, in Example 6-1, the probabilities for the 2 outcomes must depend on the unknown number of cups of coffee of each kind. Finally, it's clear that we need more background before we can assign probabilities to the events in Examples 6-8 and 6-10. We need to know more about *events*.

B. Events

An **event** is a subset A of the sample space S associated with a given random experiment. That is, A may contain one or more outcomes that belong to S, or A may contain no outcomes at all (the null set). One and only one event A of S will OCCUR—and we say that event A OCCURS if the outcome of the random experiment belongs to A.

Each outcome of the random experiment, i.e., each ELEMENT of S, is called a **simple event**. If A contains only one element of S, A is called a **simple event**. If A contains more than one element of S (i.e., more than one simple event), A is called a **compound event**.

EXAMPLE 6-11 In Example 6-3, let A be **(a)** the subset of S that is the event "the sum of the two rolls is equal to 2"; **(b)** the subset of S that is the event "the sum of the two rolls is equal to 4." List the outcomes [elements] in A for each case. In which is A a simple event? a compound event?

Solution Out of the 16 possible outcomes in S, the event A is

(a) $A = \{(1, 1)\}$

(b) $A = \{(1, 3), (2, 2), (3, 1)\}$

A is a simple event in case **(a)** and A is a compound event—made up of three simple events—in case **(b)**.

EXAMPLE 6-12 In Example 6-5, let A be the event that the integer selected contains two 3's and two 4's. Describe A.

Solution $A = \{3344, 3434, 4334, 4343, 4433, 3443\}$

C. Probability for an event

The **probability** for the event A—denoted $P(A)$—is the likelihood that an outcome of A will occur. The probability for an event is a *number* between 0 and 1. Thus an event that is *certain* to occur has a probability of 1, and an event that can *never* occur has a probability of 0.

Probability for equally likely events: For some random experiments each of the possible outcomes—the simple events—has the same chance of being observed. We call such events **equally likely events**. There is a very easy way to define probabilities associated with such experiments:

$$P(A) = \frac{k}{h}$$

where

- h is the number of outcomes in the sample space,
- k is the number of outcomes in A,
- each of the h possible outcomes is equally likely.

EXAMPLE 6-13 When you roll a 4-sided die twice, each of the 16 possible pairs of outcomes is equally likely. Give the probability for the events $A = \{(1, 1)\}$ and $A = \{(1, 3), (2, 2), (3, 1)\}$, as described in Example 6-11.

Solution There are $h = 16$ outcomes in S. If $A = \{(1, 1)\}, k = 1$; and if $A = \{(1, 3), (2, 2), (3, 1)\}, k = 3$. So

$$\textbf{(a)}\ P(A) = \frac{k}{h} = \frac{1}{16} \qquad \textbf{(b)}\ P(A) = \frac{k}{h} = \frac{3}{16}$$

Out of the 16 possible outcomes in this random experiment, the probability that you'll get a sum of 2 on the two rolls is $1/16$ and the probability that you'll get a sum of 4 on the two rolls is $3/16$.

EXAMPLE 6-14 Let A be the event that a heart is selected from a standard deck of playing cards and let B be the event that an ace is selected. Find **(a)** $P(A)$ and **(b)** $P(B)$.

Solution

(a) There are $k = 13$ hearts in the deck of $h = 52$ cards, so

$$P(A) = \frac{13}{52} = \frac{1}{4}$$

The probability of selecting a heart is one in four, or $1/4 = .25 = 25\%$.

(b) There are 4 aces in the deck, so

$$P(B) = \frac{4}{52} = \frac{1}{13}$$

EXAMPLE 6-15 Let A in Example 6-9 be the event that a junior *or* senior is selected. Find $P(A)$.

Solution Notice that A is a compound event, where $A = \{\text{Junior, Senior}\}$. And there are 42 juniors and 16 seniors among the 100 students. So

$$P(A) = \frac{42 + 16}{100} = \frac{58}{100} = .58$$

EXAMPLE 6-16 Give the probability for the event $A = \{3344, 3434, 4334, 4343, 4433, 3443\}$ in the sample space $S = \{0000, 0001, 0002, \ldots, 9999\}$ described in Example 6-12.

Solution The number of outcomes in A is 6, and the number of possible 4-digit integers between 0000 and 9999 is 10,000; so

$$P(A) = \frac{6}{10{,}000}$$

Remember that each of the h events must be equally likely. If the lottery were "fixed" so that some digits occurred more frequently than others, you couldn't figure the probability for A this way.

EXAMPLE 6-17 Give the probability for selecting a cup of regular coffee (*with* caffeine) when $S = \{\text{regular coffee, decaffeinated coffee}\}$ as in Example 6-1.

Solution Let $A = \{\text{regular coffee}\}$. In order to solve this problem, you must know the total number of cups of coffee on the buffet table, h, and the number of those cups that have caffeine, k. Then,

$$P(A) = \frac{k}{h}$$

note: In Examples 6-13 to 6-16, it was relatively easy to determine the values of k and h. For many experiments, we'll need special counting techniques to determine the values of k and h. These techniques are given in Section 6-3.

Relative frequency concept of probability: To define $P(A)$ for experiments in which the outcomes are not necessarily equally likely, we consider the repetition of the experiment n times. Then we let $\#(A)$ denote the number of times an outcome belongs to A, so that $\#(A)/n$ denotes the proportion of times that A has occurred. That is, $\#(A)/n$ is the relative frequency of the occurrence of A. As n becomes large, we expect $\#(A)/n$ to approach some number p between 0 and 1 and stabilize there. It's this number p that is equal to the probability $P(A)$ for event A.

EXAMPLE 6-18 For the roll of a fair 6-sided die, where $S = \{1, 2, 3, 4, 5, 6\}$, if $A = \{1, 2\}$, then $P(A) = (1 + 1)/6 = 2/6 = 1/3$ by our formula. How would you use the relative frequency concept of probability to test this statement?

Solution You would test this empirically by rolling a fair 6-sided die a large number of times and counting the number of times that A occurs. That is, if $\#(A)$ is the number of times that the outcome is a 1 or a 2 and if n is the number of trials, then

$$p = \frac{\#(A)}{n} \approx \frac{1}{3}$$

where n is large. In other words, if you roll the die often enough, you will roll a 1 or a 2 about one-third of the time, so that $p \approx 1/3 = P(A)$.

EXAMPLE 6-19 Describe how you could apply the relative frequency concept of probability using the Success/Failure experiment described in Example 6-8.

Solution Let event $A = \{\text{Success}\}$—so that A occurs if there is at least one match. You can't calculate $P(A)$ right now, so you have to find p experimentally by rolling a 6-sided die 6 times and seeing whether A occurs— was side 1 observed on roll 1, or side 2 observed on roll 2, ..., or side 6 observed on roll 6? If at least one match occurred, the trial was a success. Repeat the six rolls a large number (n) of times, counting the number of successes $\#(A)$. It should be true that A occurs about two-thirds of the time; i.e., $P(A) = 2/3$, so $p \approx 2/3$. [See Example 6-43.]

6-2. Properties of Probability

Given a random experiment with sample space S, let A, B, and C be subsets of S so that A, B, and C are events associated with this experiment.

A. Relationships between events: Definitions

- The **intersection** of two events A and B is denoted by $A \cap B$ and is equal to those outcomes that belong to both A AND B.
- The **union** of two events A and B is denoted by $A \cup B$ and is equal to those outcomes that belong to A OR to B OR to both A and B.
- The **empty** or **null set** is an event that contains NO outcomes of the random experiment. We use $\varnothing$ to denote the empty set.
- If A and B have no outcomes in common—that is, if $A \cap \mathrm{B} = \varnothing$—then A and B are **mutually exclusive events**.
- The **complement** of an event A, denoted A', is the set of outcomes in S that do not belong to A. Thus $A \cup A'$ is the entire sample space S, and $A \cap A'$ is the null set $\varnothing$.

EXAMPLE 6-20 A 12-sided die is rolled once. The sample space is $S = \{1, 2, 3, \ldots, 12\}$. Let $A = \{3, 6, 9, 12\}$, $B = \{4, 8, 12\}$, and $C = \{2, 6, 10\}$. Find **(a)** $A \cap B$, **(b)** $B \cup C$, **(c)** $A \cup B$, **(d)** $B \cap C$, and **(e)** A'.

Solution

(a) The intersection of A and B includes the outcomes common to both A and B. So $A \cap B = \{12\}$.
(b) The union of B and C includes the outcomes that belong to B or to C or to both B and C. So $B \cup C = \{2, 4, 6, 8, 10, 12\}$.
(c) $A \cup B = \{3, 4, 6, 8, 9, 12\}$.
[Notice that outcomes common to both sets are listed only once in the union set.]
(d) The intersection of B and C is $B \cap C = \varnothing$, the empty set.
(e) The complement of A is the set of points in S that don't belong to A. So $A' = \{1, 2, 4, 5, 7, 8, 10, 11\}$.
[Notice that $A \cup A' = S$ and $A \cap A' = \varnothing$.]

EXAMPLE 6-21 Assuming that a 6-sided die is rolled twice, consider the following events:

$$A = \{\text{sum is } 4\} = \{(1,3), (2,2), (3,1)\}$$

$$B = \{\text{greatest of the two rolls is equal to } 3\} = \{(1,3), (2,3), (3,1), (3,2), (3,3)\}$$

$$C = \{\text{sum is } 11\} = \{(5,6), (6,5)\}$$

$$D = \{\text{sum is } \leq 6\} = \{(1,1), (1,2), (1,3), (1,4), (1,5), (2,1), (2,2), (2,3), (2,4), (3,1), (3,2), (3,3), (4,1), (4,2), (5,1)\}$$

Define the following sets: **(a)** $A \cap B$, **(b)** $A \cup B$, **(c)** $A \cap C$, **(d)** $A \cup C$, **(e)** D'.

Solution Use the following figure to display S and the events A, B, C, and D.

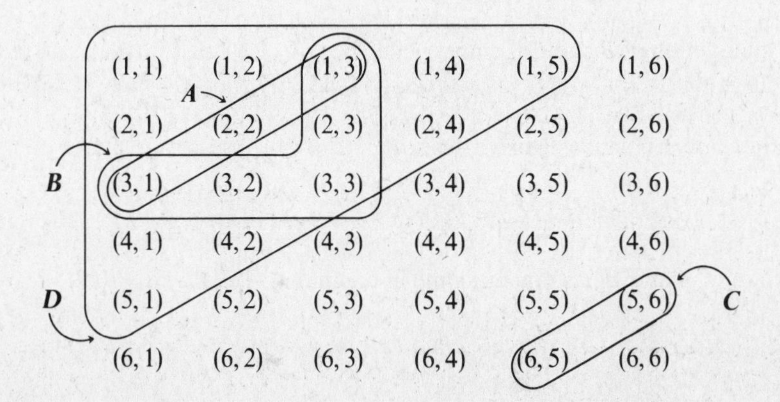

From the figure you can see that

(a) $A \cap B = \{(1,3), (3,1)\}$
(b) $A \cup B = \{(1,3), (2,2), (2,3), (3,1), (3,2), (3,3)\}$
(c) $A \cap C = \varnothing$
(d) $A \cup C = \{(1,3), (2,2), (3,1), (5,6), (6,5)\}$
(e) $D' = \{\text{sum} > 6\} = \{(1,6), (2,5), (2,6), (3,4), (3,5), (3,6), (4,3), (4,4), (4,5), (4,6), (5,2), (5,3), (5,4), (5,5), (5,6), (6,1), (6,2), (6,3), (6,4), (6,5), (6,6)\}$

B. Probabilities for unions and complements of events

- When A and B are mutually exclusive events—i.e., $A \cap B = \varnothing$—then the probability that A or B will occur is equal to the sum of the probabilities that each will occur individually:

$$P(A \cup B) = P(A) + P(B) \qquad \textbf{(6.1)}$$

- If A and B are NOT mutually exclusive events, the probability for the union of the two events is equal to the sum of their individual probabilities minus the probability of their intersection:

$$P(A \cup B) = P(A) + P(B) - P(A \cap B) \qquad \textbf{(6.2)}$$

- The empty set contains no outcomes of an experiment, so its probability is equal to zero:

$$P(\varnothing) = 0 \qquad \textbf{(6.3)}$$

note: Formula (6.2) reduces to formula (6.1) when A and B are mutually exclusive, $A \cap B = \varnothing$, since $P(A \cap B) = 0$.

Because S is the set of all possible outcomes for a random experiment, so that $P(S) = 1$, it's a "sure bet" that the experiment will result in an event of S. And since $A \cup A' = S$, we have

$$P(S) = 1$$

$$P(A \cup A') = 1$$

But from formula (6.1), since $A \cap A' = \varnothing$,

$$P(A \cup A') = P(A) + P(A')$$

then

$$P(A) + P(A') = 1$$

Thus,

$$P(A) = 1 - P(A') \quad \text{or} \quad P(A') = 1 - P(A) \qquad \textbf{(6.4)}$$

In other words:

- The probability that A will occur is 1 minus the probability that A won't occur.

EXAMPLE 6-22 For the experiment in Example 6-20, —where $S = \{1, 2, 3, \ldots, 12\}$ and $A = \{3, 6, 9, 12\}$, $B = \{4, 8, 12\}$, $C = \{2, 6, 10\}$—find **(a)** $P(B \cap C)$, **(b)** $P(B \cup C)$, **(c)** $P(A \cap B)$, **(d)** $P(A \cup B)$, **(e)** $P(A)$, and **(f)** $P(A')$.

Solution

(a) Because the intersection of B and C is empty—i.e., $B \cap C = \varnothing$—use formula (6.3) to find $P(B \cap C)$:

$$P(B \cap C) = P(\varnothing) = 0$$

(b) Because B and C are mutually exclusive—i.e., $B \cap C = \varnothing$—use formula (6.1):

$$P(B \cup C) = P(B) + P(C) = \frac{3}{12} + \frac{3}{12} = \frac{6}{12}$$

(c) The intersection of A and B has one outcome in common—i.e., $A \cap B = \{12\}$—so

$$P(A \cap B) = \frac{1}{12}$$

(d) Because A and B aren't mutually exclusive—i.e., $A \cap B = \{12\}$, use formula (6.2) to find $P(A \cup B)$:

$$P(A \cup B) = P(A) + P(B) - P(A \cap B)$$
$$= \frac{4}{12} + \frac{3}{12} - \frac{1}{12} = \frac{6}{12}$$

(e) A contains four outcomes, so

$$P(A) = \frac{4}{12}$$

(f) From formula (6.4),

$$P(A') = 1 - P(A) = 1 - \frac{4}{12} = \frac{8}{12}$$

EXAMPLE 6-23 For the experiment in Example 6-21, find **(a)** $P(A \cap B)$, **(b)** $P(A \cup B)$, **(c)** $P(A \cap C)$, **(d)** $P(A \cup C)$, **(e)** $P(D)$, **(f)** $P(D')$.

Solution

(a) There are two outcomes common to A and B—$A \cap B = \{(1, 3), (3, 1)\}$—and 36 possible outcomes in S, so

$$P(A \cap B) = \frac{2}{36}$$

(b) Because A and B aren't mutually exclusive, use (6.2).

$$P(A \cup B) = P(A) + P(B) - P(A \cap B)$$
$$= \frac{3}{36} + \frac{5}{36} - \frac{2}{36} = \frac{6}{36}$$

(c) A and C have no points in common—$A \cap C = \varnothing$—so from (6.4),

$$P(A \cap C) = P(\varnothing) = 0$$

(d) $A \cap C = \varnothing$, so from (6.1),

$$P(A \cup C) = P(A) + P(C)$$
$$= \frac{3}{36} + \frac{2}{36} = \frac{5}{36}$$

(e) The number of outcomes in D is 15. Thus

$$P(D) = \frac{15}{36}$$

(f) From (6.4),

$$P(D') = 1 - P(D) = 1 - \frac{15}{36} = \frac{21}{36}$$

EXAMPLE 6-24 In Example 6-10, let A be the event that 2 or more rolls of a pair of 6-sided dice are required to observe a sum of 7. Find $P(A)$.

Solution $S = \{1, 2, 3, 4, 5, \ldots\}$ and $A = \{2, 3, 4, 5, \ldots\}$. To find $P(A)$, you first find $P(A')$. $P(A')$ is the probability that you'll get a 7 on the first roll of a pair of 6-sided dice, so $P(A') = 6/36$. (Take a look at Example 6-21 and count the number of outcomes that have a sum of 7.) Thus

$$P(A) = 1 - P(A') = 1 - \frac{6}{36} = \frac{30}{36}$$

6-3. Methods of Enumeration

A. Multiplication principle

When we have two or more experiments in combination, we can use the multiplication principle.

Multiplication principle: If an experiment E_1 has n_1 possible outcomes and if, for each of these, an experiment E_2 has n_2 possible outcomes, then the *combined* procedure E_1 followed by E_2 has $n_1 n_2$ possible outcomes.

The multiplication principle can be extended to more than two experiments.

EXAMPLE 6-25 At a local cafeteria your meat choices are ham (H) or chicken (C) or swiss steak (S). *And*, with each meat choice, you may choose potatoes, either baked (B) or mashed (M). (**a**) Determine the number of possible meat and potato dinner combinations and (**b**) list them.

Solution

(**a**) Let E_1 be the meat choice with $n_1 = 3$ outcomes, and let E_2 be the potato choice with $n_2 = 2$ outcomes. Thus by the multiplication principle the number of meat-and-potato combinations $E_1 E_2$ has $n_1 n_2 = 3 \cdot 2 = 6$ outcomes.

(**b**) A *tree diagram* will help you list the outcomes. On the first branches list the meats—H, C, S—and on the second branches list the potatoes—B or M.

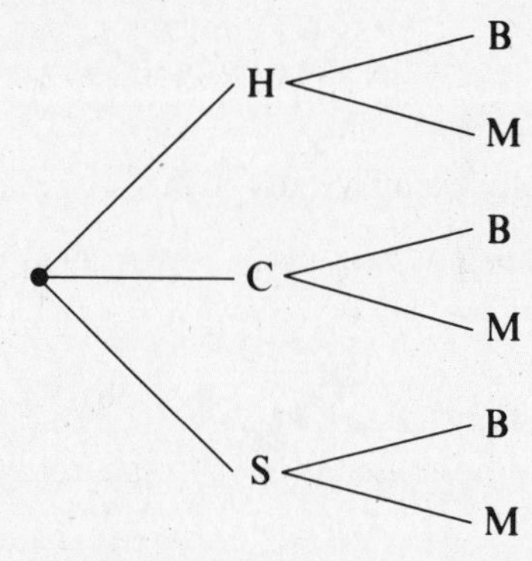

Using the tree diagram, you can easily determine that the 6 outcomes, reading from the top, are

(H, B) (H, M) (C, B) (C, M) (S, B) (S, M)

B. Sampling with replacement

If we select an object at random from a given set of n objects, note a characteristic of the object, then replace it, we're **sampling with replacement**. And, if we select r objects, one at a time, and keep track of the order in which we select the objects, we have an **ordered sample**.

- If we sample with replacement, we can use the multiplication principle to determine the number of possible ordered samples of size r that we can select from a set of n objects:

NUMBER OF ORDERED SAMPLES—SAMPLING WITH REPLACEMENT

$$\underbrace{n \cdot n \cdot n \cdots n}_{r \text{ times}} = n^r$$

EXAMPLE 6-26 From a bowl that contains ten slips of paper numbered 0, 1, 2, 3, 4, 5, 6, 7, 8, 9, you select two slips at random, one at a time, with replacement. How many ordered samples are possible?

Solution The bowl contains $n = 10$ slips of paper, and $r = 2$ were selected one at a time, so

$$n^r = 10^2 = 100$$

note: This experiment simulates the random choice of a 2-digit integer. The answer indicates that the number of 2-digit integers from 00 to 99 is 100, which makes intuitive sense.

EXAMPLE 6-27 Assume that you roll a 4-sided die three times. **(a)** How many ordered outcomes are possible? **(b)** If A is the event that all three rolls are equal, how many outcomes belong to A? **(c)** Find $P(A)$.

Solution

(a) Rolling a die is equivalent to sampling with replacement—you may come up with any of the four outcomes on each of the three rolls. Thus the number of possible ordered samples is $4 \cdot 4 \cdot 4 = 4^3 = 64$.

(b) On the first roll, four outcomes—1 or 2 or 3 or 4—are possible. If the rolls are equal, rolls two and three must match the first roll because only one outcome is possible. Then, by the multiplication principle, the number of outcomes in A is $4 \cdot 1 \cdot 1 = 4$.

(c) For equally likely outcomes with $k = 4$ and $h = 64$,

$$P(A) = \frac{k}{h} = \frac{4}{64} = \frac{1}{16}$$

C. Sampling without replacement

If we select an object at random from a given set of n objects, note a characteristic of the object, but don't replace it, we're **sampling without replacement**.

- The number of possible ordered samples of size r that we can select from a set of n objects, $r \leq n$, when we sample without replacement, is

NUMBER OF ORDERED SAMPLES—SAMPLING WITHOUT REPLACEMENT

$$\frac{n!}{(n-r)!} = n(n-1)(n-2)\cdots(n-r+1) \qquad \textbf{(6.5)}$$

- ***remark:*** $n!$, read "n factorial," is equal to the product of the first n positive integers. For example,

$$3! = 3 \cdot 2 \cdot 1 = 6$$

$$7! = 7 \cdot 6 \cdot 5 \cdot 4 \cdot 3 \cdot 2 \cdot 1 = 5040$$

Notice the following definitions:

$$1! = 1$$

$$0! = 1$$

EXAMPLE 6-28 A bowl contains three slips of paper numbered 1, 2, 3. **(a)** Find the number of possible ordered samples of size 3, sampling without replacement, and **(b)** list them. **(c)** If A is the event that the numbers are in ascending or descending order, what is $P(A)$?

Solution

(a) Since $n = 3$ and $r = 3$, the number of ordered samples is

$$\frac{n!}{(n-r)!} = \frac{3!}{0!} = 3 \cdot 2 \cdot 1 = 6$$

(b) To list the ordered samples you can construct the following tree diagram:

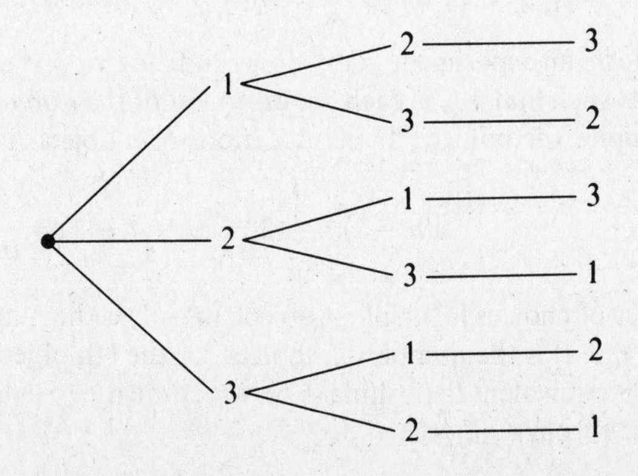

The 6 outcomes, from the top, are

$$(1, 2, 3) \quad (1, 3, 2) \quad (2, 1, 3) \quad (2, 3, 1) \quad (3, 1, 2) \quad (3, 2, 1)$$

(c) Event $A = \{(1, 2, 3), (3, 2, 1)\}$. With $k = 2$ and $h = 6$ equally likely events,

$$P(A) = \frac{2}{6} = \frac{1}{3}$$

EXAMPLE 6-29 A bowl contains 4 slips of paper numbered 1, 2, 3, 4. **(a)** Find the number of ordered samples of size 2 that can be selected from this bowl and **(b)** list them. **(c)** Let A be the event that the sum of the outcomes is 5. List the outcomes in A and **(d)** find $P(A)$.

Solution

(a) $n = 4$ and $r = 2$, so

$$\frac{4!}{(4-2)!} = \frac{4!}{2!} = 4 \cdot 3 = 12$$

(b) You can use the following tree diagram to obtain a listing:

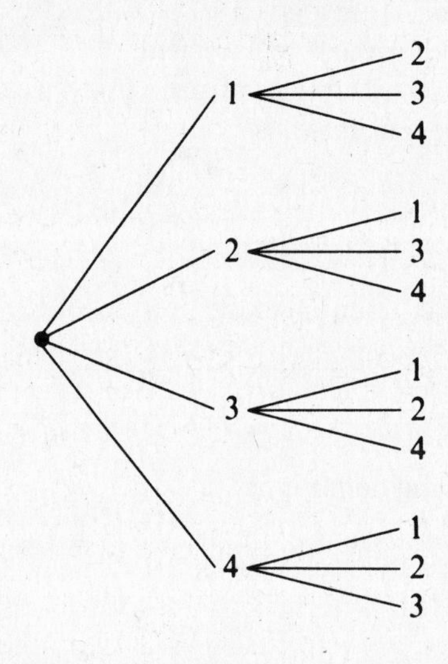

The 12 outcomes are

$$(1,2) \quad (1,3) \quad (1,4) \quad (2,1) \quad (2,3) \quad (2,4) \quad (3,1) \quad (3,2) \quad (3,4) \quad (4,1) \quad (4,2) \quad (4,3)$$

(c) $A = \{(1,4), (2,3), (3,2), (4,1)\}$.
(d) With $k = 4$ and $h = 12$ equally likely events, $P(A) = 4/12$.

D. Permutations

- Each ordered arrangement of a given set of n objects is called a **permutation** of these n objects.

The number of permutations of n objects is $n!$. Thus, if you have 3 numbers, $3! = 6$ permutations of those numbers are possible:

$$(1, 2, 3), (1, 3, 2), (2, 1, 3), (2, 3, 1), (3, 1, 2) \; (3, 2, 1)$$

If we're going to perform an experiment many times and have r positions to fill using objects selected from a set of n objects such that $r \leq n$, each arrangement of the r objects is called a **permutation of n objects taken r at a time**. The number of permutations of n objects taken r at a time is

PERMUTATIONS OF n OBJECTS r AT A TIME

$$n(n-1)(n-2)\cdots(n-r+1) = \frac{n!}{(n-r)!} \qquad \textbf{(6.6)}$$

where n is the number of choices for the first object, $(n - 1)$ is the number of choices for the second object, . . . , and $(n - r + 1)$ is the number of choices for the rth object.

note: Formula (6.6) is equivalent to formula (6.5), which we use to determine the number of ordered samples of size r from n objects.

EXAMPLE 6-30 Let $S = \{a, b, c, d\}$. List all the permutations of these letters taken three at a time.

Solution First determine the number of permutations of $n = 4$ things taken $r = 3$ at a time:

$$\frac{n!}{(n-r)!} = \frac{4!}{(4-3)!} = 4 \cdot 3 \cdot 2 = 24$$

Then list the permutations:

(a, b, c)	(a, c, b)	(b, a, c)	(b, c, a)	(c, a, b)	(c, b, a)
(a, b, d)	(a, d, b)	(b, a, d)	(b, d, a)	(d, b, a)	(d, a, b)
(a, c, d)	(a, d, c)	(c, a, d)	(c, d, a)	(d, a, c)	(d, c, a)
(b, c, d)	(b, d, c)	(c, b, d)	(c, d, b)	(d, b, c)	(d, c, b)

E. Combinations

When we select a sample of r objects out of a set of n objects and ignore the order of selection, we have an **unordered sample** of size r.

- An unordered sample of r objects taken out of a set of n objects is called a **combination of n objects taken r at a time.**

The number of combinations of n objects taken r at a time is

COMBINATIONS OF n OBJECTS r AT A TIME

$$\binom{n}{r} = \frac{n!}{r!(n-r)!} \tag{6.7}$$

EXAMPLE 6-31 Let $S = \{a, b, c, d, e\}$. **(a)** Find the number of combinations of these letters taken three at a time and **(b)** list them.

Solution

(a) The number of combinations of $n = 5$ objects taken $r = 3$ at a time is

$$\binom{n}{r} = \frac{n!}{r!(n-r)!} = \binom{5}{3} = \frac{5!}{3!2!} = \frac{5 \cdot 4 \cdot 3 \cdot 2 \cdot 1}{3 \cdot 2 \cdot 1 \cdot 2 \cdot 1} = 10$$

(b) The combinations are

(a, b, c)	(a, b, d)	(a, b, e)	(a, c, d)	(a, c, e)
(a, d, e)	(b, c, d)	(b, c, e)	(b, d, e)	(c, d, e)

EXAMPLE 6-32 Select 5 cards at random from a deck of playing cards. **(a)** How many different combinations are there? Find the probability of each of the following events: **(b)** $A = \{$all 5 cards are hearts$\}$; **(c)** $B = \{$3 clubs and 2 diamonds$\}$; **(d)** $C = \{$1 black and 4 red cards$\}$.

Solution

(a) Out of 52 cards, the number of 5-card combinations is

$$\binom{52}{5} = \frac{52!}{5!47!} = \frac{52 \cdot 51 \cdot 50 \cdot 49 \cdot 48 \cdot 47!}{5 \cdot 4 \cdot 3 \cdot 2 \cdot 1 \cdot 47!} = 2{,}598{,}960$$

note: This is the number of possible 5-card poker hands.

(b) The 5 hearts are selected out of the 13 hearts and 0 cards are selected out of the other 39 cards, so the number of outcomes in A is

$$k = \binom{13}{5}\binom{39}{0} = \left(\frac{13 \cdot 12 \cdot 11 \cdot 10 \cdot 9 \cdot 8!}{5 \cdot 4 \cdot 3 \cdot 2 \cdot 1 \cdot 8!}\right)\left(\frac{39!}{0!39!}\right) = (1287)(1) = 1287$$

Thus

$$P(A) = \frac{1287}{2{,}598{,}960} = .0005$$

(c) The 3 clubs are selected out of 13 cards, the 2 diamonds are selected out of 13 cards, and 0 cards are selected out of the remaining 26 cards. Thus the number of outcomes in B is

$$k = \binom{13}{3}\binom{13}{2}\binom{26}{0} = (286)(78)(1) = 22{,}308$$

Thus

$$P(B) = \frac{22{,}308}{2{,}598{,}960} = .0086$$

(d) The black card is selected out of 26 black cards and the 4 red cards are selected out of 26 red cards, so the number of outcomes in C is

$$\binom{26}{1}\binom{26}{4} = (26)(14{,}950) = 388{,}700$$

and

$$P(C) = \frac{388{,}700}{2{,}598{,}960} = .1496$$

6-4. Probabilities for Intersections of Events

A. Conditional probability

- The **conditional probability** $P(B \mid A)$ that event B will occur, given that event A has already occurred, is defined by

$$P(B \mid A) = \frac{P(A \cap B)}{P(A)} \qquad \textbf{(6.8)}$$

provided that $P(A) > 0$.

Think of the event A that has already occurred as creating the new sample space for which we now want to calculate $P(B)$. Think of a group of 20 children, consisting of 10 girls and 10 boys. You want to know the probability for selecting 2 girls at random from this group. You can see that the probability for selecting the first girl—call this event A—is 10/20. This is *not*, however, the probability for selecting the second girl—event B. The occurrence of event A has created a new sample space—a group of 19 children, consisting of 9 girls and 10 boys. You'll calculate the probability of the event B for this new sample space. Thus, the probability for event B, given that event A has occurred, is 9/19.

EXAMPLE 6-33 Suppose that $P(A) = .5$, $P(B) = .7$, and $P(A \cap B) = .3$, as shown on the Venn diagram in Fig. 6-1. Find $P(B \mid A)$.

Solution Given that A has already occurred, you're effectively restricting the sample space to A. So, .3 of the .5 probability also belongs to B; i.e., $P(A \cap B) = .3$. Now that A has already occurred, the probability that B will also occur is $.3/.5 = .6$. You get the same probability when you substitute the values into formula (6.8):

$$P(B \mid A) = \frac{P(A \cap B)}{P(A)} = \frac{.3}{.5} = .6$$

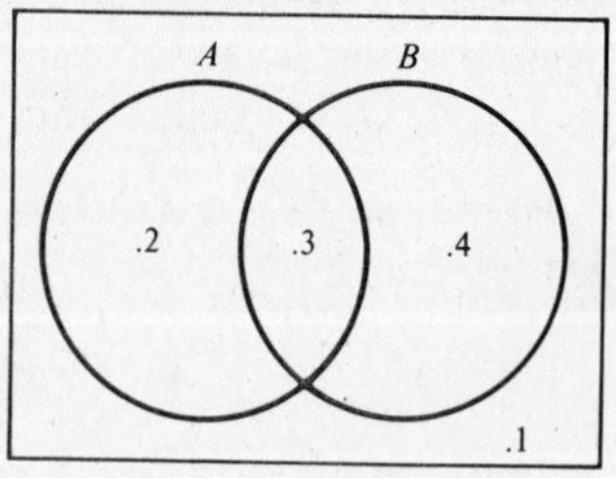

Figure 6-1

EXAMPLE 6-34 You roll a 4-sided die twice (see Example 6-3). Let A be the event that the sum of the rolls is 4 or 5 and let B be the event that the sum of the rolls is 4. Find **(a)** $P(A)$, **(b)** $P(B)$, **(c)** $P(A \cap B)$, and **(d)** $P(B \mid A)$.

Solution Notice that each of the $h = 16$ outcomes in the sample space is equally likely.

(a) The number of outcomes in A is $k = 7$, so

$$P(A) = \frac{7}{16}$$

(b) The number of outcomes in B is $k = 3$, so

$$P(B) = \frac{3}{16}$$

(c) Since B is a subset of A,

$$P(A \cap B) = P(B) = \frac{3}{16}$$

(d) Using formula (6.8),

$$P(B \mid A) = \frac{P(A \cap B)}{P(A)} = \frac{3/16}{7/16} = \frac{3}{7}$$

There are times when the nature of an experiment makes it easy to calculate $P(B \mid A)$ without using formula (6.8).

EXAMPLE 6-35 A candy jar contains 100 jelly beans, but only 10 of your favorite black ones. Let A be the event that you select a black jelly bean at random from the jar and eat it. Let B be the event that a second jelly bean you select at random is also black. Find $P(B \mid A)$.

Solution Since you have already selected a black jelly bean from the jar, there are now 99 jelly beans in the jar and nine of these are black. So,

$$P(B \mid A) = \frac{9}{99}$$

B. Multiplication rule for probabilities

If we multiply both sides of formula (6.8) by $P(A)$, we have the **multiplication rule for probabilities.**

MULTIPLICATION RULE FOR PROBABILITIES

$$P(A \cap B) = P(A)P(B \mid A) \qquad \textbf{(6.9)}$$

We can write the equivalent form of formula (6.9) as

$$P(A \cap B) = P(B)P(A \mid B)$$

EXAMPLE 6-36 An urn contains 4 red, 3 white, and 2 blue balls. You select two balls at random, one at a time, without replacement. Let $A = \{\text{red on first draw}\}$ and $B = \{\text{white on second draw}\}$. Find $P(A \cap B) = P(\text{RW})$—the probability for drawing a red ball on the first draw and a white ball on the second draw.

Solution To find $P(A \cap B)$, you need to find the probability for selecting a red the first time, $P(A) = 4/9$; then multiply $P(A)$ by the probability for selecting a white the second time, given that you've already selected a red, $P(B \mid A) = 3/(9 - 1) = 3/8$:

$$P(A \cap B) = P(A)P(B \mid A) = \frac{4}{9} \cdot \frac{3}{8} = \frac{1}{6}$$

EXAMPLE 6-37 For the experiment in Example 6-36, let BB denote the event blue on the first draw and blue on the second draw. Let WR denote the event white on the first draw and red on the second draw. Find **(a)** $P(\text{BB})$, **(b)** $P(\text{WR})$, and **(c)** $P(\text{one red and one blue})$.

Solution

(a) BB is a compound event consisting of two events—blue on the first draw, and then blue on the second draw given blue on the first draw. So you figure the probability for the first draw—i.e., 2/9—and then the probability for the second draw given that the first has already occurred—i.e., 1/8. Then from formula (6.9),

$$P(\text{BB}) = \frac{2}{9} \cdot \frac{1}{8} = \frac{1}{36}$$

(**b**) The event WR means white on the first draw and red on the second draw given white on the first draw. Thus

$$P(\mathrm{WR}) = \frac{3}{9} \cdot \frac{4}{8} = \frac{1}{6}$$

(**c**) The event {one red and one blue} can occur in two mutually exclusive ways: red then blue (RB), or blue then red (BR). Thus, using formula (6.1),

$$P(\text{one red and one blue}) = P(\mathrm{RB}) + P(\mathrm{BR}) = \frac{4}{9} \cdot \frac{2}{8} + \frac{2}{9} \cdot \frac{4}{8} = \frac{1}{9} + \frac{1}{9} = \frac{2}{9}$$

EXAMPLE 6-38 Draw three cards from a standard deck of playing cards, one at a time, without replacement. Let HCD denote heart on the first draw, club on the second draw, diamond on the third draw. Find (**a**) $P(\mathrm{HCD})$ and (**b**) $P(\mathrm{SSS})$, the probability for drawing three spades.

Solution

(**a**) Extending the multiplication rule (6.9), you have

$$P(\mathrm{HCD}) = P(\mathrm{H})P(\mathrm{C} \mid \mathrm{H})P(\mathrm{D} \mid \mathrm{HC}) = \frac{13}{52} \cdot \frac{13}{51} \cdot \frac{13}{50}$$

(**b**) The probability for drawing three spades is $P(\mathrm{SSS}) = \dfrac{13}{52} \cdot \dfrac{12}{51} \cdot \dfrac{11}{50}$

C. Independent events

We say that A and B are **independent events** when the probability for event B isn't affected by whether or not event A has occurred. We can state this definition in either of two formulas:

- A and B are independent events if

$$P(B \mid A) = P(B) \qquad \textbf{(6.10)}$$

Then by the multiplication rule (6.9) and formula (6.10), we can write $P(A \cap B) = P(A)P(B \mid A) = P(A)P(B)$, so

- A and B are independent events if

$$P(A \cap B) = P(A)P(B) \qquad \textbf{(6.11)}$$

note: Events that aren't independent are **dependent events**.

EXAMPLE 6-39 An urn contains four balls numbered 1, 2, 3, and 4. Select one ball at random from the urn, then put it back. Let $A = \{1, 2\}$ and $B = \{1, 3\}$. Show that A and B are independent events.

Solution It's clear that $P(A) = 2/4$ and $P(B) = 2/4$. Since $A \cap B = \{1\}$, $P(A \cap B) = 1/4$. Now from (6.8),

$$P(B \mid A) = \frac{P(A \cap B)}{P(A)} = \frac{1/4}{2/4} = \frac{1}{2} = P(B)$$

Whether A has occurred or not, the value of $P(B)$ remains the same, so A and B are independent events.

The second definition of independent events (6.11) also shows that A and B are independent events:

$$P(A)P(B) = \frac{2}{4} \cdot \frac{2}{4} = \frac{1}{4} = P(A \cap B)$$

Sometimes the trials of an experiment are independent because of the nature of the experiment. If an experiment consists of two trials and the outcome of the first trial doesn't affect the outcome of the second trial, we say that these are **independent trials**. If A is an outcome of the first trial and B is an outcome of the second trial, and the trials are independent, then

$$P(A \cap B) = P(A)P(B)$$

EXAMPLE 6-40 Consider two successive rolls of a 4-sided die. The sample space is

(1, 1)	(1, 2)	(1, 3)	(1, 4)
(2, 1)	(2, 2)	(2, 3)	(2, 4)
(3, 1)	(3, 2)	(3, 3)	(3, 4)
(4, 1)	(4, 2)	(4, 3)	(4, 4)

(A encloses the first two rows; B encloses the last three columns.)

where A is the event that a 1 or 2 is the outcome on the first roll, and B is the event that a 2 or 3 or 4 is the outcome on the second roll. Show that A and B are independent events.

Solution It's common sense that successive rolls of a fair die must be independent, but you can SHOW that A and B are independent as follows:

$$P(A)P(B) = \frac{8}{16} \cdot \frac{12}{16} = \frac{2}{4} \cdot \frac{3}{4} = \frac{6}{16} = P(A \cap B)$$

[*Count* 'em—there are 6 out of 16 outcomes in the intersection.]

We can extend the idea of two independent events and two independent trials (6.11) to *several* independent events or trials.

EXAMPLE 6-41 Suppose that in a certain college, $A = \{\text{student is from Iowa}\}$, $B = \{\text{student is a sophomore}\}$, and $C = \{\text{student is a female}\}$. If $P(A) = 7/10$, $P(B) = 1/3$, and $P(C) = 1/2$, and if A, B, and C are independent, find $P(A \cap B \cap C)$.

Solution Extend formula (6.11) for independent events so that

$$P(A \cap B \cap C) = P(A)P(B)P(C) = \frac{7}{10} \cdot \frac{1}{3} \cdot \frac{1}{2} = \frac{7}{60}$$

EXAMPLE 6-42 Roll a tetrahedron four times. Find the probability for observing a 1 on each roll.

Solution Because the trials are independent,

$$P(\{1, 1, 1, 1\}) = \frac{1}{4} \cdot \frac{1}{4} \cdot \frac{1}{4} \cdot \frac{1}{4} = \frac{1}{256}$$

EXAMPLE 6-43 When you roll a 6-sided die 6 times, you can say that a match occurs if side k is rolled on the kth roll. If you define "Success" as the possibility that at least one match will occur during the 6 rolls and "Failure" as the possibility that no matches will occur, what is $P(\text{Success})$?

Solution If $A = \{\text{Success}\}$, then $A' = \{\text{Failure}\}$. You can find $P(A')$ and subtract this answer from 1 to get $P(A)$. A' occurs if 1 is NOT rolled on the first roll, 2 is not rolled on the second roll, etc. These rolls are independent, so

$$P(A') = \frac{5}{6} \cdot \frac{5}{6} \cdot \frac{5}{6} \cdot \frac{5}{6} \cdot \frac{5}{6} \cdot \frac{5}{6} = \left(\frac{5}{6}\right)^6$$

Thus

$$P(A) = P(\text{Success}) = 1 - P(A') = 1 - \left(\frac{5}{6}\right)^6 = .665$$

SUMMARY

1. The *sample space* S is the set of all possible outcomes for a random experiment.
2. A subset A of a sample space S is called an *event*.
3. Each individual outcome of a random experiment is a *simple event*.
4. A *compound event* contains more than one simple event.
5. For equally likely outcomes, $P(A)$ equals the number of outcomes k in event A divided by the number of outcomes h in the entire sample space S.

6. The relative frequency of the occurrence of A in n trials of a given experiment, $\#(A)/n$, should be close to $P(A)$ when n is large.
7. When *sampling without replacement*, you don't replace an object after it has been selected, so that particular object can't be chosen again.
8. If you keep track of the order of selection in an experiment, you have an *ordered sample*.
9. The number of ordered samples of size r when sampling with replacement from a set of n objects is n^r.
10. The number of ordered samples of size r when sampling without replacement from a set of n objects is

$$n(n-1)(n-2)\cdots(n-r+1) = \frac{n!}{(n-r)!}$$

This is called the *number of permutations* of n objects taken r at a time.

11. The number of permutations of n objects is $n!$.
12. The number of unordered samples of size r when sampling without replacement from a set of n objects is

$$\binom{n}{r} = \frac{n!}{r!(n-r)!}$$

This is called the *number of combinations* of n objects taken at r at a time.

13. The *conditional probability* for event B, given that event A has occurred, is $P(B|A) = P(A \cap B)/P(A)$.
14. The *multiplication rule for probabilities* is $P(A \cap B) = P(A)P(B|A)$.
15. If $P(A \cap B) = P(A)P(B)$ or if $P(B|A) = P(B)$, then A and B are *independent events*.

RAISE YOUR GRADES

Can you . . . ?

☑ describe a random experiment
☑ define sample space
☑ define simple and compound events
☑ find probabilities for equally likely outcomes
☑ define the relative frequency concept of probability
☑ find $P(A \cup B)$, $P(A \cap B)$, $P(A')$ given S and sets A and B
☑ explain the difference between sampling with and without replacement
☑ evaluate $\binom{n}{r}$ for different values of n and r
☑ make a tree diagram
☑ find conditional probabilities
☑ apply the multiplication rule for probabilities
☑ define independent events using two different formulas

RAPID REVIEW

1. A ____________ event contains more than one simple event.
2. If a sample space S contains 100 equally likely outcomes and event A contains 16 of these outcomes, then $P(A) =$ ____________.
3. If a pair of 6-sided dice is rolled and the outcome is the sum of these dice, then the sample space is $S =$ ____________.
4. In question 3, if $A = \{11\}$, then $P(A) =$ ____________.
5. In question 3, if $B = \{7\}$ and $P(B) = 1/6$, then $P(B') =$ ____________.
6. If $A = \varnothing$, then $P(A) =$ ____________.
7. If event A and event B have no common outcomes, then A and B are ____________ ____________.

8. If $P(A) = .2$, $P(B) = .3$, and $A \cap B = \varnothing$, then $P(A \cup B) =$ __________.
9. If $P(A) = .6$, $P(B) = .7$, and $P(A \cap B) = .4$, then $P(A \cup B) =$ __________.
10. The number of possible permutations using the letters H, O, P, E is __________.
11. The number of 3-letter code words that can be formed using the letters in H, O, P, E, C is __________.
12. The number of 3-card hands that can be selected from a deck of 10 cards is __________.
13. If $P(A) = .7$ and $P(B \mid A) = .8$, then $P(A \cap B) =$ __________.
14. An urn contains 3 red balls and 7 blue balls. Two balls are selected at random, one at a time. $P(RB) =$ **(a)** __________ if the balls are selected without replacement, and $P(RB) =$ **(b)** __________ if the balls are selected with replacement.

Answers **(1)** compound **(2)** .16 **(3)** $\{2, 3, 4, 5, 6, 7, 8, 9, 10, 11, 12\}$ **(4)** 2/36 **(5)** 5/6 **(6)** 0 **(7)** mutually exclusive **(8)** .5 **(9)** .9 **(10)** 24 **(11)** 60 **(12)** 120 **(13)** .56 **(14)** **(a)** 21/90 **(b)** 21/100

SOLVED PROBLEMS

Math Review

PROBLEM 6-1 Evaluate **(a)** 3!, **(b)** 4!, **(c)** 8!.

Solution

(a) 3!, read "3 factorial," is the product of the first 3 positive integers. Thus

$$3! = 3 \cdot 2 \cdot 1 = 6$$

(b) Similarly, 4! is the product of the first 4 positive integers.

$$4! = 4 \cdot 3 \cdot 2 \cdot 1 = 24$$

(c)

$$8! = 8 \cdot 7 \cdot 6 \cdot 5 \cdot 4 \cdot 3 \cdot 2 \cdot 1 = 40{,}320$$

PROBLEM 6-2 Evaluate **(a)** $\dfrac{10!}{8!}$, **(b)** $\dfrac{20!}{17!}$, **(c)** $\dfrac{52!}{47!}$.

Solution

(a) Since $8 = 10 - 2$, you know that $n = 10$ and $r = 2$ and you can use formula (6.5):

$$\frac{n!}{(n-r)!} = \frac{10!}{(10-2)!} = \frac{10 \cdot 9 \cdot \cancel{8!}}{\cancel{8!}} = 10 \cdot 9 = 90$$

or, simply

$$\frac{10!}{8!} = \overbrace{10 \cdot 9}^{r=2} = 90$$

(b) Again, if you use formula (6.5), $n = 20$ and $17 = 20 - 3$, so $r = 3$. Thus

$$\frac{20!}{17!} = \frac{20 \cdot 19 \cdot 18 \cdot 17!}{17!} = \overbrace{20 \cdot 19 \cdot 18}^{r=3} = 6840$$

(c)

$$\frac{52!}{47!} = \frac{52 \cdot 51 \cdot 50 \cdot 49 \cdot 48 \cdot 47!}{47!} = 52 \cdot 51 \cdot 50 \cdot 49 \cdot 48 = 311{,}875{,}200$$

note: Answer (c) represents the number of ordered samples of five cards that can be selected from a standard deck of playing cards when sampling without replacement.

PROBLEM 6-3 Evaluate (**a**) $\binom{8}{5}$, (**b**) $\binom{8}{3}$, (**c**) $\binom{10}{4}$, (**d**) $\binom{10}{6}$.

Solution Use formula (6.7)

$$\binom{n}{r} = \frac{n!}{r!(n-r)!}$$

(**a**) Let $n = 8$ and $r = 5$:

$$\binom{8}{5} = \frac{8!}{5!3!} = \frac{8 \cdot 7 \cdot 6 \cdot \cancel{5} \cdot \cancel{4} \cdot \cancel{3} \cdot \cancel{2} \cdot \cancel{1}}{\cancel{5} \cdot \cancel{4} \cdot \cancel{3} \cdot \cancel{2} \cdot \cancel{1} \cdot 3 \cdot 2 \cdot 1} = \frac{8 \cdot 7 \cdot 6}{3 \cdot 2 \cdot 1} = 56$$

(**b**)

$$\binom{8}{3} = \frac{8!}{3!5!} = \frac{8!}{5!3!} = \binom{8}{5} = 56$$

by part (**a**). [Notice that $\binom{n}{r} = \binom{n}{n-r}$.]

(**c**)

$$\binom{10}{4} = \frac{10!}{4!6!} = \frac{10 \cdot 9 \cdot 8 \cdot 7 \cdot 6!}{4!6!} = \frac{10 \cdot 9 \cdot 8 \cdot 7}{4 \cdot 3 \cdot 2 \cdot 1} = 10 \cdot 3 \cdot 7 = 210$$

(**d**)

$$\binom{10}{6} = \binom{n}{r} = \binom{n}{n-r} = \binom{10}{4} = 210$$

The Concept of Probability

PROBLEM 6-4 A certain statistics class has 3 freshmen, 24 sophomores, 15 juniors, and 2 seniors. You perform an experiment such that one student is selected at random and the student's class is determined. (**a**) Define the sample space. (**b**) Let A = {freshman, sophomore} and give the value of $P(A)$.

Solution

(**a**) Because you're interested in the student's class, the sample space is

$$S = \{\text{freshman, sophomore, junior, senior}\}$$

(**b**) The number of freshmen and sophomores is $k = 3 + 24 = 27$. Because each student has the same chance of being selected out of the 44 students in the class, $h = 44$ and

$$P(A) = \frac{k}{h} = \frac{27}{44} = .61$$

PROBLEM 6-5 A box of 50 crocus bulbs contains 25 bulbs for purple crocuses, 15 bulbs for white crocuses, and 10 bulbs for yellow crocuses. Select one bulb at random. (**a**) Define the sample space in terms of the color of the flower. (**b**) Let B = {yellow}. If you plant the bulb that was selected, what is the probability that it will yield a yellow flower—that is, give $P(B)$?

Solution

(**a**) Because you're interested in the color of the flower,

$$S = \{\text{purple, white, yellow}\}$$

(**b**) Out of the 50 bulbs, 10 will produce yellow crocuses. So

$$P(B) = \frac{10}{50} = .2$$

PROBLEM 6-6 A 20-sided die is an icosahedron with 20 faces, each of which is an equilateral triangle. The faces are numbered from 0 to 9, with each of these digits appearing on two faces. Roll this 20-sided die once. (**a**) Define the sample space S. (**b**) Let A = {0, 1, 2} and give the value of $P(A)$.

Solution

(**a**) This die is a random-digit generator and the sample space is

$$S = \{0, 1, 2, 3, 4, 5, 6, 7, 8, 9\}$$

(b) The number of outcomes that belong to A is 6—two 0's, two 1's, and two 2's. The number of faces on the die is 20. Thus

$$P(A) = \frac{6}{20} = .30$$

PROBLEM 6-7 A typical roulette wheel used in a casino has 38 slots numbered $1, 2, 3, \ldots, 36, 0, 00$. The 0 and 00 slots are colored green. Half of the remaining slots are red and the other half are black. Also, half of the integers between 1 and 36 inclusive are odd and half are even, with 0 and 00 defined as neither odd nor even. A ball is rolled around the wheel and ends up in one of the slots. **(a)** Assume that you're interested only in the color of the outcome. Define S. **(b)** Let $A = \{\text{red}\}$ and give the value of $P(A)$. **(c)** Assume that you're interested in the probabilities of certain compound events; in particular, let $B = \{0, 00\}$ and let $C = \{14, 15, 17, 18\}$. Give the values of $P(B)$ and $P(C)$.

Solution

(a) There are three possible colors, so

$$S = \{\text{red, black, green}\}$$

(b) Each of the 38 slots has the same chance of occurring, so each slot has a probability of 1/38. Of the 38 possible outcomes, 18 are red. Thus

$$P(A) = P(\text{red}) = \frac{18}{38} = .474$$

(c) Because all the outcomes are equally likely,

$$P(B) = \frac{2}{38} = .053$$

$$P(C) = \frac{4}{38} = .105$$

PROBLEM 6-8 Urn A contains 3 red balls numbered 1, 2, and 3. Urn B contains 3 blue balls numbered 1, 2, and 3. Select one ball at random from each urn. **(a)** Describe S by listing the 9 pairs of numbers (x, y) where x is the number on the red ball and y is the number on the blue ball. **(b)** Let A be the event that the sum of the outcomes is equal to 4. Find $P(A)$.

Solution

(a) The 9 points in the S are

(1,1) (1,2) (1,3) ← A

(2,1) (2,2) (2,3)

(3,1) (3,2) (3,3)

(b) Each of the 9 outcomes is equally likely. There are 3 outcomes whose sum is 4. Thus

$$P(A) = \frac{3}{9} = .33$$

PROBLEM 6-9 Take two "decks" of 10 cards—say, the ace to ten of clubs and the ace to ten of diamonds. Shuffle each deck, and then compare the two decks to determine whether a particular number occupies the same position in each deck. Call this a match. You can do this by checking whether the top cards in each deck are the same, then checking the second cards, etc. Call the experiment a success if at least one match occurs in the ten positions. Explain how you could use the relative frequency concept of probability to estimate $p = P(\text{Success})$.

Solution Repeat this experiment a large number of times and keep track of the number of trials in which at least one success was observed. The relative frequency of successes should be close to .632, which is much easier to show experimentally than to prove theoretically!

Properties of Probability

PROBLEM 6-10 Urn A contains three red balls numbered 1, 2, 3. Urn B contains four blue balls numbered 1, 2, 3, 4. Select a red ball at random from Urn A and then a blue ball at random from Urn B. The sample space S contains the following points (x, y), where x is the red ball number and y is the blue ball number:

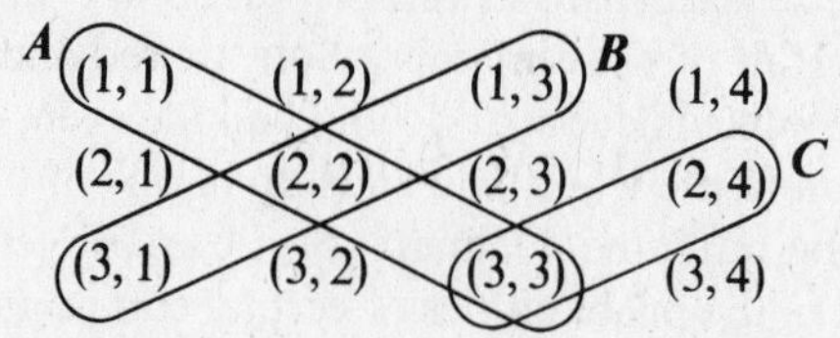

Let $A = \{\text{outcomes are equal}\}$, $B = \{\text{sum equals } 4\}$, and $C = \{\text{sum equals } 6\}$. List the outcomes in **(a)** $A \cup B$, **(b)** $A \cup C$, **(c)** $B \cap C$, **(d)** $A \cap C$, **(e)** A'.

Solution The events are shown on the sample space.

(a) $A \cup B$ is the set of points in A or in B or in both A and B. Thus

$$A \cup B = \{1, 1), (2, 2), (3, 3), (1, 3), (3,1)\}$$

(b) $A \cup C$ is the set of points in A or in C or in both A and C. Thus

$$A \cup C = \{(1, 1), (2, 2), (3, 3), (2, 4)\}$$

(c) $B \cap C$ is the set of points in both B and C. There are no such points, So $B \cap C$ is the empty set:

$$B \cap C = \varnothing$$

(d) $A \cap C$ is the set of points in both A and C. Thus

$$A \cap C = \{(3, 3)\}$$

(e) A' is the set of points in S that don't belong to A, so

$$A' = \{(1, 2), (1, 3), (1, 4), (2, 1), (2, 3), (2, 4), (3, 1), (3, 2), (3, 4)\}$$

PROBLEM 6-11 If $P(A) = .3$, $P(B) = .6$, and $A \cap B = \varnothing$, find $P(A \cup B)$.

Solution Because A and B are mutually exclusive, you use formula (6.1):

$$P(A \cup B) = P(A) + P(B) = .3 + .6 = .9$$

PROBLEM 6-12 If $P(A) = .6$, $P(B) = .3$, and $P(A \cap B) = .2$, find $P(A \cup B)$.

Solution Because A and B aren't mutually exclusive, you must subtract $P(A \cap B)$ so the outcomes in *both* A and B aren't counted twice. By formula (6.2),

$$P(A \cup B) = P(A) + P(B) - P(A \cap B) = .6 + .3 - .2 = .7$$

You may want to draw a Venn diagram similar to Fig. 6-1 to help you visualize this result.

PROBLEM 6-13 A field of beans is planted in "hills," with three seeds per hill. For each hill, 0, 1, 2, or 3 seeds will germinate. Suppose that the probabilities for three of these possible outcomes are $P(\{0\}) = 1/64$, $P(\{1\}) = 9/64$, $P(\{2\}) = 27/64$. Give the value of $P(\{3\})$.

Solution If you let $A = \{3\}$, then the complement of A is $A' = \{0, 1, 2\}$. By Formula (6.4):

$$P(A) = 1 - P(A') = 1 - P(\{0, 1, 2\}) = 1 - [P(\{0\}) + P(\{1\}) + P(\{2\})]$$

$$= 1 - \left(\frac{1}{64} + \frac{9}{64} + \frac{27}{64}\right) = 1 - \frac{37}{64} = \frac{27}{64}$$

PROBLEM 6-14 An 8-sided die is an octahedron that has equilateral triangles as its eight faces. These eight faces are numbered from 1 to 8. Roll a fair octahedron once. Each of the possible outcomes in $S = \{1, 2, 3, 4, 5, 6, 7, 8\}$ has the same chance of occurring. Let $A = \{2, 4, 6, 8\}$, $B = \{3, 6\}$, $C = \{2, 5, 7\}$, and $D = \{1, 3, 5, 7\}$. **(a)** Give the values of $P(A)$, $P(B)$, $P(C)$, and $P(D)$. **(b)** For the events A, B, C, and D, list all pairs of events that are mutually exclusive. **(c)** Give the values of $P(A \cap B)$, $P(B \cap C)$, and $P(C \cap D)$. **(d)** Give the values of $P(A \cup B)$, $P(B \cup C)$, and $P(C \cup D)$. **(e)** Give the value of $P(B')$.

Solution

(a) Because each of the eight outcomes has the same chance of occurring, the probability for each event is the number of outcomes in the event divided by 8:

$$P(A) = \frac{4}{8}, \qquad P(B) = \frac{2}{8}, \qquad P(C) = \frac{3}{8}, \qquad P(D) = \frac{4}{8}$$

(b) The pairs of events with no points in common—the mutually exclusive events—are

A and D $\qquad$ B and C

(c)

$$P(A \cap B) = P(\{6\}) = \frac{1}{8}$$

$$P(B \cap C) = P(\varnothing) = 0 \qquad \text{by formula (6.4)}$$

$$P(C \cap D) = P(\{5, 7\}) = \frac{2}{8}$$

(d) $A \cap B = \{6\}$, so these are not mutually exclusive events. Therefore, use formula (6.2):

$$P(A \cup B) = \frac{4}{8} + \frac{2}{8} - \frac{1}{8} = \frac{5}{8}$$

Since B and C are mutually exclusive events, you can use formula (6.1):

$$P(B \cup C) = \frac{2}{8} + \frac{3}{8} = \frac{5}{8}$$

By formula (6.2),

$$P(C \cup D) = \frac{3}{8} + \frac{4}{8} - \frac{2}{8} = \frac{5}{8}$$

(e) By formula (6.3),

$$P(B') = 1 - P(B) = 1 - \frac{2}{8} = \frac{6}{8}$$

PROBLEM 6-15 The formula for the probability for the union of two events can be extended to give a formula for the probability for three events. Given three events A, B, and C,

$$P(A \cup B \cup C) = P(A) + P(B) + P(C) - P(A \cap B) - P(A \cap C) - P(B \cap C) + P(A \cap B \cap C) \qquad \textbf{(6.12)}$$

To illustrate formula (6.12), roll a 12-sided die. The possible outcomes are the integers 1 to 12 inclusive. Let A be the event that the outcome is even, let B be the event that the outcome is a multiple of 3, and let C be the event that the outcome is a multiple of 4. **(a)** List the outcomes in each of the events A, B, and C. **(b)** Give the values of $P(A)$, $P(B)$, $P(C)$, $P(A \cap B)$, $P(A \cap C)$, $P(B \cap C)$, and $P(A \cap B \cap C)$. **(c)** Use your answers to part **(b)** with formula (6.12) to find $P(A \cup B \cup C)$, and check this result by finding the number of events in the set $A \cup B \cup C$.

Solution

(a) $A = \{2, 4, 6, 8, 10, 12\}$, $B = \{3, 6, 9, 12\}$, and $C = \{4, 8, 12\}$.
(b) Each of these probabilities is the number of outcomes in the event divided by 12:

$$P(A) = \frac{6}{12}, \qquad P(B) = \frac{4}{12}, \qquad P(C) = \frac{3}{12}, \qquad P(A \cap B) = P(\{6, 12\}) = \frac{2}{12}$$

$$P(A \cap C) = P(C) = \frac{3}{12}, \qquad P(B \cap C) = P(\{12\}) = \frac{1}{12}, \qquad P(A \cap B \cap C) = P(\{12\}) = \frac{1}{12}$$

(c) By formula (6.12),

$$P(A \cup B \cup C) = \frac{6}{12} + \frac{4}{12} + \frac{3}{12} - \frac{2}{12} - \frac{3}{12} - \frac{1}{12} + \frac{1}{12} = \frac{14}{12} - \frac{6}{12} = \frac{8}{12}$$

To check this result, you can determine that $A \cup B \cup C = \{2, 3, 4, 6, 8, 9, 10, 12\}$ and thus $P(A \cup B \cup C) = 8/12$.

note: Counting the number of outcomes may seem easier than using formula (6.12)—and in this case it is, but only because you're working with relatively small, simple sets in this problem.

Methods of Enumeration

PROBLEM 6-16 When you order a sandwich at the deli, you may choose from 6 breads, 7 meats, and 4 cheeses. If you select one from each, how many different sandwiches could you make?

Solution Let E_1 be the bread choice with $n_1 = 6$ outcomes, E_2 be the meat choice with $n_2 = 7$ outcomes, and E_3 be the cheese choice with $n_3 = 4$ outcomes. Then by the multiplication principle, the number of choices is

$$n_1 n_2 n_3 = 6 \cdot 7 \cdot 4 = 168$$

PROBLEM 6-17 A 6-sided die is rolled three times. Give the probability that all three rolls are equal.

Solution The roll of a die is an example of sampling with replacement, so the number of ordered rolls of this die is

$$6 \cdot 6 \cdot 6 = 6^3 = 216$$

If A is the event that all three rolls are equal, then

$$A = \{(1, 1, 1), (2, 2, 2), (3, 3, 3), (4, 4, 4), (5, 5, 5), (6, 6, 6)\}$$

Because each of the 216 outcomes is equally likely,

$$P(A) = \frac{6}{216} = .0278$$

PROBLEM 6-18 For Valentine's Day your Dutch uncle gives you a wooden shoe that contains ten tulip bulbs: six for red tulips and four for yellow tulips. You take two bulbs at random and plant them. Think of this as selecting the bulbs one at a time without replacement. You want to know the probability for the colors of the tulips that will come up. Let $A = \{\text{two red tulips}\} = \{\text{RR}\}$, $B = \{\text{two yellow tulips}\} = \{\text{YY}\}$, and $C = \{\text{one red tulip and one yellow tulip}\} = \{\text{RY, YR}\}$. Find **(a)** $P(A)$, **(b)** the probability that both tulips are the same color, and **(c)** $P(C)$.

Solution

(a) Use formula (6.5) to find the number of ordered samples of size $r = 2$ out of $n = 10$ outcomes:

$$\frac{n!}{(n-r)!} = \frac{10!}{(10-2)!} = \frac{10!}{8!} = 10(9) = 90$$

Then let E_1 be the first red tulip with $n_1 = 6$ outcomes and E_2 be the second red tulip with $n_2 = 5$ outcomes, and use the multiplication principle to find the number of the 90 samples that will produce two red tulips:

$$6 \cdot 5 = 30$$

Now, since each of the 90 outcomes is equally likely,

$$P(A) = P(\text{RR}) = \frac{30}{90} = \frac{1}{3}$$

(b) In the same way, you find that the number of samples that will produce two yellow tulips is $4 \cdot 3 = 12$. Events A and B are mutually exclusive, so you can use formula (6.1) to find the probability that both tulips will be the same color:

$$P(A \cup B) = P(\text{RR}) + P(\text{YY}) = \frac{30}{90} + \frac{12}{90} = \frac{42}{90} = \frac{7}{15}$$

(c) Since $C' = A \cup B$, you can use formula (6.4) to find $P(C)$:

$$P(C) = 1 - P(C') = 1 - \frac{7}{15} = \frac{8}{15}$$

Or, since $P(\text{RY}) = (6 \cdot 4)/90$ and $P(\text{YR}) = (4 \cdot 6)/90$, and RY and YR are mutually exclusive—i.e., $\text{RY} \cap \text{YR} = \varnothing$—you could also find $P(C)$ in the following way:

$$P(C) = P(\text{RY}) + P(\text{YR}) = \frac{24}{90} + \frac{24}{90} = \frac{48}{90} = \frac{8}{15}$$

PROBLEM 6-19 **(a)** How many two-letter codes can you form using two different letters from the vowels A, E, I, O, U? **(b)** List all of these permutations.

Solution

(a) Use formula (6.6) with $n = 5$ and $r = 2$:

$$\frac{n!}{(n-r)!} = \frac{5!}{(5-2)!} = 5 \cdot 4 = 20$$

(b) The 20 permutations are

AE	AI	AO	AU	EA	EI	EO	EU	IA	IE
IO	IU	OA	OE	OI	OU	UA	UE	UI	UO

PROBLEM 6-20 You have a package of gumdrops containing 55 pieces of candy. Twenty-two gumdrops are black and the rest are either yellow, orange, red, or green. Select three gumdrops at random and don't replace them (in fact, you may eat them). **(a)** How many different combinations of three gumdrops are possible? **(b)** How many combinations of size 3 contain all black gumdrops? **(c)** Give the probability for selecting three black gumdrops.

Solution

(a) Using formula (6.7) with $n = 55$ and $r = 3$, we have

$$\binom{n}{r} = \frac{n!}{r!(n-r)!} = \binom{55}{3} = \frac{55!}{3!52!} = \frac{55 \cdot 54 \cdot 53}{3 \cdot 2 \cdot 1} = 26{,}235$$

(b) Since there are $n = 22$ black gumdrops, the number of combinations of $r = 3$ black gumdrops is

$$\binom{22}{3} = \frac{22!}{3!19!} = \frac{22 \cdot 21 \cdot 20}{3 \cdot 2 \cdot 1} = 1540$$

(c) Using the quotient of the answers to parts **(a)** and **(b)**, you find

$$P(\text{BBB}) = \frac{1540}{26{,}235} = .0587$$

PROBLEM 6-21 You select 13 cards out of a deck of 52 playing cards. **(a)** Find the number of simple events. Find the probability of the following events: **(b)** A: 6 red and 7 black cards, **(c)** B: 4 clubs, 3 diamonds, 3 hearts, and 3 spades.

Solution

(a) The number of combinations of 52 cards taken 13 at a time is

$$h = \binom{52}{13} = \frac{52!}{13!39!} = 635{,}013{,}559{,}600$$

note: This is the number of possible bridge hands. Simplifying the factorials can be a difficult task without a good calculator.

(b) The number of outcomes in A is the number of combinations of 26 red cards taken 6 at a time multiplied by the number of combinations of 26 black cards taken 7 at a time:

$$k = \binom{26}{6}\binom{26}{7} = (230,230)(657,800) = 151{,}445{,}294{,}000$$

Thus

$$P(A) = \frac{151{,}445{,}294{,}000}{635{,}013{,}559{,}600} = .2385$$

(c) The number of outcomes in B is

$$k = \binom{13}{4}\binom{13}{3}\binom{13}{3}\binom{13}{3} = (715)(286)(286)(286) = 16{,}726{,}464{,}040$$

Thus

$$P(B) = \frac{16{,}726{,}464{,}040}{635{,}013{,}559{,}600} = .0263$$

PROBLEM 6-22 If a collection of n objects is composed of r objects of one kind and $(n - r)$ objects of another kind—for example, r red tulips and $(n - r)$ yellow tulips—then the number of distinguishable permutations of these n objects is

$$\binom{n}{r} = \frac{n!}{r!(n-r)!} \qquad \textbf{(6.13)}$$

Formula (6.13) is the same as formula (6.7), the combination of n things taken r at a time.

Suppose that a fair coin is flipped 6 times and the outcome on each flip is denoted by H = heads or T = tails. Give the number of possible distinguishable outcomes for **(a)** two heads and four tails, and for **(b)** three heads and three tails. **(c)** List the outcomes in part **(a)**.

Solution

(a) Use formula (6.13) with $n = 6$ and $r = 2$ to find the number of possible outcomes:

$$\binom{6}{2} = \frac{6!}{2!4!} = \frac{6 \cdot 5 \cdot 4!}{2 \cdot 1 \cdot 4!} = \frac{6 \cdot 5}{2} = 15$$

(b) By (6.13):

$$\binom{6}{3} = \frac{6!}{3!3!} = \frac{6 \cdot 5 \cdot 4 \cdot 3!}{3 \cdot 2 \cdot 1 \cdot 3!} = \frac{6 \cdot 5 \cdot 4}{3 \cdot 2 \cdot 1} = 20$$

(c) The 15 distinguishable permutations of two H's and four T's are

HHTTTT HTHTTT HTTHTT HTTTHT HTTTTH
THHTTT THTHTT THTTHT THTTTH TTHHTT
TTHTHT TTHTTH TTTHHT TTTHTH TTTTHH

Probabilities for Intersections of Events

PROBLEM 6-23 If $P(A) = .4$, $P(B) = .7$, and $P(A \cap B) = .2$, find $P(B \mid A)$.

Solution Since you want to find the probability of B given the condition that A has already occurred, you use formula (6.8) for conditional probability:

$$P(B \mid A) = \frac{P(A \cap B)}{P(A)} = \frac{.2}{.4} = .5$$

PROBLEM 6-24 Fifty-four college students were classified by sex F or M and by their choice of whole milk (W) or skim milk (S). The classifications are recorded in the following table:

	M	F	Totals
W	18	13	31
S	6	17	23
Totals	24	30	54

If you select one student at random, what is the probability that the student will be (**a**) a female, (**b**) a female who drinks skim milk; (**c**) a female given that the student selected drinks skim milk?

Solution

(**a**) Because you're selecting randomly, each student is equally likely to be chosen, so

$$P(F) = \frac{30}{54}$$

(**b**) From the table you can see that the intersection of F and S—the number of females who drink skim milk—is 17, so

$$P(F \cap S) = \frac{17}{54}$$

(**c**) You want to determine the probability that the student will be a female if you already know that the selected student drinks skim milk, so you use formula (6.8):

$$P(F \mid S) = \frac{P(F \cap S)}{P(S)} = \frac{17/54}{23/54} = \frac{17}{23}$$

Or, since you're concerned only with those students who drink skim milk, you can eliminate the W column in the table from consideration. The total of skim milk drinkers is 23 and 17 of these are female, so $P(F \mid S) = 17/23$.

PROBLEM 6-25 The genes that determine the eye color—red and white—of fruit flies are (R, W). The offspring receive one eye-color gene from each parent. (**a**) Define the sample space for the eye-color genes of the fruit fly offspring, assuming that each of the four possible outcomes is equally likely. (**b**) If a fruit fly offspring ends up with either two red genes or one red and one white gene, its eyes will look red. Given that an offspring's eyes look red, what's the conditional probability that it has two red genes for eye color?

Solution

(**a**) $S = \{(\text{R}, \text{R}), (\text{R}, \text{W}), (\text{W}, \text{R}), (\text{W}, \text{W})\}$

(**b**) Let $A = \{\text{red eyes}\} = \{(\text{R}, \text{R}), (\text{R}, \text{W}), (\text{W}, \text{R})\}$. Let $B = \{(\text{R}, \text{R})\}$. Then

$$P(B \mid A) = \frac{P(A \cap B)}{P(A)} = \frac{1/4}{1/3} = \frac{1}{3}$$

One-third of the fruit fly offspring with red eyes have two red eye-color genes.

PROBLEM 6-26 You have a box that contains 6 Golden Delicious apples and 4 Red Delicious apples. Two apples are selected at random from the box, one at a time, without replacement. What is the probability (**a**) that both apples will be Golden Delicious? (**b**) that one apple will be Golden and the other Red?

Solution First define the sets: $A = \{\text{Golden Delicious on 1st draw}\}$, $B = \{\text{Golden Delicious on 2nd draw}\}$, $C = \{\text{Red Delicious on 1st draw}\}$, $D = \{\text{Red Delicious on 2nd draw}\}$.

(**a**) The selection of two Golden Delicious is $A \cap B$. Use the multiplication rule (6.9) to find

$$P(A \cap B) = P(A)P(B \mid A) = \frac{6}{10} \cdot \frac{5}{9} = \frac{30}{90} = \frac{1}{3}$$

(**b**) The selection of one apple of each kind is the union of the mutually exclusive events $A \cap D$ and $C \cap B$. So the probability that one apple of each kind is selected is

$$P(A \cap D) + P(C \cap B) = P(A)P(D \mid A) + P(C)P(B \mid C) = \frac{6}{10} \cdot \frac{4}{9} + \frac{4}{10} \cdot \frac{6}{9} = \frac{48}{90} = \frac{8}{15}$$

PROBLEM 6-27 Given that A and B are independent events and $P(A) = .4$, $P(B) = .5$, find the values of (**a**) $P(A \cap B)$, (**b**) $P(A \cup B)$, and (**c**) $P(B \mid A)$.

Solution

(a) Use formula (6.11):

$$P(A \cap B) = P(A)P(B) = (.4)(.5) = .2$$

(b) From formula (6.2):

$$P(A \cup B) = .4 + .5 - .2 = .7$$

(c) From formula (6.10):

$$P(B \mid A) = P(B) = .5$$

You can also determine $P(B \mid A)$ by using formula (6.8):

$$P(B \mid A) = \frac{P(A \cap B)}{P(A)} = \frac{.2}{.4} = .5$$

PROBLEM 6-28 Roll a 4-sided die five times and call a roll of $\{1\}$ a success and a roll of $\{2, 3, 4\}$ a failure. Denote success with S and failure with F. Find **(a)** $P(\text{SSFFF})$, the probability for success on the first two rolls and failure on the last three rolls; **(b)** $P(\text{FSFSF})$; and **(c)** $P(2 \text{ successes}, 3 \text{ failures})$.

Solution

(a) Because these trials are independent, $P(S) = 1/4$ on each trial and $P(F) = 3/4$ on each trial. Now use the multiplication principle for independent events:

$$P(\text{SSFFF}) = \frac{1}{4}\cdot\frac{1}{4}\cdot\frac{3}{4}\cdot\frac{3}{4}\cdot\frac{3}{4} = \frac{27}{1024}$$

(b) Similarly,

$$P(\text{FSFSF}) = \frac{3}{4}\cdot\frac{1}{4}\cdot\frac{3}{4}\cdot\frac{1}{4}\cdot\frac{3}{4} = \frac{27}{1024}$$

(c) Notice that the answers to parts **(a)** and **(b)** are equal. In fact *each* permutation of 2 successes and 3 failures will have this same probability. Now you need to know the number of permutations of 2 S's and 3 F's, so use (6.13):

$$\binom{n}{r} = \frac{n!}{r!(n-r)!} = \binom{5}{2} = \frac{5!}{2!3!} = 10$$

Then you multiply the *number* of permutations times the probability for *each* permutation.

$$P(2 \text{ S's}, 3 \text{ F's}) = \frac{5!}{2!3!}\left(\frac{1}{4}\right)^2\left(\frac{3}{4}\right)^3 = 10\left(\frac{27}{1024}\right) = .264$$

PROBLEM 6-29 Roll a fair 6-sided die eight times and let success $= \{1, 2\}$ and failure $= \{3, 4, 5, 6\}$. Find **(a)** $P(\text{SSSFFFFF})$, **(b)** $P(\text{FSFSFSFF})$, and **(c)** $P(3 \text{ S's}, 5 \text{ F's})$.

Solution Because the trials are independent,

(a)
$$P(\text{SSSFFFFF}) = \frac{1}{3}\cdot\frac{1}{3}\cdot\frac{1}{3}\cdot\frac{2}{3}\cdot\frac{2}{3}\cdot\frac{2}{3}\cdot\frac{2}{3}\cdot\frac{2}{3} = \left(\frac{1}{3}\right)^3\left(\frac{2}{3}\right)^5 = \left(\frac{1}{27}\right)\left(\frac{32}{243}\right) = \frac{32}{6561} = .00488$$

(b)
$$P(\text{FSFSFSFF}) = \frac{2}{3}\cdot\frac{1}{3}\cdot\frac{2}{3}\cdot\frac{1}{3}\cdot\frac{2}{3}\cdot\frac{1}{3}\cdot\frac{2}{3}\cdot\frac{2}{3} = \left(\frac{1}{3}\right)^3\left(\frac{2}{3}\right)^5 = .00488$$

(c) The number of distinguishable permutations of 3 S's and 5 F's is $\frac{8!}{3!5!}$ and each of these outcomes has the same probability. Thus

$$P(3 \text{ S's}, 5 \text{ F's}) = \frac{8!}{3!5!}\left(\frac{1}{3}\right)^3\left(\frac{2}{3}\right)^5 = \frac{8\cdot7\cdot6\cdot5!}{3\cdot2\cdot1\cdot5!}(.00488) = .273$$

PROBLEM 6-30 Roll a 12-sided die (a dodecahedron) three times. The possible outcomes on each roll are the integers 1 through 12. Find the probability **(a)** that the outcome on each roll will be ≤ 5 and **(b)** that the outcome on two rolls will be ≤ 5 and on one roll will be > 5.

Solution Let success $= \text{S} = \{1, 2, 3, 4, 5\}$ and failure $= \text{F} = \{6, 7, 8, 9, 10, 11, 12\}$, so that $P(\text{success}) = P(\text{S}) = 5/12$ on each roll and $P(\text{failure}) = P(\text{F}) = 7/12$.

(a)
$$P(\text{SSS}) = \frac{5}{12} \cdot \frac{5}{12} \cdot \frac{5}{12} = \left(\frac{5}{12}\right)^3 = \frac{125}{1728} = .072$$

(b)
$$P(\text{SSF}) + P(\text{SFS}) + P(\text{FSS}) = 3\left(\frac{5}{12}\right)^2\left(\frac{7}{12}\right) = .304$$

Supplementary Exercises

PROBLEM 6-31 Roll a fair 8-sided die two times. Let A be the event that the outcome on the first roll is 3 or 4 and let B be the event that the outcome on the second roll is 5, 6, or 7. **(a)** Define the sample space with a display, and circle the outcomes that belong to A and to B. **(b)** Give the values of $P(A)$ and $P(B)$. **(c)** Use your answers in part **(b)** to find $P(A \cap B)$, assuming that A and B are independent. Is this a valid assumption?

Answer

(a)

				B	B	B		
1, 1	1, 2	1, 3	1, 4	1, 5	1, 6	1, 7	1, 8	
2, 1	2, 2	2, 3	2, 4	2, 5	2, 6	2, 7	2, 8	
3, 1	3, 2	3, 3	3, 4	3, 5	3, 6	3, 7	3, 8	A
4, 1	4, 2	4, 3	4, 4	4, 5	4, 6	4, 7	4, 8	A
5, 1	5, 2	5, 3	5, 4	5, 5	5, 6	5, 7	5, 8	
6, 1	6, 2	6, 3	6, 4	6, 5	6, 6	6, 7	6, 8	
7, 1	7, 2	7, 3	7, 4	7, 5	7, 6	7, 7	7, 8	
8, 1	8, 2	8, 3	8, 4	8, 5	8, 6	8, 7	8, 8	

(b) $P(A) = 1/4$, $P(B) = 3/8$ **(c)** 3/32; yes

PROBLEM 6-32 Roll a 12-sided die once. Find the probability that the outcome is a 9, given that the outcome is divisible by 3.

Answer 1/4

PROBLEM 6-33 In the game of craps, a shooter rolls a pair of dice. The shooter wins on the first roll if the sum of the dice is 7 or 11, and he loses if the sum is 2, 3, or 12. If he rolls a 4, 5, 6, 8, 9, or 10, that number becomes his "point." In order to win, the shooter must roll his point before he rolls a 7—if he rolls a 7 before he rolls his point, he loses. Assume that the dice are fair and each roll has a probability of 1/36. **(a)** Find the probability that the shooter wins on the first roll. **(b)** Given that the shooter rolls an 8 on his first roll, what is the probability that he will roll an 8 before a 7 and thus win? Note that at this point you are only interested in 7's and 8's, so you must find $P(8 \mid [7 \text{ or } 8])$. **(c)** Find the probability that the shooter will roll an 8 on the first roll and then win—i.e., $P(8)P(8 \mid [7 \text{ or } 8])$. **(d)** Find the total probability that a shooter will win in the game of craps. [*Hint:* The answer to **(d)** includes the answers to **(a)** and **(c)** plus the probability of rolling a 4, 5, 6, 9 or 10 on the first roll and winning. You calculate these latter 5 probabilities just as you did in **(c)**.]

Answer **(a)** 8/36 **(b)** 5/11 **(c)** (5/36)(5/11) **(d)** .49293

PROBLEM 6-34 At Izzy's Deli, a sandwich customer may choose from 7 bread, 8 meat, 5 cheese, and 13 extras such as lettuce, mustard, and alfalfa sprouts. **(a)** If you choose 1 bread, 1 meat, and 1 cheese, how many different sandwiches are possible? (**b**) You may choose from 0 to 13 of the 13 extras. How many possibilities are there? (**c**) If you combine 1 bread, 1 meat, and 1 cheese selection with from 0 to 13 extras, how many sandwiches are possible?

Answer (**a**) 280 (**b**) $2^{13} = 8192$ (for each extra, there are 2 choices, yes or no) (**c**) 2,293,760

PROBLEM 6-35 In a multilevel marketing scheme, the distributor is at the top. There are three people below the distributor at the first level. Below *each* of these three, there are three people at the second level. Below *each* of these nine at the second level there are three people at the third level, etc. For a seven-level scheme, give (**a**) the number of people at each level and (**b**) the total number of people, including the distributor.

Answer (**a**) 1st level: 3; 2nd level: 9; 3rd level: 27; 4th level: 81; 5th level: 243; 6th level: 729; 7th level: 2187 (**b**) 3280

PROBLEM 6-36 A dish contains 60 rainbow-colored mints: 17 pink, 18 yellow, 15 white, and 10 green. If 4 mints are selected, one at a time without replacement, give the probability for selecting (**a**) mints colored pink, yellow, white, green, in that order, and (**b**) 1 mint of each color. (**c**) If 8 mints are selected, one at a time with replacement, give the probability for selecting 3 white mints and 5 colored mints.

Answer (**a**) $\left(\frac{17}{60}\right)\left(\frac{18}{59}\right)\left(\frac{15}{58}\right)\left(\frac{10}{57}\right)$ (**b**) $(4!)\left(\frac{17}{60}\right)\left(\frac{18}{59}\right)\left(\frac{15}{58}\right)\left(\frac{10}{57}\right)$ (**c**) $\left(\frac{8!}{3!5!}\right)\left(\frac{1}{4}\right)^3\left(\frac{3}{4}\right)^5$

PROBLEM 6-37 Draw two cards from a standard deck of playing cards. What's the probability for drawing two cards of the same color if you select the cards (**a**) without replacement? (**b**) with replacement?

Answer (**a**) 25/51 (**b**) 1/2

PROBLEM 6-38 A jar contains the following numbers of yellow (Y) and orange (O) rock candy pieces (R) and mints (M).

	R	M	Totals
Y	22	12	34
O	28	8	36
Totals	50	20	70

Select one piece of candy at random from the jar. Find (**a**) $P(\text{Y})$, (**b**) $P(\text{M} \cap \text{Y})$, (**c**) $P(\text{M})$, and (**d**) $P(\text{M} \mid \text{Y})$.

Answer (**a**) 34/70 (**b**) 12/70 (**c**) 20/70 (**d**) 12/34

PROBLEM 6-39 Consider the following three events:

$A = \{$an orange piece of candy is selected from a package of rock candy pieces$\}$
$B = \{$an orange piece of candy is selected from a package of mints$\}$
$C = \{$an orange piece of candy is selected from a package of jelly beans$\}$

Assume that A, B, and C are mutually independent and that $P(A) = .5$, $P(B) = .2$, and $P(C) = .25$. If one piece of candy is selected from each of the three packages, find the probability that exactly two of the three pieces of candy will be orange.

Answer .2

PROBLEM 6-40 If A and B are independent events, $P(A) = .7$, and $P(B) = .2$, give the values of (**a**) $P(A \cap B)$, (**b**) $P(A \cup B)$, and (**c**) $P(B \mid A)$.

Answer (**a**) .14 (**b**) .76 (**c**) .2

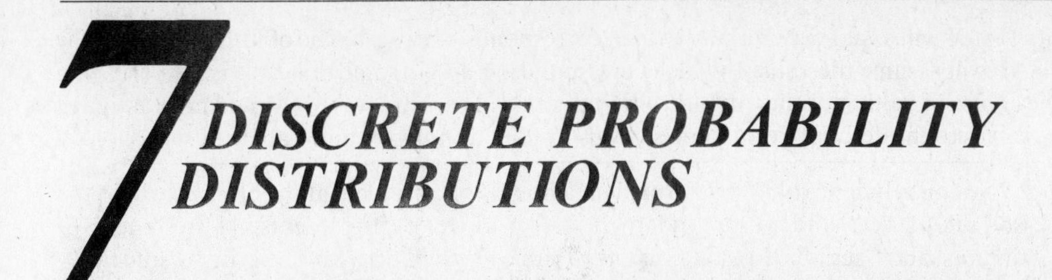

7 DISCRETE PROBABILITY DISTRIBUTIONS

THIS CHAPTER IS ABOUT

- ☑ **Discrete Random Variables**
- ☑ **Probability Functions**
- ☑ **Distribution Functions**
- ☑ **Mean, Variance, and Standard Deviation of a Random Variable**

There is a theoretical probability distribution associated with a random experiment. When the set of possible outcomes is discrete (finite or countable), we define a random variable of the discrete type along with probabilities associated with the possible outcomes of the experiment. The mean, variance, and standard deviation are defined for this probability distribution.

7-1. Discrete Random Variables

There are two major classes of probability distributions: those for which the set of possible outcomes for the random experiment is a discrete set of points, and those for which the set of possible outcomes is an interval or a union of intervals—a continuous set of points. This chapter will concentrate on those distributions associated with random experiments that have a discrete set of points as the set of possible outcomes.

In most of the examples in Chapter 6 the sample space contains a finite set of points (see Examples 6-1 to 6-9). In Example 6-10 there are as many outcomes as there are positive integers in the sample space. For such sample spaces we say that the sample space is *countable* (or countably infinite).

- If the sample space is either finite or countable, we say that it is *discrete* and the associated random experiment is of the *discrete type.*

The outcome of an experiment may be either numerical or descriptive—for example, success or failure, heads or tails, male or female. There are advantages in working with quantitative rather than qualitative outcomes, so

- When the outcomes of a random experiment are qualitative, we use a **random variable** that assigns a numerical value to each outcome of a random experiment.

A random variable associated with a random experiment of the discrete type is called, aptly, a **discrete-type random variable**, or a random variable of the discrete type. We denote a random variable with a letter near the end of the alphabet, most commonly X, Y, or Z, although other letters are also used.

EXAMPLE 7-1 Perform a random experiment by flipping a fair coin once. Denote the outcome of the experiment with the random variable X and assign a numerical value to each possibility. If the outcome is tails, let $X = 0$, and if the outcome is heads, let $X = 1$. What are the probabilities associated with the outcomes of the random variable X?

Solution The probability that tails will come up is equal to $1/2$ and the probability that heads will come up is equal to $1/2$, so

$$P(X = 0) = \frac{1}{2}, \qquad P(X = 1) = \frac{1}{2}$$

The probability that X will assume the value 0 (which corresponds to an outcome of tails) is 1/2 and the probability that X will assume the value 1 (which corresponds to an outcome of heads) is also 1/2. You can see that the choice of values for the random variable is entirely arbitrary—you could have assigned any numerical values to the descriptive outcomes, heads or tails, although 0 and 1 are the usual choices.

EXAMPLE 7-2 Not one student volunteers to explain a homework problem at the blackboard, so the statistics professor selects one student at random from her class of 50 students; 11 freshmen, 19 sophomores, 14 juniors, and 6 seniors. Let the random variable X equal the year of the lucky student—i.e., 1, 2, 3, or 4 for a freshman, sophomore, junior, or senior, respectively. **(a)** Define the probabilities associated with all the possible outcomes of X. **(b)** What's the probability that a freshman or a sophomore will be selected? What's the probability that the student will *not* be a freshman?

Solution

(a) Since each student in the class has the same chance of being selected,

$$P(X = 1) = \frac{11}{50} = .22 \qquad P(X = 2) = \frac{19}{50} = .38$$

$$P(X = 3) = \frac{14}{50} = .28 \qquad P(X = 4) = \frac{6}{50} = .12$$

(b)

$$P(X \le 2) = P(X = 1) + P(X = 2) = \frac{30}{50} = .6$$

$$P(X \ge 2) = 1 - P(X = 1) = \frac{39}{50} = .78$$

EXAMPLE 7-3 Flip a fair coin three times. Let H denote heads and T denote tails. If $Y =$ the number of heads that come up, what are the probabilities for flipping 0, 1, 2, or 3 heads?

Solution The sample space for 3 tosses of a fair coin is

$$S = \{\text{HHH, HHT, HTH, THH, HTT, THT, TTH, TTT}\}$$

Each of the 8 outcomes has the same probability—i.e., 1/8—so you have

$$P(Y = 0) = P(\{\text{TTT}\}) = \frac{1}{8}$$

$$P(Y = 1) = P(\{\text{HTT, THT, TTH}\}) = \frac{3}{8}$$

$$P(Y = 2) = P(\{\text{HHT, HTH, THH}\}) = \frac{3}{8}$$

$$P(Y = 3) = P(\{\text{HHH}\}) = \frac{1}{8}$$

7-2. Probability Functions

Given a random variable X of the discrete type, we're interested in defining the probabilities for the possible outcomes of X, $P(X = x)$. We do this with a function,

$$f(x) = P(X = x)$$

called the probability function of X. We can define $f(x)$ with a formula, a table, or by listing the values of $f(x)$ for each value of x. Although we usually use $f(x)$ for the probability function, we can also use $g(x)$. If the random variable is Y we could use $f(y) = P(Y = y)$ or $g(y) = P(Y = y)$ for the probability function.

EXAMPLE 7-4 Define the probability function for the flip of a fair coin as given in Example 7-1.

Solution Since $X = 0$ when the outcome is tails and $X = 1$ when the outcome is heads, you have

$$f(x) = P(X = x)$$

$$f(0) = P(X = 0) = \frac{1}{2}$$

$$f(1) = P(X = 1) = \frac{1}{2}$$

The value of the probability function at 0 (X takes the value 0) is 1/2, and the value of the probability function at 1 ($X = 1$) is also 1/2.

We can also define the probability function by using a table, particularly when there are many possible values for X.

EXAMPLE 7-5 Use a table to define the probability function for the selection of a student as described in Example 7-2.

Solution In the x column, you list each value of the random variable X followed in column $f(x)$ by the probability for its occurrence:

x	$f(x)$
1	11/50 = .22
2	19/50 = .38
3	14/50 = .28
4	6/50 = .12
	$\sum f(x) = 50/50 = 1.00$

EXAMPLE 7-6 Use a table to define the probability function $g(y)$ for the random variable Y defined in Example 7-3.

Solution

y	$g(y)$
0	1/8
1	3/8
2	3/8
3	1/8
	$\sum g(y) = 8/8 = 1$

Because a probability function assigns probabilities, it's always true that $0 \leq f(x) \leq 1$ for each value of x that the random variable X assumes. That is, the probability for any outcome of a random variable must lie somewhere at or between the extremes of never occurring $[f(x) = 0]$ and always occurring $[f(x) = 1]$. Also, the sum of all the values of $f(x)$ must equal one ($\sum f(x) = 1$) which indicates that one of the values of X must occur.

EXAMPLE 7-7 Roll a 4-sided die twice. Let Y equal the sum of the two outcomes, so that the possible values of Y are 2, 3, 4, 5, 6, 7, 8. Define the probability function $g(y)$, **(a)** by using a table, and **(b)** by determining a formula that describes the probabilities.

Solution

(a)

y	$g(y)$
2	1/16
3	2/16
4	3/16
5	4/16
6	3/16
7	2/16
8	1/16
	$\sum g(y) = 16/16$

(b) A formula that can be used to define this probability function is

$$g(y) = \frac{4 - |y - 5|}{16}, \quad y = 2, 3, 4, 5, 6, 7, 8$$

This formula may be difficult for you to derive, but you can check to see that the formula holds for each value of y.

You can use two types of graphs to depict the probability function for a discrete-type variable: a bar graph or a probability histogram. A **bar graph** for the probability function $f(x)$ of the random variable X has a vertical line segment of length $f(x)$ at each possible value x of the random variable X. When X takes on integer values, a **probability histogram** of $f(x)$ has a rectangle of height $f(x)$ and a base of length one centered at x for each possible value of X.

EXAMPLE 7-8 Draw **(a)** a bar graph and **(b)** a probability histogram for the probability function defined in Example 7-5.

Solution

(a) Label the x-axis with the values of X, and the y-axis with the values of the probability function $f(x)$. Then draw a line whose length corresponds to $f(x)$ for each value of X. See Fig. 7-1a.

(b) Label the x- and y-axes as in Fig. 7-1a, but draw the lines so that each value of X is centered in a rectangle of height $f(x)$. See Fig. 7-1b.

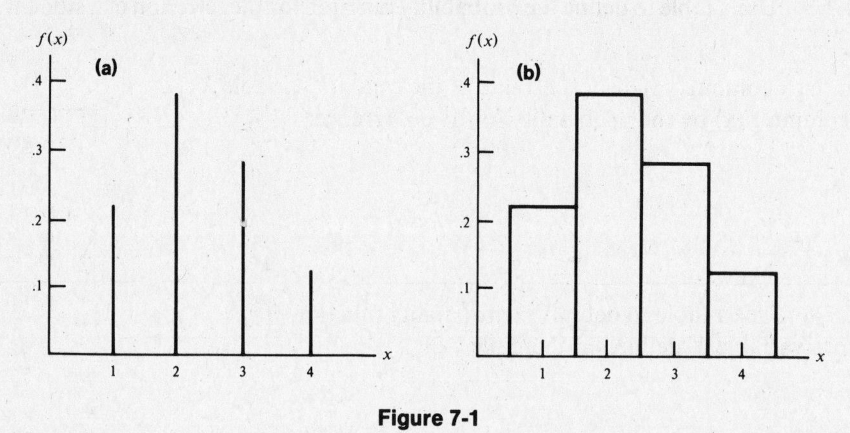

Figure 7-1

EXAMPLE 7-9 Draw **(a)** a bar graph and **(b)** a probability histogram for the probability function defined in Example 7-7.

Solution **(a)** See Fig. 7-2a. **(b)** See Fig. 7-2b.

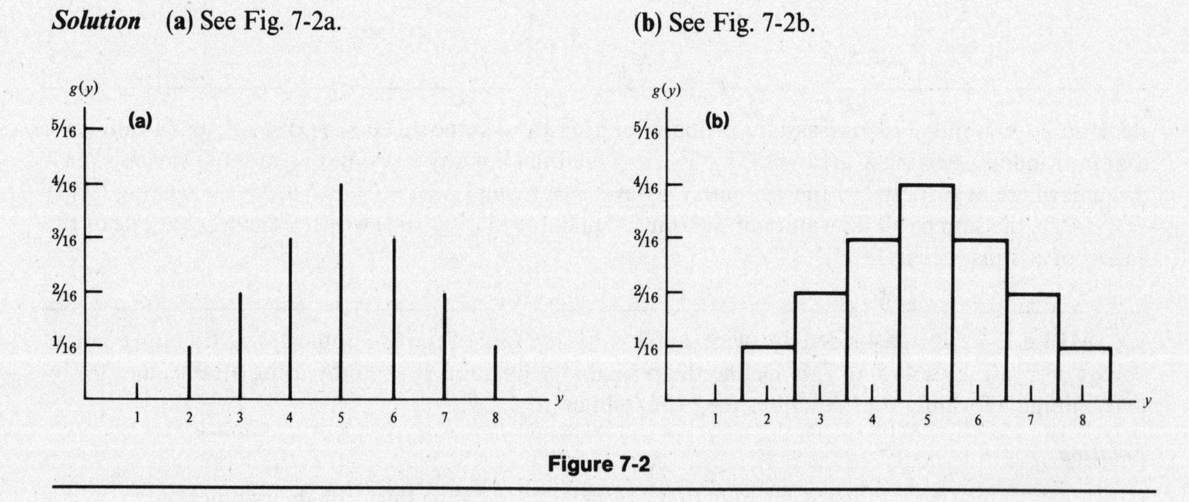

Figure 7-2

7-3. Distribution Functions

The probability function gives the probability for an individual value of X. If we want to know the probability for not one, but several outcomes of X, we can use the **distribution function** of the random variable X. This function gives the probability that X is less than or equal to some number x. It is defined by

$$F(x) = P(X \leq x)$$

The distribution function is also called the **cumulative distribution function**, because it gives the cumulative probabilities for a distribution. When the probability function is $f(x)$, we use $F(x)$ for the distribution function. Similarly, we use $G(y)$ for the distribution function when the probability function is $g(y)$. Notice that we define the distribution function for all real numbers—not just for possible values of the random variable.

EXAMPLE 7-10 (**a**) Define the distribution function for the distribution in Example 7-5, and (**b**) draw a graph of this function.

Solution

(**a**) Because $F(x) = P(X \leq x)$, you have, e.g.,

$$F(2) = P(X \leq 2) = f(1) + f(2) = .22 + .38 = .60$$

$$F(3) = P(X \leq 3) = f(1) + f(2) + f(3) = .88$$

Also,

$$F(2.7) = P(X \leq 2.7) = f(1) + f(2) = .60$$

since 1 and 2 are the only values of X that are less than or equal to 2.7. In general we find the values of $F(x)$ by adding together all the individual probabilities for X for stipulated values x given in the right column, so

$$F(x) = P(X \leq x) = \begin{cases} 0, & x < 1 \\ .22, & 1 \leq x < 2 \\ .60, & 2 \leq x < 3 \\ .88, & 3 \leq x < 4 \\ 1.00, & 4 \leq x \end{cases}$$

(**b**) See Fig. 7-3.

Notice that because the distribution function is cumulative, the graph never decreases as x increases. Also notice the jumps in the graph that occur at $x = 1, 2, 3$, and 4, the possible outcomes of X. The size of the jump at x is the probability that the random variable X equals x–at $x = 2$, there is a jump from .22 to .60, a difference of .38, which equals the probability that $X = 2$.

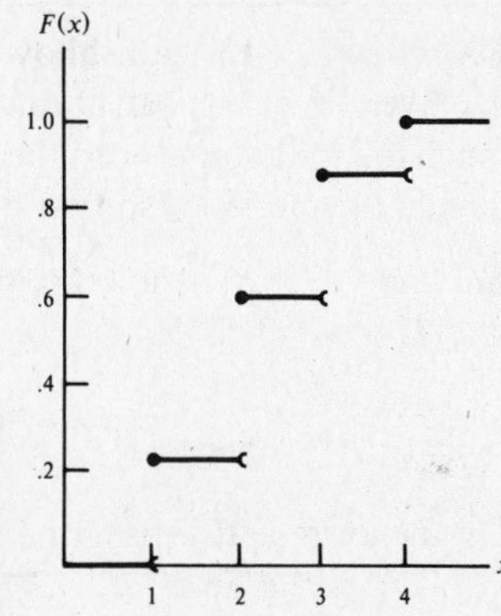

Figure 7-3

EXAMPLE 7-11 Draw the graph of the distribution function for the experiment described in Example 7-7.

Solution See Fig. 7-4.

Since the random variable is Y and the probability function is $g(y)$, we use $G(y)$ for the distribution function. We have

$$G(y) = P(Y \leq y) = \begin{cases} 0, & y < 2 \\ 1/16, & 2 \leq y < 3 \\ 3/16, & 3 \leq y < 4 \\ 6/16, & 4 \leq y < 5 \\ 10/16, & 5 \leq y < 6 \\ 13/16, & 6 \leq y < 7 \\ 15/16, & 7 \leq y < 8 \\ 1, & 8 \leq y \end{cases}$$

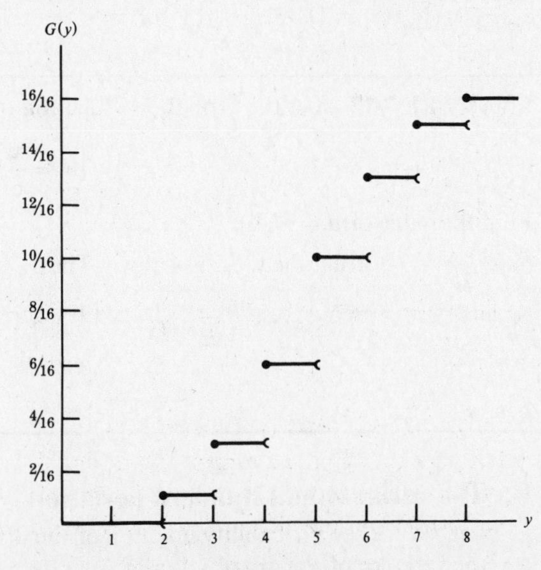

Figure 7-4

7-4. Mean, Variance, and Standard Deviation of a Random Variable

In Sections 2-1 and 3-1, we defined the sample mean, the sample variance, and the sample standard deviation of empirical distributions, which give us a measure of the center and the spread of an actual sample set of measurements. In this section we give definitions for summary measures of central tendency and variability for the underlying distribution from which the sample was taken.

A. The mean of X

Let $f(x)$ be the probability function for the discrete-type random variable X. The mean of X (or the mean of the distribution of X) is a weighted average achieved by summing the products of possible outcomes and the probabilities of these outcomes:

DISTRIBUTION MEAN

$$\mu = E(X) = \sum xf(x) \tag{7.1}$$

where the summation is over all possible values of X. $E(X)$ is the **expected value** of X, which is the same as the mean. Formula (7.1) states that the mean or expected value of X is the sum of all the products of x times the probability that x will be observed.

note: We denote the sample mean with the Roman letter $\bar{x}$ and the mean of the underlying distribution with the Greek letter μ (mu). This is true in general: Roman letters denote values for a sample while Greek letters denote the projected population counterparts.

EXAMPLE 7-12 A fishbowl at the county fair contains three balls: two labeled \$1 and one labeled \$4. A player draws one ball at random from the bowl and wins the amount on the label, either \$1 or \$4. How much (on the average) can the player expect to win? How much should the player pay to play if this is to be a fair game—i.e., so that he would break even after playing all day?

Solution Let X denote the possible wins and $f(x)$ the probability function for X, so that $f(1) = 2/3$ and $f(4) = 1/3$. Then

$$\mu = E(X) = \sum xf(x) = 1f(1) + 4f(4) = 1\left(\frac{2}{3}\right) + 4\left(\frac{1}{3}\right) = 2$$

On the average the player can expect to win \$2, so \$2 would be a fair charge to play the game.

You can construct a table to help in calculating the mean. The first column (x) gives the possible values of X; the second column $[f(x)]$ gives the probability function at each value of x; and the third column $xf(x)$ gives the value of x times the probability that x will be observed. The sum of the values of $xf(x)$ gives the mean of X. Remember that the total of the values in the second column must equal one $(\sum f(x) = 1)$.

EXAMPLE 7-13 Let the probability function of X be defined by

$$f(x) = \frac{x}{6}, \quad x = 1, 2, 3$$

Find the value of $\mu = E(X)$.

Solution Construct the table as shown. Thus,

$$\mu = E(X) = \sum xf(x) = \frac{14}{6} = 2.33$$

x	$f(x)$	$xf(x)$
1	1/6	1(1/6) = 1/6
2	2/6	2(2/6) = 4/6
3	3/6	3(3/6) = 9/6
	6/6 = 1	$\mu = 14/6$

B. The variance and standard deviation of X

Let $f(x)$ be the probability function of the discrete-type random variable X. The **variance of** X (or the variance of the distribution of X) is defined by

DISTRIBUTION VARIANCE

$$\sigma^2 = E[(X - \mu)^2] = \sum (x - \mu)^2 f(x) \tag{7.2}$$

where $E[(X - \mu)^2]$ is the expected value of the squared difference between the values of the random variable X and the mean, and the summation is over all possible values of X.

A working formula that simplifies calculations for the variance is

DISTRIBUTION VARIANCE (alternate form)

$$\sigma^2 = \sum x^2 f(x) - \mu^2 \tag{7.3}$$

Because the variance gives us the squared difference between X and μ, we take the square root of the variance to obtain the **standard deviation of** X.

$$\sigma = \sqrt{\sigma^2} \tag{7.4}$$

Notice that we denote the population variance and standard deviation with the Greek letter σ (sigma).

EXAMPLE 7-14 Let $f(x) = 1/3$, $x = 4, 5, 6$, be the probability function of the random variable X. Find **(a)** the mean, **(b)** the variance, and **(c)** the standard deviation.

Solution

(a) The mean is

$$\mu = 4\left(\frac{1}{3}\right) + 5\left(\frac{1}{3}\right) + 6\left(\frac{1}{3}\right) = \frac{15}{3} = 5$$

(b) If you use formula (7.2),

$$\begin{aligned}\sigma^2 &= \sum (x - \mu)^2 f(x) \\ &= (4 - 5)^2\left(\frac{1}{3}\right) + (5 - 5)^2\left(\frac{1}{3}\right) + (6 - 5)^2\left(\frac{1}{3}\right) = \frac{1}{3} + 0 + \frac{1}{3} = \frac{2}{3}\end{aligned}$$

If you use (7.3), you don't have to calculate $(x - \mu)^2$ for each value of X.

$$\sigma^2 = \sum x^2 f(x) - \mu^2 = 16\left(\frac{1}{3}\right) + 25\left(\frac{1}{3}\right) + 36\left(\frac{1}{3}\right) - 25 = \frac{77}{3} - \frac{75}{3} = \frac{2}{3}$$

(c) From formula (7.4), the standard deviation is

$$\sigma = \sqrt{\sigma^2} = \sqrt{2/3}$$

EXAMPLE 7-15 Let $g(y) = 1/3$, $y = 3, 5, 7$, be the probability function for the random variable Y. Find **(a)** the mean, **(b)** the variance, and **(c)** the standard deviation.

Solution

(a) The mean is

$$\mu = 3\left(\frac{1}{3}\right) + 5\left(\frac{1}{3}\right) + 7\left(\frac{1}{3}\right) = 5$$

(b) The variance from (7.2) is

$$\sigma^2 = (3 - 5)^2\left(\frac{1}{3}\right) + (5 - 5)^2\left(\frac{1}{3}\right) + (7 - 5)^2\left(\frac{1}{3}\right) = \frac{8}{3}$$

or from (7.3) is

$$\sigma^2 = 9\left(\frac{1}{3}\right) + 25\left(\frac{1}{3}\right) + 49\left(\frac{1}{3}\right) - 25 = \frac{8}{3}$$

(c) The standard deviation is

$$\sigma = \sqrt{8/3} = 2\sqrt{2/3}$$

note: The standard deviation in Example 7-15 is twice as large as the standard deviation in Example 7-14, which indicates that the probability is more spread out in Example 7-15. You can also see that this is the case if you compare the values for X and Y.

EXAMPLE 7-16 Roll a pair of 4-sided dice, and let X equal the sum of the dice. Find **(a)** the mean, **(b)** the variance, and **(c)** the standard deviation.

Solution First construct a table as shown. [You won't be able to complete the last column until you solve part **(a)**.]

x	$f(x)$	$xf(x)$	$x^2f(x)$	$(x-\mu)^2f(x)$
2	1/16	2/16	4/16	9/16
3	2/16	6/16	18/16	8/16
4	3/16	12/16	48/16	3/16
5	4/16	20/16	100/16	0/16
6	3/16	18/16	108/16	3/16
7	2/16	14/16	98/16	8/16
8	1/16	8/16	64/16	9/16
	16/16	80/16	440/16	40/16

(a) You can determine the mean by (7.1):

$$\mu = \frac{80}{16} = 5$$

(b) From (7.2),

$$\sigma^2 = \sum (x-\mu)^2 f(x) = \frac{40}{16} = 2.5$$

or from (7.3),

$$\sigma^2 = \sum x^2 f(x) - \mu^2 = \frac{440}{16} - 5^2 = \frac{40}{16} = 2.5$$

(c) $$\sigma = \sqrt{2.5} = 1.58$$

EXAMPLE 7-17 Roll a pair of 4-sided dice, and let Y equal the largest number on the two dice. **(a)** Use a table to define the probability function of Y. Find **(b)** the mean, **(c)** the variance, and **(d)** the standard deviation of Y.

Solution

(a) First describe the sample space with an array of all the possible outcomes in the random experiment where $y = 1, 2, 3$, and 4. Then you can use the array to construct the table as shown. The first two columns define the probability function of Y.

$y = 1$	1, 1	1, 2	1, 3	1, 4
$y = 2$	2, 1	2, 2	2, 3	2, 4
$y = 3$	3, 1	3, 2	3, 3	3, 4
$y = 4$	4, 1	4, 2	4, 3	4, 4

y	$f(y)$	$yf(y)$	$y^2f(y)$
1	1/16	1/16	1/16
2	3/16	6/16	12/16
3	5/16	15/16	45/16
4	7/16	28/16	112/16
	16/16	50/16	170/16

(b) $$\mu = \frac{50}{16} = 3.125$$

(c) Use (7.3) to avoid the extra calculations:

$$\sigma^2 = \frac{170}{16} - \left(\frac{50}{16}\right)^2 = \frac{220}{256} = .859$$

(d) $$\sigma = \sqrt{.859} = .927$$

SUMMARY

1. The number of possible outcomes for a discrete-type random variable must be either finite or countable.
2. A *random variable* X assigns a numerical value to each outcome of a random experiment.
3. The *probability function* $f(x)$ defines the probabilities for the possible outcomes of X.
4. The *distribution function* $F(x)$ defines cumulative probabilities for the distribution of X.
5. The *distribution function* is $F(x) = P(X \le x)$.
6. The *mean of* X is $\mu = \sum xf(x)$.
7. The *variance of* X is $\sigma^2 = \sum (x-\mu)^2 f(x) = \sum x^2 f(x) - \mu^2$.
8. The *standard deviation of* X is $\sigma = \sqrt{\sigma^2}$.

RAISE YOUR GRADES

Can you . . . ?

☑ describe a discrete-type random experiment
☑ give the definition of a random variable

☑ give the properties of a probability function
☑ define a distribution function for a given probability function
☑ draw a bar graph
☑ draw a probability histogram
☑ graph a distribution function
☑ calculate a mean of a random variable X
☑ calculate a variance of a random variable X
☑ calculate a standard deviation of a random variable X
☑ describe the difference between a sample mean, variance, and standard deviation and a distribution mean, variance, and standard deviation

RAPID REVIEW

1. If 10 dice are rolled and the number of 5's is counted, the number of outcomes in the sample space is ____________.
2. If you count the number of rolls of a die before a 5 is observed, the sample space is ____________.
3. If a sample space is finite or countable, we say it is ____________.
4. A ____________ ____________ assigns a numerical value to each outcome of a random experiment.
5. A ____________ ____________ defines the probabilities for the possible outcomes of X.
6. If the possible values of X are 1, 2, and 3, and $f(1) = 1/3$ and $f(2) = 1/2$, then $f(3) =$ ____________.
7. Two ways to graph a probability function for a discrete-type random variable are with a **(a)** ____________ ____________ and with a **(b)** ____________ ____________.
8. Cumulative probabilities for X are defined by the ____________ ____________.
9. If $f(2) = 3/4$ and $f(5) = 1/4$, then $\mu =$ ____________.
10. If $f(1) = 1/4$, $f(2) = 1/2$, and $f(3) = 1/4$, then $\mu =$ **(a)** ____________ and $\sigma^2 =$ **(b)** ____________.
11. If $\sigma^2 = 81$ then $\sigma =$ ____________.

Answers **(1)** finite (or 11) **(2)** countable **(3)** discrete **(4)** random variable **(5)** probability function **(6)** 1/6 **(7)** **(a)** bar graph **(b)** probability histogram **(8)** distribution function **(9)** 11/4 **(10)** **(a)** 2 **(b)** 1/2 **(11)** 9

SOLVED PROBLEMS

Discrete Random Variables

PROBLEM 7-1 There are 4 cups of decaffeinated coffee on a table and 16 cups of coffee with caffeine. Select one cup of coffee at random. Let $X = 0$ if the coffee is decaffeinated, and let $X = 1$ if the coffee has caffeine. Define the probabilities associated with X.

Solution Assuming that the 20 outcomes are equally likely,

$$P(X = 0) = \frac{4}{20} = \frac{1}{5} \qquad P(X = 1) = \frac{16}{20} = \frac{4}{5}$$

PROBLEM 7-2 Select a letter at random from the letters in the two words HOPE COLLEGE. Let the random variable X assign to each letter the position that it occupies in the alphabet (A = 1, B = 2, etc.). **(a)** List the possible values that X can equal. **(b)** Define the probabilities associated with X.

Solution

(a) The possible values of X are $S = \{3, 5, 7, 8, 12, 15, 16\}$.

(b) The probabilities associated with X are

$$P(X = 3) = P(\{C\}) = 1/11 \qquad P(X = 5) = P(\{E\}) = 3/11$$

$$P(X = 7) = P(\{G\}) = 1/11 \qquad P(X = 8) = P(\{H\}) = 1/11$$

$$P(X = 12) = P(\{L\}) = 2/11 \qquad P(X = 15) = P(\{O\}) = 2/11$$

$$P(X = 16) = P(\{P\}) = 1/11$$

PROBLEM 7-3 You see an empty parked car with its lights on and you stop to turn the lights off. Call your attempt a failure if the doors of the car are locked and a success if the doors are unlocked. Define a random variable Y for this "experiment."

Solution It's usual to let $Y = 0$ for a failure and $Y = 1$ for a success.

PROBLEM 7-4 In the casino game of chuck-a-luck, it's possible to lose one dollar or win one, two, or three dollars on a one-dollar bet. What's a natural way to define a random variable for this game?

Solution If you let Y equal the payoff, then $Y = -1$ if a dollar is lost and $Y = 1$, $Y = 2$, or $Y = 3$ if one, two, or three dollars are won, respectively.

PROBLEM 7-5 Suppose that you've been buying Freakies cereal and would like to complete the collection of all 7 freakies (See Example 6-4). When you have collected all of the freakies but Hamhose, let X equal the number of boxes of cereal that you must purchase to obtain Hamhose. Describe the possible values of X.

Solution Hamhose could be in the first box that you purchase, or the second, or the third, etc. Thus the possible values for X are

$$S = \{1, 2, 3, \ldots\}$$

This is an example of a sample space that is countable but not finite.

Probability Functions

PROBLEM 7-6 To satisfy your sweet tooth, you buy a bag of candy. After you have eaten all you want, there are 12 mints (M) and 8 butterscotch drops (B) left in the bag. You generously offer the leftovers to your little sister, who selects a piece of candy from the bag and eats it, and then selects a second piece. Let X equal the number of mints that she selects. **(a)** Define the probability function of X. **(b)** Draw a probability histogram for the probability function of X.

Solution

(a) If you let $\{X = 0\} = \{BB\} = \{2 \text{ butterscotch drops}\}$, $\{X = 1\} = \{BM, MB\} = \{1 \text{ mint and } 1 \text{ butterscotch}\}$, $\{X = 2\} = \{MM\} = \{2 \text{ mints}\}$, the probability function of X is

$$f(0) = P(X = 0) = P(BB) = \frac{8}{20} \cdot \frac{7}{19} = \frac{14}{95}$$

$$f(1) = P(X = 1) = P(BM) + P(MB)$$

$$= \frac{8}{20} \cdot \frac{12}{19} + \frac{12}{20} \cdot \frac{8}{19} = \frac{48}{95}$$

$$f(2) = P(X = 2) = P(MM) = \frac{12}{20} \cdot \frac{11}{19} = \frac{33}{95}$$

Notice that the sum of these probabilities is one.

(b) See Fig. 7-5.

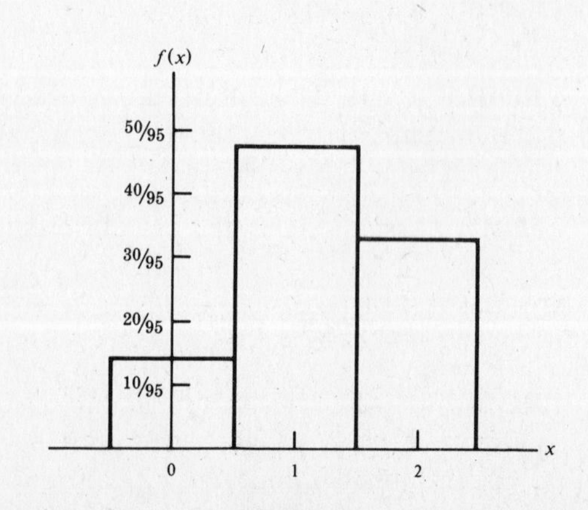

Figure 7-5

PROBLEM 7-7 A 20-sided die can be used to generate the digits 0, 1, 2, 3, 4, 5, 6, 7, 8, 9 at random. Let X equal the outcome when this die is rolled once. **(a)** Define the probability function of X. Draw **(b)** a bar graph and **(c)** a probability histogram for the probability function of X. **(d)** Letting Y equal the outcome when this die is rolled twice to generate a two-digit integer, define the probability function of Y.

Solution

(a) The probability function of X is

$$f(x) = P(X = x) = \frac{1}{10}, \quad x = 0, 1, 2, 3, 4, 5, 6, 7, 8, 9$$

(b) See Fig. 7-6a.
(c) See Fig. 7-6b.
(d) When you roll the die once, there are 10 possible outcomes, so if you roll the die twice, there will be $10^2 = 100$ equally possible outcomes. So

$$g(y) = P(Y = y) = \frac{1}{100}, \quad y = 00, 01, 02, \ldots, 99$$

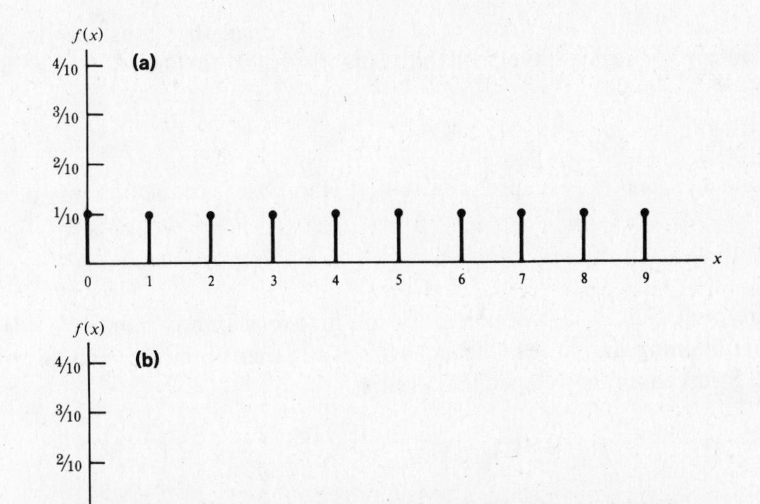

Figure 7-6

PROBLEM 7-8 There are six cartons of yogurt in the refrigerator: two cherry (C) and four strawberry (S). Select two cartons at random without replacement, and let Y equal the number of cartons of strawberry yogurt that you select. **(a)** Define the probability function of Y. **(b)** Draw a bar graph of the probability function.

Solution

(a) You might select no strawberry yogurts, one strawberry yogurt, or two strawberry yogurts, so the probability function of Y is

$$g(0) = P(Y = 0) = P(\text{CC}) = \frac{2}{6} \cdot \frac{1}{5} = \frac{1}{15}$$

$$g(1) = P(Y = 1) = P(\text{CS}) + P(\text{SC})$$

$$= \frac{2}{6} \cdot \frac{4}{5} + \frac{4}{6} \cdot \frac{2}{5} = \frac{8}{15}$$

$$g(2) = P(Y = 2) = P(\text{SS}) = \frac{4}{6} \cdot \frac{3}{5} = \frac{6}{15}$$

(b) See Fig. 7-7.

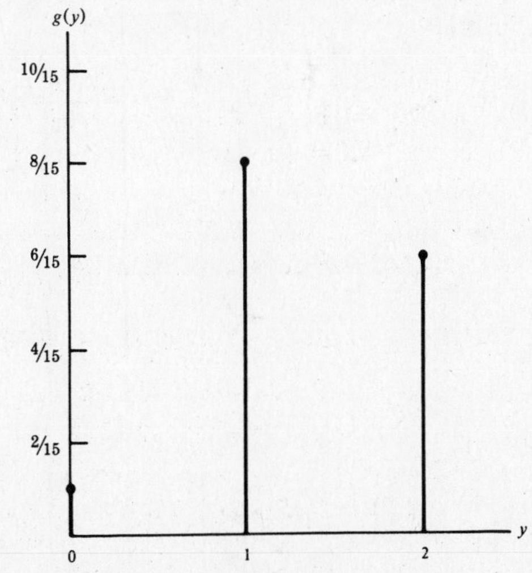

Figure 7-7

PROBLEM 7-9 The probability function of the random variable Y that gives the payoff for chuck-a-luck in Problem 7-4 is

$$g(-1) = P(Y = -1) = \left(\frac{1}{6}\right)^0\left(\frac{5}{6}\right)^3 = \frac{125}{216} = .579$$

$$g(1) = P(Y = 1) = 3\left(\frac{1}{6}\right)^1\left(\frac{5}{6}\right)^2 = \frac{75}{216} = .347$$

$$g(2) = P(Y = 2) = 3\left(\frac{1}{6}\right)^2\left(\frac{5}{6}\right)^1 = \frac{15}{216} = .069$$

$$g(3) = P(Y = 3) = \left(\frac{1}{6}\right)^3\left(\frac{5}{6}\right)^0 = \frac{1}{216} = .005$$

Draw a bar graph for this probability function.

Solution See Fig. 7-8.

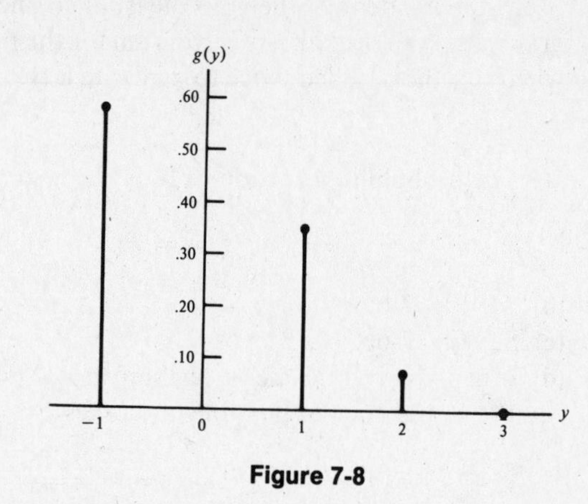

Figure 7-8

PROBLEM 7-10 Let X equal the number of boxes of cereal that must be purchased to obtain Hamhose (see Problem 7-5). **(a)** Give the probability function of X. **(b)** Draw a probability histogram for this distribution.

Solution

(a) Hamhose is one of 7 freakies. The probability that Hamhose is in the first box of cereal is

$$f(1) = P(X = 1) = \frac{1}{7}$$

The probability that Hamhose is obtained for the first time in box number 5, for example, is the product of not obtaining Hamhose in the first 4 boxes and then obtaining Hamhose in box number 5. Because these 5 purchases are independent events,

$$f(5) = P(X = 5) = \frac{6}{7}\cdot\frac{6}{7}\cdot\frac{6}{7}\cdot\frac{6}{7}\cdot\frac{1}{7} = \left(\frac{6}{7}\right)^4\left(\frac{1}{7}\right)$$

In general, the probability that Hamhose is first obtained in box number x is

$$f(x) = P(X = x) = \left(\frac{6}{7}\right)^{x-1}\left(\frac{1}{7}\right), \quad x = 1, 2, 3, 4, \ldots$$

defined for all positive integers.

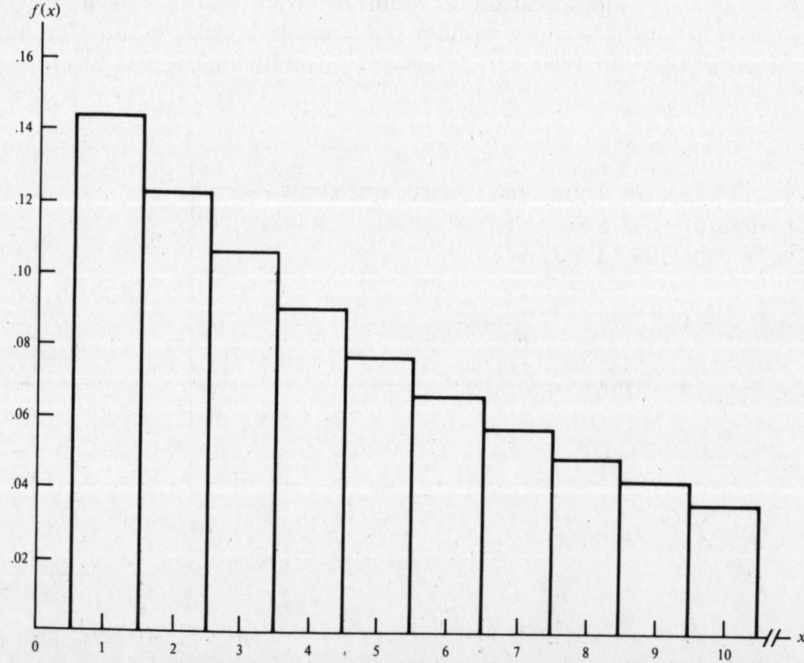

Figure 7-9

(b) The probabilities for the first ten integers are as shown in the tabulation. These are graphed in Fig. 7-9.

x	$f(x)$	x	$f(x)$
1	.143	6	.066
2	.122	7	.057
3	.105	8	.049
4	.090	9	.042
5	.077	10	.036

note: The probabilities associated with large integers are small. However, $P(X = k) > 0$ for each positive integer k.

Distribution Functions

PROBLEM 7-11 Let the probability function of X be defined by $f(x) = x/10$, $x = 1, 2, 3, 4$. **(a)** Graph the probability function as a bar graph. **(b)** Define and **(c)** graph the distribution function of X.

Solution

(a) See Fig. 7-10a.

(b) The distribution function of X gives $P(X \leq x)$, the probability that an observed value of X is less than or equal to a number x, so define

$$F(x) = \begin{cases} 0, & x < 1 \\ 1/10, & 1 \leq x < 2 \\ 3/10, & 2 \leq x < 3 \\ 6/10, & 3 \leq x < 4 \\ 10/10, & 4 \leq x \end{cases}$$

(c) Label the x-axis with the possible values of X, and the y-axis with the probabilities of those values, as in Fig. 7-10b.

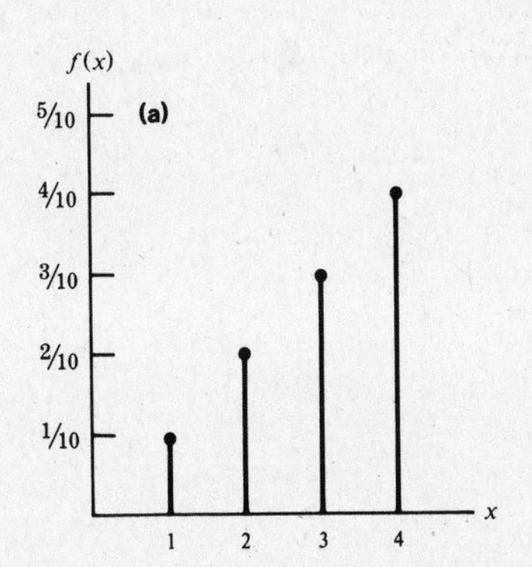

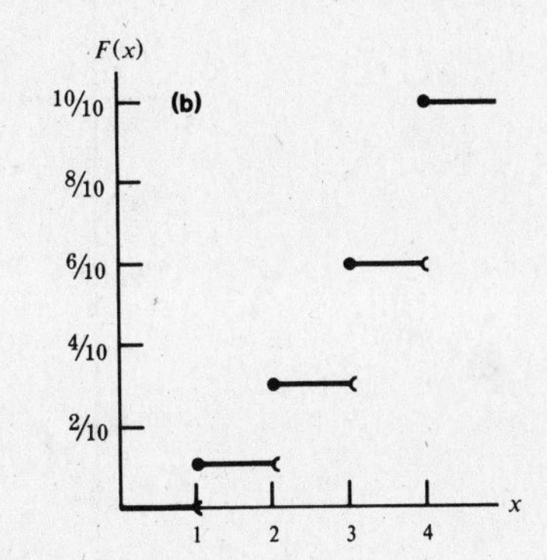

Figure 7-10

PROBLEM 7-12 Let the probability function of X be defined by $f(x) = x/9$, $x = 2, 3, 4$. **(a)** Draw a bar graph for this probability function. **(b)** Define and **(c)** graph the distribution function of X.

Solution

(a) See Fig. 7-11a.

(b) The distribution function is defined by

$$F(x) = \begin{cases} 0, & x < 2 \\ 2/9, & 2 \leq x < 3 \\ 5/9, & 3 \leq x < 4 \\ 9/9, & 4 \leq x \end{cases}$$

(c) See Fig. 7-11b.

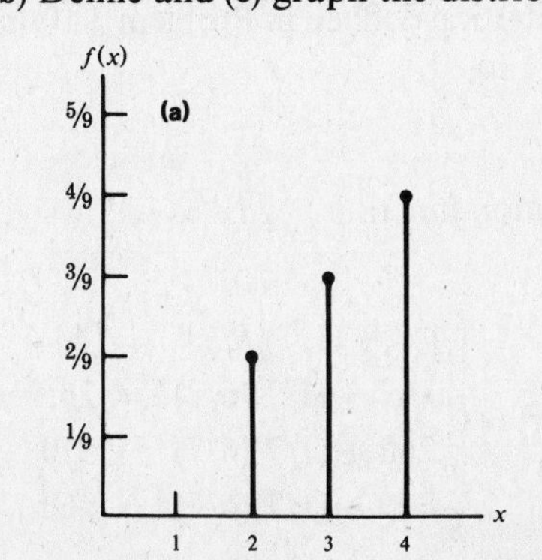

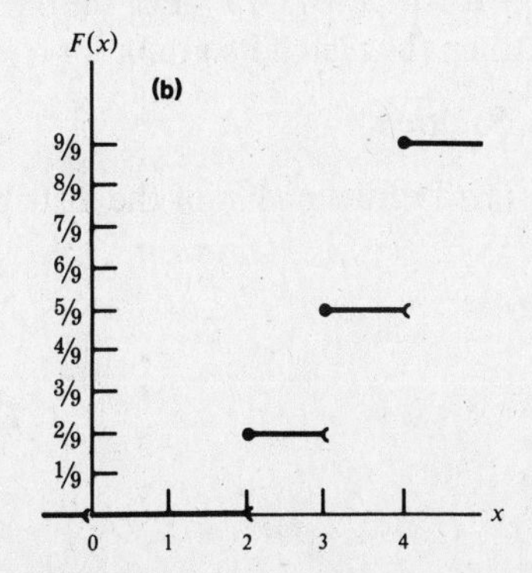

Figure 7-11

PROBLEM 7-13 For the distribution that is defined in Problem 7-8, (**a**) define and (**b**) graph the distribution function.

Solution

(**a**) Since the probability function is denoted by $g(y)$, use $G(y)$ for the distribution function. You have

$$G(y) = \begin{cases} 0, & y < 0 \\ 1/15, & 0 \le y < 1 \\ 9/15, & 1 \le y < 2 \\ 15/15, & 2 \le y \end{cases}$$

(**b**) See Fig. 7-12.

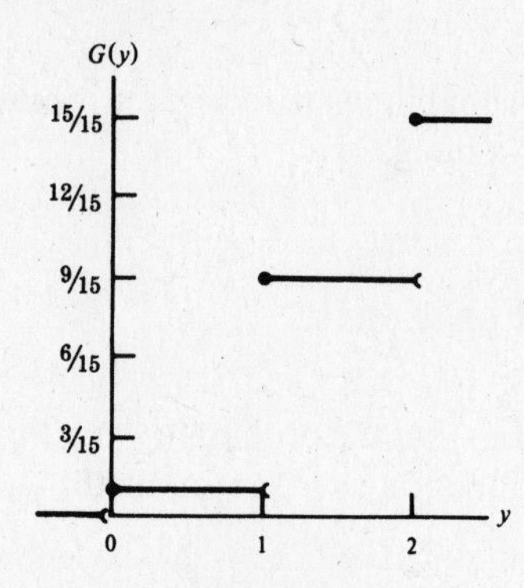

Figure 7-12

PROBLEM 7-14 Sketch the graph for the distribution function of Y (the payoff in chuck-a-luck) as described in Problem 7-9.

Solution See Fig. 7-13.

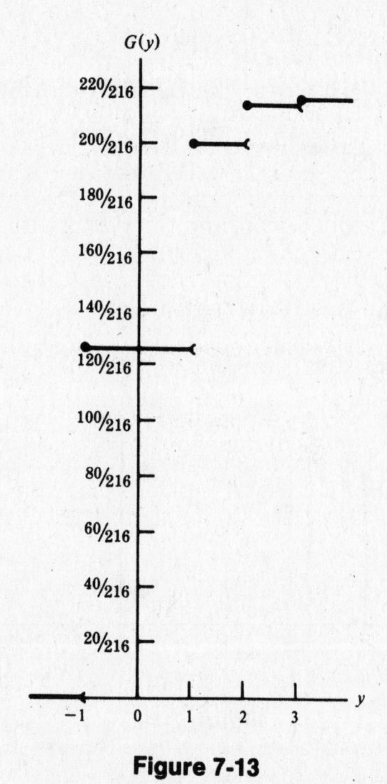

Figure 7-13

PROBLEM 7-15 For the distribution defined in Problem 7-10, (**a**) define the distribution function and then (**b**) sketch its graph for $x = 1$ to 11.

Solution

(**a**) The definition of the distribution function begins as follows:

$$F(x) = \begin{cases} 0, & x < 1 \\ 1/7 + (1/7)(6/7)^0 = .143 & 1 \le x < 2 \\ .143 + (1/7)(6/7)^1 = .265 & 2 \le x < 3 \\ .265 + (1/7)(6/7)^2 = .370 & 3 \le x < 4 \\ .370 + (1/7)(6/7)^3 = .460 & 4 \le x < 5 \\ \vdots & \vdots \end{cases}$$

We can also define this distribution function in a way that requires a knowledge of the sum of a geometric series:

$$F(x) = \begin{cases} 0, & x < 1 \\ 1 - (6/7)^k, & k \leq x < k + 1 \end{cases}$$

where k is any positive integer. This formula simplifies the evaluation of $F(x)$ when x is large.

(b) See Fig. 7-14.

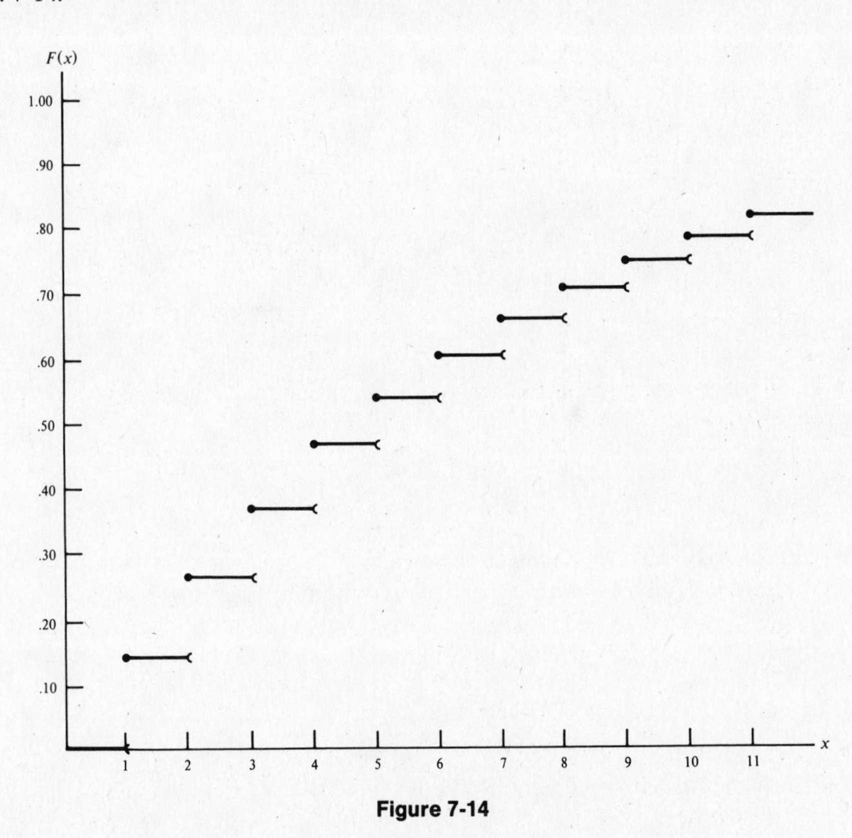

Figure 7-14

Mean, Variance, and Standard Deviation of a Random Variable

PROBLEM 7-16 Let the probability function of X be defined by $f(x) = (5 - x)/10, x = 1, 2, 3, 4$. Find **(a)** the mean, **(b)** the variance using formula (7.2), **(c)** the variance using formula (7.3), and **(d)** the standard deviation.

Solution Construct the table as shown.

x	$f(x)$	$xf(x)$	$x^2f(x)$	$(x - \mu)^2f(x)$
1	4/10	4/10	4/10	4/10
2	3/10	6/10	12/10	0/10
3	2/10	6/10	18/10	2/10
4	1/10	4/10	16/10	4/10
	10/10	20/10	50/10	10/10

(a) Use formula (7.1) to determine the mean:

$$\mu = \sum xf(x) = \frac{20}{10} = 2$$

(b) From formula (7.2), the variance is

$$\sigma^2 = \sum (X - \mu)^2 f(x) = \frac{10}{10} = 1$$

(c) If you use formula (7.3), you don't have to calculate the final column:

$$\sigma^2 = \sum x^2 f(x) - \mu^2 = \frac{50}{10} - 2^2 = 5 - 4 = 1$$

(d) Use formula (7.4) to determine the standard deviation:

$$\sigma = \sqrt{\sigma^2} = \sqrt{1} = 1$$

PROBLEM 7-17 Find (**a**) the mean, (**b**) the variance, and (**c**) the standard deviation for the probability distribution defined in Problem 7-1.

Solution

(**a**) The mean is

$$\mu = 0\left(\frac{1}{5}\right) + 1\left(\frac{4}{5}\right) = \frac{4}{5}$$

(**b**) From (7.3),

$$\sigma^2 = 0^2\left(\frac{1}{5}\right) + 1^2\left(\frac{4}{5}\right) - \left(\frac{4}{5}\right)^2 = \frac{4}{5} - \left(\frac{4}{5}\right)^2$$

Factoring out $\frac{4}{5}$, you get

$$\left(\frac{4}{5}\right)\left(1 - \frac{4}{5}\right) = \left(\frac{4}{5}\right)\left(\frac{1}{5}\right) = \frac{4}{25}$$

Notice that if you let

$$p = P(X = 1) = P(\text{Success})$$
$$q = 1 - p = P(X = 0) = P(\text{Failure})$$

then

$$\sigma^2 = p(1 - p) = pq$$

This gives you an easy way to calculate the variance for experiments that have two possible outcomes—success and failure—with respective probabilities p and $1 - p = q$.

(**c**) Then from formula (7.4) the standard deviation is

$$\sigma = \sqrt{4/25} = 2/5$$

PROBLEM 7-18 If the probability function of X is

$$f(x) = \frac{1}{10}, \qquad x = 0, 1, 2, 3, 4, 5, 6, 7, 8, 9$$

(see Problem 7-6), find (**a**) the mean, (**b**) the variance using formula (7.2), (**c**) the variance using formula (7.3), and (**d**) the standard deviation.

Solution

x	$f(x)$	$xf(x)$	$x^2f(x)$	$(x-\mu)^2f(x)$
0	1/10	0/10	0/10	20.25/10
1	1/10	1/10	1/10	12.25/10
2	1/10	2/10	4/10	6.25/10
3	1/10	3/10	9/10	2.25/10
4	1/10	4/10	16/10	.25/10
5	1/10	5/10	25/10	.25/10
6	1/10	6/10	36/10	2.25/10
7	1/10	7/10	49/10	6.25/10
8	1/10	8/10	64/10	12.25/10
9	1/10	9/10	81/10	20.25/10
	10/10	45/10	285/10	82.50/10

(**a**) $\mu = \frac{45}{10} = 4.5$

(**b**) From formula (7.2),

$$\sigma^2 = \frac{82.50}{10} = 8.25$$

(**c**) From formula (7.3),

$$\sigma^2 = \frac{285}{10} - (4.5)^2 = 8.25$$

(**d**) $\sigma = \sqrt{8.25} = 2.87$

PROBLEM 7-19 For the probability distribution in Problem 7-8, find (**a**) the mean, (**b**) the variance, and (**c**) the standard deviation.

Solution Construct the table as shown.

y	$g(y)$	$yg(y)$	$y^2g(y)$
0	1/15	0/15	0/15
1	8/15	8/15	8/15
2	6/15	12/15	24/15
	15/15	20/15	32/15

(**a**) $$\mu = \frac{20}{15} = 1.33$$

(**b**) $$\sigma^2 = \frac{32}{15} - \left(\frac{20}{15}\right)^2 = \frac{16}{45} = .356$$

(**c**) $$\sigma = \sqrt{.356} = .597$$

PROBLEM 7-20 The probability function of X is defined by the given tabulation. Find (**a**) the mean, (**b**) the variance, and (**c**) the standard deviation.

x	$f(x)$
1	.1
2	.6
3	.3

Solution First you construct a table as shown.

x	$f(x)$	$xf(x)$	$x^2f(x)$
1	.1	.1	.1
2	.6	1.2	2.4
3	.3	.9	2.7
	1.0	2.2	5.2

(**a**) $$\mu = 2.2$$

(**b**) $$\sigma^2 = 5.2 - (2.2)^2 = .36$$

(**c**) $$\sigma = \sqrt{.36} = .6$$

PROBLEM 7-21 Let Y be the payoff for the casino game of chuck-a-luck. On the average, how much can a bettor expect to win per bet?

Solution The probability function (payoff function) as given in Problem 7-9 is

$$g(-1) = \frac{125}{216}, \qquad g(1) = \frac{75}{216}, \qquad g(2) = \frac{15}{216}, \qquad g(3) = \frac{1}{216}$$

So the expected payoff is

$$\mu = (-1)\left(\frac{125}{216}\right) + (1)\left(\frac{75}{216}\right) + (2)\left(\frac{15}{216}\right) + (3)\left(\frac{1}{216}\right) = -\frac{17}{216} = -.0787$$

That is, the bettor can expect to lose, on the average, about 8 cents for each $1 bet.

PROBLEM 7-22 In the casino game of roulette (see Problem 6-7), let X equal the payoff when a $1 bet is placed on red. Then the probability function of X is

$$f(-1) = \frac{20}{38}, \quad f(1) = \frac{18}{38}$$

That is, the bettor can expect to lose one dollar 20/38 of the time and win one dollar 18/38 of the time. Find $\mu = E(X)$.

Solution On the average, the bettor can expect to win

$$\mu = (-1)\left(\frac{20}{38}\right) + (1)\left(\frac{18}{38}\right) = -\frac{2}{38} = -.0526$$

So, on the average, a bettor can expect to lose about 5 cents on each $1 bet.

PROBLEM 7-23 In the casino game of craps, the probability for winning is .49293 (see Problem 6-33). (**a**) If you let X equal the payoff function for a $1 bet, what's the probability function of X? (**b**) Find the mean of X. That is, on the average, how much can a bettor expect to win per bet?

Solution

(**a**) Let $f(1) = P(\text{winning})$ and $f(-1) = P(\text{losing})$, so that

$$P(\text{losing}) = 1 - P(\text{winning}) = 1 - f(1) = 1 - .49293 = .50707$$

Thus, the probability function of X is

$$f(-1) = .50707 \text{ and } f(1) = .49293$$

(b)
$$\mu = (-1)(.50707) + (1)(.49293) = -.01414$$

Thus a bettor can expect to lose 1.4 cents on the average for each \$1 bet that is placed.

Supplementary Problems

PROBLEM 7-24 A bag contains 2 gold coins and 2 silver coins. You select two coins at random from the bag, one at a time, without replacement. Let X equal the number of gold coins in the sample. Define the probabilities associated with X.

Answer $P(X = 0) = 1/6$, $P(X = 1) = 2/3$, $P(X = 2) = 1/6$

PROBLEM 7-25 The probability function of X is defined by the table as shown. Graph this probability function as **(a)** a bar graph and as **(b)** a probability histogram.

x	$f(x)$
1	.3
2	.5
3	.2

Answer **(a)** See Fig. 7-15a. **(b)** See Fig. 7-15b.

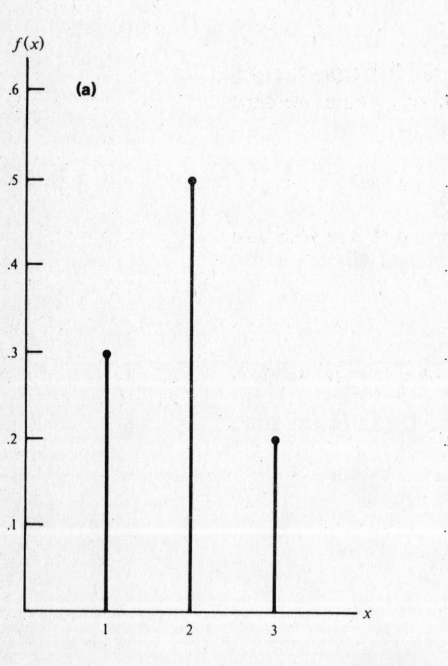

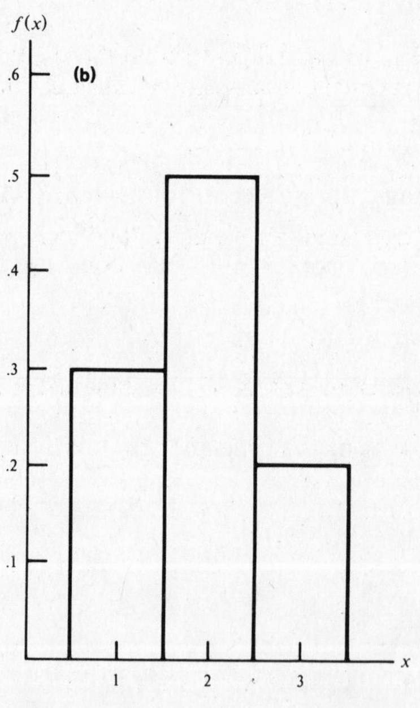

Figure 7-15

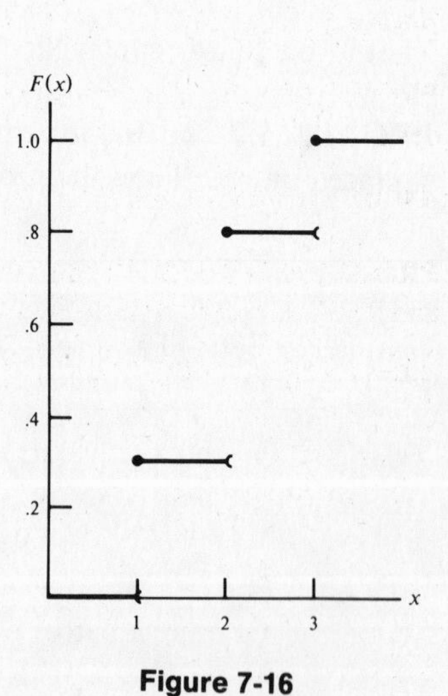

Figure 7-16

PROBLEM 7-26 For the distribution in Problem 7-25, **(a)** define and **(b)** graph the distribution function.

Answer

(a)
$$F(x) = \begin{cases} 0, & x < 1 \\ .3, & 1 \le x < 2 \\ .8, & 2 \le x < 3 \\ 1.0, & 3 \le x \end{cases}$$

(b) See Fig. 7-16.

PROBLEM 7-27 Urn A contains 3 red balls numbered 1, 2, and 3. Urn B contains 3 blue balls numbered 1, 2, and 3. Select one ball at random from each urn. Let X equal the sum of the two balls that are selected. (**a**) Define the probability function of X, and (**b**) draw a bar graph for this probability function. (**c**) Define the distribution function, and (**d**) sketch the graph of the distribution function.

Answer

(**a**)
$$f(x) = \begin{cases} 1/9, & x = 2 \\ 2/9, & x = 3 \\ 3/9, & x = 4 \\ 2/9, & x = 5 \\ 1/9, & x = 6 \end{cases}$$

(**c**)
$$F(x) = \begin{cases} 0, & x < 2 \\ 1/9, & 2 \le x < 3 \\ 3/9, & 3 \le x < 4 \\ 6/9, & 4 \le x < 5 \\ 8/9, & 5 \le x < 6 \\ 9/9, & 6 \le x \end{cases}$$

(**b**) See Fig. 7-17a.

(**d**) See Fig. 7-17b.

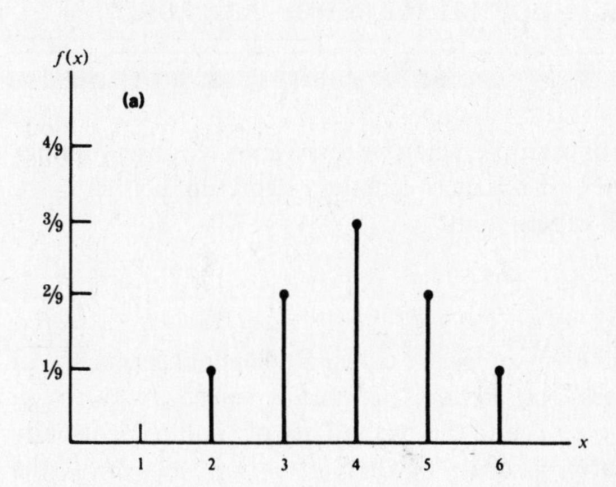

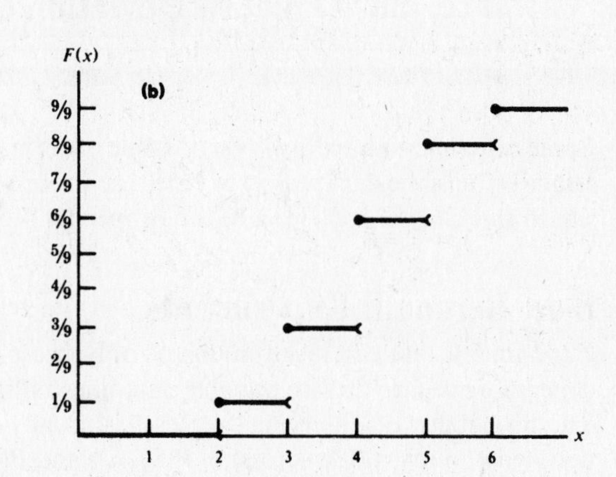

Figure 7-17

PROBLEM 7-28 An urn contains 4 balls labeled \$5, \$5, \$10, and \$20. One ball is selected at random from the urn. Let X equal the payoff according to the label on the ball. Find (**a**) the mean, (**b**) the variance, and (**c**) the standard deviation of X.

Answer (**a**) $\mu = 10$, (**b**) $\sigma^2 = 37.5$, (**c**) $\sigma = 6.12$

PROBLEM 7-29 For the distribution in Problem 7-25, find (**a**) the mean, (**b**) the variance, and (**c**) the standard deviation.

Answer (**a**) 1.9 (**b**) .49 (**c**) .7

PROBLEM 7-30 Roll a pair of 6-sided dice, and let X equal the sum of the dice. Find (**a**) the mean, (**b**) the variance, and (**c**) the standard deviation of X.

Answer (**a**) 7 (**b**) 5.833 (**c**) 2.415

THE BINOMIAL DISTRIBUTION

THIS CHAPTER IS ABOUT

- ☑ **Bernoulli Experiments**
- ☑ **Binomial Probabilities**
- ☑ **The Mean, Variance, and Standard Deviation of a Binomial Random Variable**

Some experiments have only two possible outcomes. For example, when we toss a coin, we observe either heads or tails; the next person to enter the room will be either male or female; a lightbulb is either on or off. In this chapter we'll look at the properties of such experiments.

8-1. Bernoulli Experiments

Experiments that can result in one of only two possible outcomes are called **Bernoulli experiments.** In general, we refer to the two possible outcomes as "success" and "failure," denoted as S and F, respectively. The probability of observing "success" is usually denoted p, and the probability of "failure" is usually denoted q. We write $P(\text{success}) = P(\text{S}) = p$ and $P(\text{failure}) = P(\text{F}) = q$, where $p + q = 1$ since one of the two outcomes *must* occur.

EXAMPLE 8-1 Explain how to interpret each of the following as a Bernoulli experiment: **(a)** the flip of a coin, **(b)** the random selection of a female or male rat from a cage, **(c)** the hatching of an incubated egg.

Solution

(a) The flip of a coin results in only two possible outcomes, heads or tails. If the coin is fair and we call heads success, then we can write

$$p = P(\text{heads}) = P(\text{S}) = .5$$

$$q = 1 - p = P(\text{tails}) = P(\text{F}) = .5$$

Notice that we arbitrarily chose heads as success.

(b) If we randomly select a rat from a cage, we get either a female or a male rat. Let's call choosing a female rat a success. The value of $p = P(\text{success})$ depends on how many female and male rats are in the cage. If out of 10 rats in the cage, there are 7 females, $p = P(\text{female}) = 7/10 = .7$ and $q = 1 - p = .3$.

(c) An egg that's being incubated either does or doesn't hatch. Let's call an egg that does hatch a success. The value of $p = P(\text{success})$ would depend on several factors and would probably have to be estimated using the proportion of eggs that hatched in a similar setting as a "best guess."

8-2. Binomial Probabilities

A. Properties of binomial probability experiments

A **binomial experiment** is a random experiment that satisfies the following properties:

1. A Bernoulli (success/failure) experiment is carried out n times.
2. The n trials are independent (see Section 6-4C). The outcome on any trial doesn't affect the outcome on any other trial.

3. On each trial, $p = P(\text{success})$ and $q = 1 - p = P(\text{failure})$; that is, the probabilities of success and failure don't change from trial to trial.
4. A random variable X "tallies" the number of successes observed in these n trials. X must equal an integer between 0 and n, inclusive, since out of n trials, we may observe anywhere from 0 to n successes.

Thus, we can think of a binomial experiment as a *sequence* of n Bernoulli trials, in which we're interested in determining the number of successes that occur.

B. The binomial probability function

The **binomial probability function** is given by

BINOMIAL PROBABILITY FUNCTION

$$f(x) = P(X = x)$$

$$= \binom{n}{x} p^x(1-p)^{n-x}, \quad x = 0, 1, \ldots, n$$

$$= \frac{n!}{x!(n-x)!} p^x(1-p)^{n-x}, \quad x = 0, 1, \ldots, n \qquad \textbf{(8.1)}$$

This defines the probability that the random variable X takes on the value x as x varies from zero to n. To determine the probability that we obtain $X = x$ successes in n trials, we only need to substitute the appropriate values of n, x, and p into the above formula.

The coefficients

$$\binom{n}{x} = \frac{n!}{x!(n-x)!} \qquad \textbf{(8.2)}$$

are called **binomial coefficients.** We can easily find values of $f(x) = P(X = x)$ for varying values of x and p by using the binomial probability table (Table 2 in the Appendix).

C. The binomial distribution

If n is the number of trials and p is the probability of success on each trial, then n and p are the *parameters* for the **binomial distribution** of X. We use the notation that the distribution of X is $b(n, p)$.

We call this distribution *binomial* because of the relationship between the binomial probability function and the expansion of the binomial $[p + (1 - p)]^n$. Notice that the two terms of the binomial are p, the probability of success, and $(1 - p)$, the probability of failure. It's true that

$$\sum_{x=0}^{n} f(x) = \sum_{x=0}^{n} P(X = x)$$

$$= \sum_{x=0}^{n} \frac{n!}{x!(n-x)!} p^x(1-p)^{n-x}$$

$$= [p + (1-p)]^n = 1$$

EXAMPLE 8-2 You have to take a surprise ten-question, multiple-choice test. For each question there are four possible answers, but only one is correct. You don't know the answer to any of the questions, so you have to guess on each one. Let X equal the number of answers you get correct if you guess on each question. What is the distribution of X, and what is the probability function of X?

Solution Because you guess either right (success) or wrong (failure) on each of the ten questions, you know that X has a binomial distribution where the number of trials is $n = 10$ and the probability of success (i.e., guessing a correct answer) is $p = \frac{1}{4}$. You can completely describe the distribution of X as $b(10, \frac{1}{4})$. The probability function of X is

$$f(x) = \frac{n!}{x!(n-x)!} p^x(1-p)^{n-x}, \quad x = 0, 1, 2, \ldots, n$$

$$= \frac{10!}{x!(10-x)!} \left(\frac{1}{4}\right)^x \left(\frac{3}{4}\right)^{10-x}, \quad x = 0, 1, 2, \ldots, 10$$

EXAMPLE 8-3 What's the probability of guessing exactly two correct answers on the multiple-choice test described in Example 8-2?

Solution You find the probability of guessing exactly two correct answers by substituting $x = 2$ into the probability function:

$$f(2) = \frac{10!}{2!8!}\left(\frac{1}{4}\right)^2\left(\frac{3}{4}\right)^8 = .2816$$

EXAMPLE 8-4 Use the binomial probability table to find the probability of guessing two correct answers on your quiz (Example 8-2).

Solution A part of the binomial probability table (see the Appendix, Table 2) is reproduced here. You know that $n = 10$, $x = 2$, and $p = \frac{1}{4} = .25$, so $f(2) = .2816$.

You'll find the binomial probability table very helpful in computing these probabilities, so you'll seldom need to do the calculations "by hand."

		p		
n	*x*	.20	.25	.30
10	0	.1074	.0563	.0282
	1	.2684	.1877	.1211
	2	.3020	.2816	.2335
	3	.2013	.2503	.2668
	4	.0881	.1460	.2001
	5	.0264	.0584	.1029
	6	.0055	.0162	.0368
	7	.0008	.0031	.0090
	8	.0001	.0004	.0014
	9	.0000	.0000	.0001
	10	.0000	.0000	.0000

EXAMPLE 8-5 What's the probability of guessing correctly **(a)** on two, three, *or* four questions, **(b)** on *at most* three questions, **(c)** on *at least* six questions in the test in Example 8-2?

Solution

(a) The probability of guessing correctly on two, three *or* four questions is

$$P(2 \le X \le 4) = f(2) + f(3) + f(4)$$
$$= .2816 + .2503 + .1460 = .6779$$

You can use the binomial probability table to obtain these values quickly.

(b) The probability of guessing correctly on *at most* three questions is

$$P(X \le 3) = f(0) + f(1) + f(2) + f(3)$$
$$= .0563 + .1877 + .2816 + .2503$$
$$= .7759$$

(c) The probability of guessing correctly on *at least* six questions is

$$P(X \ge 6) = f(6) + f(7) + f(8) + f(9) + f(10)$$
$$= .0162 + .0031 + .0004 + .0000 + .0000$$
$$= .0197$$

EXAMPLE 8-6 Suppose that $p = .8$ is the probability that an egg laid by a moorhen (a bird that nests in marshlands) will hatch. In a nest containing 7 eggs, what is the probability **(a)** that 6 eggs will hatch, **(b)** that 5, 6, or 7 eggs will hatch?

Solution

(a) Let X equal the number of eggs that hatch. If you assume the events are independent, the distribution of X is $b(7, .8)$, and you can compute

$$P(X = 6) = f(6)$$
$$= \frac{7!}{6!1!}(.8)^6(.2)^1$$
$$= .3670$$

You could also obtain the answer using Table 2 with $n = 7$, $p = .8$, and $x = 6$.

(b) Using the binomial probability table with $n = 7$ and $p = .8$,

$$P(X \geq 5) = f(5) + f(6) + f(7)$$
$$= .2753 + .3670 + .2097$$
$$= .8520$$

EXAMPLE 8-7 Among the gifted 7th graders who score very high on a mathematics exam, approximately 20% are either left-handed or ambidextrous. Let X equal the number of left-handed or ambidextrous students among a random sample of 12 of these gifted 7th graders. What's the probability **(a)** that at most three of these students are left-handed or ambidextrous, **(b)** more than three of these students are left-handed or ambidextrous?

Solution

(a) The distribution of X is $b(12, .20)$. Thus,

$$P(X \leq 3) = f(0) + f(1) + f(2) + f(3)$$
$$= .0687 + .2062 + .2835 + .2362 = .7946$$

(b) In part **(a)** you found the probability of at most 3 successes. Now you want to find the probability for more than 3 successes. Instead of finding $f(4) + f(5) + \cdots + f(12)$ recall that $P(X \leq 3) + P(X > 3) = 1$. It follows that

$$P(X > 3) = 1 - P(X \leq 3) = 1 - .7946 = .2054$$

D. The probability histogram

The effect of different values of p, the probability of success, on the shape of probability histograms is shown in Fig. 8-1. Notice that each of the graphs has $n = 12$ trials, and as p increases, the whole distribution shifts to the right.

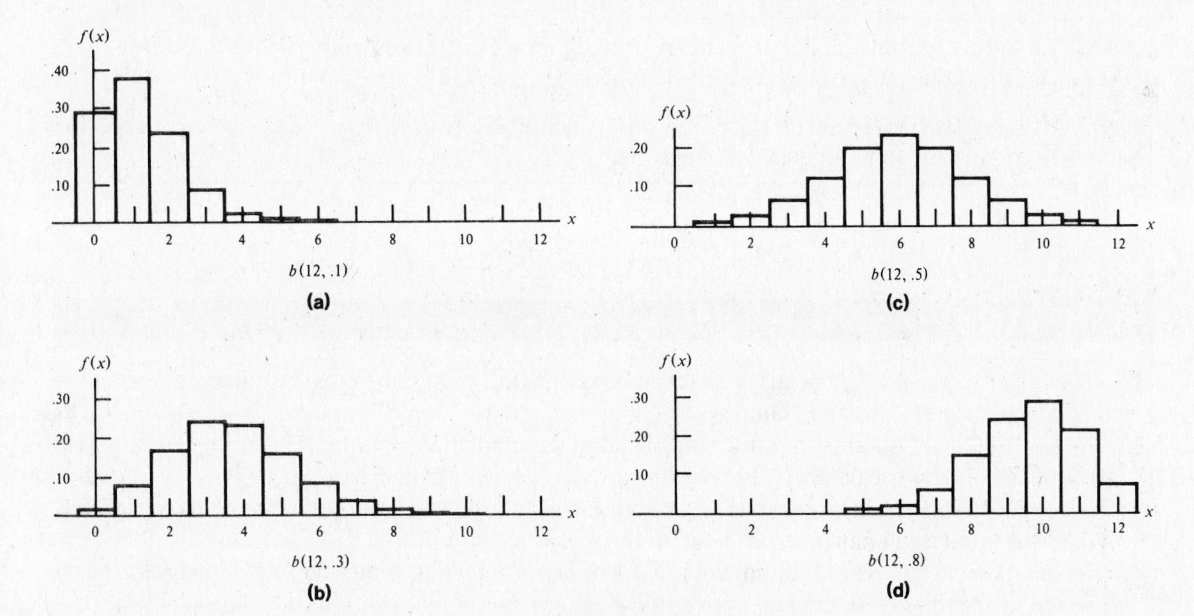

Figure 8-1

8-3. The Mean, Variance, and Standard Deviation of a Binomial Random Variable

Let X have a binomial distribution with parameters n and p; that is, X is $b(n, p)$. Then

- the mean of X is $\mu = np$
- the variance of X is $\sigma^2 = np(1 - p) = npq$
- the standard deviation of X is $\sigma = \sqrt{npq}$

To say that the mean of X is $\mu = np$ should make sense to you intuitively: If p is the probability of success on each of n trials, then, on the average, we expect p of the n trials to end in success. For instance, if we toss a fair coin $n = 100$ times, we expect on the average to observe $np = 100(\frac{1}{2}) = 50$ heads.

EXAMPLE 8-8 What are the mean, the variance, and the standard deviation of X, the number of correct answers on the multiple-choice list described in Example 8-2?

Solution In Example 8-2, $n = 10$ and $p = \frac{1}{4}$, so that the distribution of X is $b(10, \frac{1}{4})$. Thus,

$$\mu = np = 10\left(\frac{1}{4}\right) = 2.5$$

$$\sigma^2 = npq = 10\left(\frac{1}{4}\right)\left(\frac{3}{4}\right) = 1.875$$

$$\sigma = \sqrt{npq} = \sqrt{1.875} = 1.369$$

This indicates that by guessing alone you would expect to answer an average of 2.5 questions correctly. Of course, you could never correctly guess exactly 2.5 questions, since this is not an integer value. This simply means that if you took many such 10-question tests you would *average* 2.5 correct answers.

EXAMPLE 8-9 What are the mean, the variance, and the standard deviation of X, the number of eggs that hatch, in Example 8-6?

Solution In Example 8-6, the distribution of X is $b(7, .8)$. Thus,

$$\mu = 7(.8) = 5.6$$

$$\sigma^2 = 7(.8)(.2) = 1.12$$

$$\sigma = \sqrt{1.12} = 1.058$$

EXAMPLE 8-10 In roulette, the probability of winning with a bet on red is $p = 18/38$. If you make 1000 bets on red, how many can you expect to win on the average?

Solution In $n = 1000$ independent bets each with probability of success $p = 18/38$, you can expect to win an average of $\mu = 1000(\frac{18}{38}) = 473.68$ bets.

SUMMARY

1. A *Bernoulli experiment* (or trial) is an experiment having only two possible outcomes, generally denoted as "success" (S) and "failure" (F).
2. The probability of success in a Bernoulli trial is denoted $P(\text{S}) = p$; the probability of failure is denoted $P(\text{F}) = 1 - P(\text{S}) = q$.
3. A *binomial experiment* consists of a sequence of n Bernoulli trials, in which a random variable X tallies the number of successes observed in the n independent trials.
4. The distribution of X in a binomial experiment is $b(n, p)$: $n =$ the number of independent trials, $p =$ the probability of success on each trial.
5. The *binomial probability function* defines the probability that the random variable X takes on the value x as x varies from 0 to n:

$$f(x) = P(X = x) = \frac{n!}{x!(n-x)!} p^x(1-p)^{n-x}, \quad x = 0, 1, \ldots, n$$

6. The *mean, variance,* and *standard deviation* of a binomial random variable are $\mu = np$, $\sigma^2 = np(1-p) = npq$, and $\sigma = \sqrt{np(1-p)} = \sqrt{npq}$, respectively.
7. An increase in the value of p shifts the whole distribution of a probability histogram to the right.

RAISE YOUR GRADES

Can you . . . ?

☑ give the properties of a Bernoulli trial
☑ give the properties of a binomial experiment
☑ explain the meaning of the parameters n and p for a binomial distribution
☑ explain how the binomial distribution received its name
☑ explain the relationship between Bernoulli trials and a binomial random variable
☑ find probabilities using the binomial probability function
☑ find probabilities using the binomial probability table
☑ give the mean and variance of a binomial random variable

RAPID REVIEW

1. In a Bernoulli experiment **(a)** ______ = P(success) and **(b)** ______ = P(failure).
2. On each trial of a binomial experiment there are always **(a)** ______ possible outcomes. These are often called **(b)** ______ and **(c)** ______.
3. In a binomial experiment, $p + q$ must always equal ______.
4. In a binomial experiment the trials must be ______.
5. If X has a binomial distribution, $b(n, p)$, then its mean is $\mu =$ ______ and its variance is $\sigma^2 =$ ______.
6. In a binomial experiment, the random variable X counts the number of ______.
7. For a binomial distribution, $b(n, p)$, n and p are called ______.
8. If X has a binomial distribution, $b(25, .5)$, then $\mu =$ **(a)** ______ and $\sigma^2 =$ **(b)** ______.
9. A fair 6-sided die is rolled 12 times and X is equal to the number of times that a 5 is observed. The distribution of X is **(a)** ______. For this distribution, $n =$ **(b)** ______ and $p =$ **(c)** ______.
10. If X has a binomial distribution, $b(7, .4)$, then the probability function of X is $f(x) = P(X = x) =$ ______.

Answers **(1)** **(a)** p **(b)** $q = 1 - p$ **(2)** **(a)** two **(b)** success **(c)** failure **(3)** one **(4)** independent **(5)** **(a)** np **(b)** $np(1 - p) = npq$ **(6)** successes **(7)** parameters **(8)** **(a)** $25(.5) = 12.5$ **(b)** $25(.5)(.5) = 6.25$ **(9)** **(a)** $b(12, \frac{1}{6})$ **(b)** 12 **(c)** $\frac{1}{6}$ **(10)** $\frac{7!}{x!(7-x)!}(.4)^x(.6)^{7-x}, \quad x = 0, 1, 2, \ldots, 7$

SOLVED PROBLEMS

Math Review

PROBLEM 8-1 Evaluate **(a)** 3!, **(b)** 5!, **(c)** 0!.

Solution

(a) $3! = (3)(2)(1) = 6$
(b) $5! = (5)(4)(3)(2)(1) = 120$
(c) $0! = 1$ by definition

PROBLEM 8-2 Evaluate the binomial coefficients **(a)** $\binom{5}{3}$, **(b)** $\binom{5}{4}$, **(c)** $\binom{4}{3}$.

Solution

(a) $\binom{5}{3} = \frac{5!}{3!2!} = \frac{5\cdot 4\cdot \cancel{3\cdot 2\cdot 1}}{\cancel{3\cdot 2\cdot 1}\cdot 2\cdot 1} = \frac{5\cdot 4}{2\cdot 1} = \frac{20}{2} = 10$

(b) $\binom{5}{4} = \frac{5!}{4!1!} = \frac{5\cdot \cancel{4\cdot 3\cdot 2\cdot 1}}{\cancel{4\cdot 3\cdot 2\cdot 1}\cdot 1} = \frac{5}{1} = 5$

(c) $\binom{4}{3} = \frac{4!}{3!1!} = \frac{4\cdot \cancel{3\cdot 2\cdot 1}}{\cancel{3\cdot 2\cdot 1}\cdot 1} = \frac{4}{1} = 4$

PROBLEM 8-3 Prove that $\binom{n}{k} = \binom{n}{n-k}$.

Solution $\binom{n}{k} = \frac{n!}{k!(n-k)!} = \frac{n!}{(n-k)!k!} = \binom{n}{n-k}$

PROBLEM 8-4 You can easily evaluate binomial coefficients by looking them up in binomial coefficient tables, like Table 1 of this book's Appendix. Use Table 1 to evaluate **(a)** $\binom{5}{3}$, **(b)** $\binom{12}{10}$, **(c)** $\binom{15}{12}$, and **(d)** $\binom{18}{12}$.

Solution A part of the binomial coefficients table (see Table 1 of the Appendix) is reproduced here.

n	$\binom{n}{2}$	$\binom{n}{3}$	$\binom{n}{4}$
1	—	—	—
2	1	—	—
3	3	1	—
4	6	4	1
[5]	10	[10]	5

(a) You know that $n = 5$ (go down 5 places in the n column), and that $k = 3$ (go from the 5 in the n column over to the corresponding value in the $\binom{n}{3}$ column). The answer is 10.

(b) $\binom{12}{10} = 66$ **(c)** $\binom{15}{12} = 455$ **(d)** $\binom{18}{12} = 18{,}564$

PROBLEM 8-5 Give the expansion of $(a + b)^n$.

Solution

$$(a+b)^n = \binom{n}{0}a^n b^0 + \binom{n}{1}a^{n-1}b^1 + \binom{n}{2}a^{n-2}b^2 + \cdots + \binom{n}{k}a^{n-k}b^k + \cdots + \binom{n}{n}a^0 b^n$$

PROBLEM 8-6 Use the result of Problem 8-5 to give the expansion of $(a + b)^5$.

Solution

$$(a+b)^5 = \binom{5}{0}a^5 b^0 + \binom{5}{1}a^4 b^1 + \binom{5}{2}a^3 b^2 + \binom{5}{3}a^2 b^3 + \binom{5}{4}a^1 b^4 + \binom{5}{5}a^0 b^5$$
$$= a^5 + 5a^4b^1 + 10a^3b^2 + 10a^2b^3 + 5a^1b^4 + b^5$$

PROBLEM 8-7 The binomial coefficients are given by Pascal's triangle:

Row

Row													
1							1						
2						1		1					
3					1		2		1				
4				1		3		3		1			
5			1		4		6		4		1		
6		1		5		10		10		5		1	
7	1		6		15		20		15		6		1

. . . .

. . . .

In Pascal's triangle the $(n + 1)$st row contains the coefficients of the terms in the expansion of $(a + b)^n$. (Compare row 6 with the answer to Problem 8-6.) You obtain each entry in Pascal's triangle that isn't equal to 1 by adding together the two numbers in the row above it, as indicated by the braces. Pascal's triangle is based on the equation

$$\binom{n}{k} = \binom{n-1}{k-1} + \binom{n-1}{k}$$

Verify this equation.

Solution

$$\binom{n-1}{k-1} + \binom{n-1}{k} = \frac{(n-1)!}{(k-1)!(n-1-(k-1))!} + \frac{(n-1)!}{k!(n-1-k)!}$$

$$= \frac{(n-1)!}{(k-1)!(n-k)!}\left(\frac{k}{k}\right) + \frac{(n-1)!}{k!(n-1-k)!}\,\frac{(n-k)}{(n-k)}$$

$$= \frac{(n-1)!k}{k!(n-k)!} + \frac{(n-1)!(n-k)}{k!(n-k)!} = \frac{(n-1)!(k+n-k)}{k!(n-k)!}$$

$$= \frac{n!}{k!(n-k)!} = \binom{n}{k}$$

PROBLEM 8-8 For Pascal's triangle in Problem 8-7, give the entires that would be in the next row (row 8).

Solution Each row begins and ends with a 1, and the other entries are the sums of the two numbers above it, so the next row is

1 7 21 35 35 21 7 1

Notice that there are 8 number in this row—they're the appropriate coefficients of the terms in the expansion of $(a + b)^7$.

Bernoulli Experiments

PROBLEM 8-9 Conduct a Bernoulli experiment in which the random variable $X = 1$ if you observe success and $X = 0$ if you observe failure. Define the probability function $f(x)$ of X.

Solution The possible values of X are 0 and 1. So,

$$f(0) = P(X = 0) = P(\text{failure}) = 1 - p = q$$

$$f(1) = P(X = 1) = P(\text{success}) = p$$

We often write this compactly as

$$f(x) = \begin{cases} 1 - p, & x = 0 \\ p, & x = 1 \end{cases}$$

PROBLEM 8-10 Find the mean μ and variance σ^2 of a Bernoulli random variable X.

Solution From the definition of the mean,

$$\mu = \sum_{x=0}^{1} xf(x) = 0(1-p) + 1p = p$$

From the definition of the variance,

$$\sigma^2 = \sum_{x=0}^{1} (x-p)^2 f(x) = (0-p)^2(1-p) + (1-p)^2 p$$

$$= p^2(1-p) + p(1-p)^2$$

$$= p(1-p)[p + (1-p)]$$

$$= p(1-p) = pq$$

Alternatively, you could think of the Bernoulli distribution as a binomial with $n = 1$. Then the mean is $\mu = np = p$ and the variance is $\sigma^2 = npq = pq$.

PROBLEM 8-11 Let X be a Bernoulli random variable associated with one roll of a fair 6-sided die. Call the roll a "success" if either a 5 or 6 is observed. **(a)** Define the probability function of X. **(b)** Give the mean of X and **(c)** the variance of X.

Solution

(a) You have two chances out of six to get a 5 or a 6, so $p = 2/6 = 1/3$. So the probability function is

$$P(\text{success}) = f(1) = p = \frac{1}{3} \quad \text{and} \quad P(\text{failure}) = f(0) = q = 1 - p = \frac{2}{3}$$

or more compactly, you can write

$$f(x) = \begin{cases} 2/3, & x = 0 \\ 1/3, & x = 1 \end{cases}$$

(b) The mean of X is $\mu = np = (1)\left(\frac{1}{3}\right) = \frac{1}{3}$

(c) The variance of X is $\sigma^2 = npq = (1)\left(\frac{1}{3}\right)\left(1 - \frac{1}{3}\right) = \frac{2}{9}$

PROBLEM 8-12 In a state lottery, you select a 3-digit integer. If the state selects the same integer, you'll win. Let X equal 1 if you win (success!) and 0 if you lose. **(a)** Define the probability function of X. **(b)** Give the mean of X and **(c)** the variance of X.

Solution

(a) Because there are 1000 three-digit integers, 000–999 inclusive, the probability of winning is $p = \frac{1}{1000}$. So the probability function of X is

$$P(\text{success}) = f(1) = p = \frac{1}{1000} = .001 \quad \text{and} \quad P(\text{failure}) = f(0) = q = 1 - p = \frac{999}{1000} = .999$$

or

$$f(x) = \begin{cases} .999, & x = 0 \\ .001, & x = 1 \end{cases}$$

(b) The mean of X is $\mu = np = (1)\left(\frac{1}{1000}\right) = .001$

(c) The variance of X is $\sigma^2 = npq = (1)(.001)(.999) = .000999$

Binomial Probabilities

PROBLEM 8-13 Let the distribution of X be $b(14, .3)$. **(a)** Give the values of n and p. **(b)** Find $P(X = 4)$. **(c)** Find $P(3 \leq X \leq 5)$.

Solution

(a) The distribution of X in a binomial experiment is $b(n, p)$. So the parameters of $b(14, .3)$ are $n = 14$ and $p = .3$.

(b) You use the binomial probability function (8.1) to determine the probability that the random variable X will take on the value $x = 4$. Knowing that $n = 14$, $p = .3$, and given that $x = 4$,

$$f(x) = \frac{n!}{x!(n-x)!}p^x(1-p)^{n-x}$$

$$f(4) = \frac{14!}{4!(14-4)!}(.3)^4(1-.3)^{14-4}$$

$$= \frac{14!}{4!10!}(.3)^4(.7)^{10} = \frac{14 \cdot 13 \cdot 12 \cdot 11}{4 \cdot 3 \cdot 2 \cdot 1}(.3)^4(.7)^{10}$$

$$= (1001)(.0081)(.02825) = .2290$$

(**c**) Use the binomial probability table (see Table 2 in the Appendix) to find $P(3 \leq X \leq 5)$:

$$\begin{aligned} P(3 \leq X \leq 5) &= f(3) + f(4) + f(5) \\ &= .1943 + .2290 + .1963 \\ &= .6196 \end{aligned}$$

PROBLEM 8-14 Let the distribution of X be $b(7, .9)$. (**a**) Give the values of n and p. (**b**) Use the binomial probability table (Table 2 in the Appendix) to find the probability that X is *at least* 5. (**c**) Find the probability that X is *at most* 4.

Solution

(**a**) The distribution of X is $b(n, p) = b(7, .9)$; $n = 7$ and $p = .9$.

(**b**) $$\begin{aligned} P(X \geq 5) &= f(5) + f(6) + f(7) \\ &= .1240 + .3720 + .4783 \\ &= .9743 \end{aligned}$$

(**c**) $$\begin{aligned} P(X \leq 4) &= f(0) + f(1) + f(2) + f(3) + f(4) \\ &= .0000 + .0000 + .0002 + .0026 + .0230 \\ &= .0258 \end{aligned}$$

Notice that (except for round-off error) the answers to (**b**) and (**c**) sum to one, as they should—i.e., for $n = 7$, the number of successes *must* be either ≥ 5 or ≤ 4.

PROBLEM 8-15 Let X equal the number of successes in n independent Bernoulli trials with probability of success p on each trial. Verify that

$$\begin{aligned} f(x) &= P(X = x) \\ &= \binom{n}{x} p^x (1-p)^{n-x}, \qquad x = 0, 1, 2, \ldots, n \end{aligned}$$

Solution The probability of observing x successes (S) in a row, followed by $n - x$ failures (F), is

$$P(\underbrace{\underbrace{SS \ldots S}_{x \text{ (successes)}}\underbrace{FF \ldots F}_{n-x \text{ (failures)}}}_{n \text{ trials}}) = p^x(1-p)^{n-x}$$

because the trials are independent. Every arrangement of x S's and $n - x$ F's has this same probability, $p^x(1-p)^{n-x}$. The number of possible arrangements is $\binom{n}{x}$, namely, the number of ways of choosing x positions out of n positions for the S's. So

$$f(x) = \binom{n}{x} p^x (1-p)^{n-x}, \qquad x = 0, 1, 2, \ldots, n$$

PROBLEM 8-16 Low birth weight (LBW)—a weight of less than 2500 g—is a good indicator of a newborn baby's chances of survival. Suppose that in northern Africa the proportion of babies with LBW is $p = .10$. In a random sample of $n = 13$ newborn babies, let X equal the number of babies with LBW. (**a**) What's the distribution of X? (**b**) Use Table 2 in the Appendix to find $P(X \leq 2)$ and (**c**) $P(X \geq 3)$.

Solution

(**a**) If you treat each baby in the sample as the observation of a Bernoulli trial, then X has a binomial distribution with $n = 13$ and $p = .10$—i.e., X is $b(13, .10)$.

(**b**) Using Table 2,

$$\begin{aligned} P(X \leq 2) &= f(0) + f(1) + f(2) \\ &= .2542 + .3672 + .2448 \\ &= .8662 \end{aligned}$$

(c) Knowing that $P(X \leq 2) + P(X \geq 3)$ must equal one, you can use the answer to part **(b)** to find the answer to part **(c)**:

$$P(X \geq 3) = 1 - P(X \leq 2) = 1 - .8662 = .1338$$

The Mean, Variance, and Standard Deviation of a Binomial Random Variable

PROBLEM 8-17 Suppose that $p = .9$ is the probability that a bean seed germinates. Let X equal the number of seeds that germinate out of 12 randomly selected seeds that have been planted. **(a)** How is X distributed? **(b)** Determine the mean and variance of X. **(c)** Find the probability that *exactly* 10 out of the 12 seeds germinate. **(d)** Find the probability that *at least* 10 seeds germinate. **(e)** Find the probability that *at most* 10 seeds germinate.

Solution

(a) The distribution of X is $b(12, .9)$. That is, X has a binomial distribution where $n = 12$ is the number of trials (bean seeds) and $p = .9$ is the probability of success (germination) on each trial.

(b) The mean (μ) and variance (σ^2) of X are

$$\mu = 12(.9) = 10.8 \qquad \text{and} \qquad \sigma^2 = 12(.9)(.1) = 1.08$$

(c) $P(X = 10) = \dfrac{12!}{10!2!}(.9)^{10}(.1)^2 = (66)(.3487)(.01) = .2301$

(d) Using Table 2 in the Appendix, you find that

$$P(X \geq 10) = .2301 + .3766 + .2824 = .8891$$

(e) The appropriate probability is given by

$$P(X \leq 10) = .0000 + .0005 + .0038 + .0213 + .0852 + .2301$$
$$= .3409$$

PROBLEM 8-18 Suppose that among tree pruners, only 25% use pruners of the correct length. In a random sample of 20 tree pruners, let X equal the number whose pruners are the correct length. **(a)** How is X distributed? **(b)** Give the mean and variance of X. **(c)** Find $P(X \leq 3)$, $P(X = 5)$, and $P(X \geq 10)$.

Solution

(a) The distribution of X is $b(20, .25)$.

(b) $\mu = np = 20(.25) = 5 \qquad \text{and} \qquad \sigma^2 = np(1 - p) = 20(.25)(.75) = 3.75$

(c) Using Table 2 in the Appendix,

$$P(X \leq 3) = .0032 + .0211 + .0669 + .1339 = 0.2251$$

$$P(X = 5) = .2023$$

$$P(X \geq 10) = .0099 + .0030 + .0008 + .0002 + .0000 = .0139$$

PROBLEM 8-19 Using the definition of expectation (7.1), show that the mean of a binomial random variable is $\mu = np$.

Solution

$$\mu = E(X)$$
$$= \sum_{x=0}^{n} xf(x) = \sum_{x=1}^{n} x\left[\frac{n!}{x!(n-x)!}p^x(1-p)^{n-x}\right]$$
$$= \sum_{x=1}^{n} \frac{n!}{(x-1)!(n-x)!}p^x(1-p)^{n-x}$$
$$= np\sum_{x=1}^{n} \frac{(n-1)!}{(x-1)!(n-x)!}p^{x-1}(1-p)^{n-x}$$

Let $k = x - 1$. Then $x = k + 1$ so that

$$\mu = np \sum_{k=0}^{n-1} \frac{(n-1)!}{k!(n-1-k)!} p^k(1-p)^{n-1-k}$$
$$= np(p + 1 - p)^{n-1} = np$$

PROBLEM 8-20 In the casino game of craps, the probability of winning is $p = .49423$. If you place $n = 1000$ independent bets, how many times can you expect to win on the average?

Solution $\mu = 1000(.49423) = 494.23$

Supplementary Exercises

PROBLEM 8-21 Let X be a Bernoulli random variable associated with the flip of a fair coin, and let heads be a success. **(a)** Define the probability function of X, and **(b)** give the mean and variance of X.

Answer **(a)** $f(0) = 1/2, f(1) = 1/2$ **(b)** $\mu = 1/2, \sigma^2 = 1/4$

PROBLEM 8-22 Evaluate **(a)** 4!, **(b)** 6!, **(c)** 10!.

Answer **(a)** 24 **(b)** 720 **(c)** 3,628,800

PROBLEM 8-23 Evaluate the binomial coefficients "by hand" (i.e., without using a table): **(a)** $\binom{7}{3}$, **(b)** $\binom{8}{5}$, **(c)** $\binom{6}{4}$.

Answer **(a)** 35 **(b)** 56 **(c)** 15

PROBLEM 8-24 Use the binomial coefficients table (Table 1 in the Appendix) to evaluate **(a)** $\binom{14}{5}$, **(b)** $\binom{17}{8}$, **(c)** $\binom{13}{11}$, **(d)** $\binom{20}{15}$.

Answer **(a)** 2002, **(b)** 24,310 **(c)** 78 **(d)** 15,504

PROBLEM 8-25 Give the expansion of $(a + b)^4$.

Answer $(a + b)^4 = a^4 + 4a^3b + 6a^2b^2 + 4ab^3 + b^4$

PROBLEM 8-26 Let X be a Bernoulli random variable associated with selecting a gem stone from a bag of 10 gem stones containing 3 diamonds. Call the selection of one of the diamonds a success. **(a)** Define the probability function of X and **(b)** give the mean and variance of X.

Answer **(a)** $f(0) = 7/10, f(1) = 3/10$ **(b)** $\mu = 3/10, \sigma^2 = .21$

PROBLEM 8-27 Let the distribution of X be $b(11, .7)$. **(a)** Give the values of μ and σ^2. **(b)** Find $P(X = 9)$. **(c)** Find $P(X \geq 8)$.

Answer **(a)** $\mu = 7.7, \sigma^2 = 2.31$ **(b)** .1998 **(c)** .5696

PROBLEM 8-28 Let the distribution of X be $b(15, .10)$. **(a)** Give the values of the mean and variance of X. **(b)** Find the probability that X is at least 4. **(c)** Find the probability that X is at most 2.

Answer **(a)** 1.5, 1.35 **(b)** .0555 **(c)** .8160

PROBLEM 8-29 Flip a fair coin 8 times. Let X equal the number of heads you observe. Find **(a)** $P(X = 4)$, **(b)** $P(X \leq 3)$, **(c)** $P(X \geq 6)$.

Answer **(a)** .2734 **(b)** .3633 **(c)** .1445

PROBLEM 8-30 Roll a fair 4-sided die 7 times. Let X equal the number of times that you observe a 3. **(a)** How is X distributed? **(b)** Give the mean and variance of X. **(c)** Find $P(X = 2)$, $P(X \leq 2)$, $P(X \geq 4)$.

Answer **(a)** $b(7, .25)$ **(b)** 7/4, 21/16 **(c)** .3115, .7565, .0706

PROBLEM 8-31 Suppose that 40% of the registered voters who expect to vote in the next election plan to vote "yes" on Proposal C. In a random sample of 20 of these voters, let X equal the number who plan to vote yes. **(a)** How is X distributed? **(b)** Give the mean and variance of X. **(c)** What's the probability that at least 11 of these voters plan to vote "yes?" **(d)** What's the probability that at most 9 out of the 20 voters plan to vote "yes?"

Answer **(a)** $b(20, .4)$ **(b)** 8, 4.8 **(c)** .1276 **(d)** .7552

PROBLEM 8-32 A certain brand of dry soap is packaged in 1000-g boxes. Let p equal the probability that the amount of soap in a box selected at random weighs more than 1050 g. Suppose that $p = .20$. In a random sample of 15 boxes, let X equal the number of boxes that weigh more than 1050 g. **(a)** How is X distributed? **(b)** Give the mean, variance, and standard deviation of X. **(c)** Find $P(X < 3)$. **(d)** Find $P(X > 7)$.

Answer **(a)** $b(15, .20)$ **(b)** 3, 2.4, 1.55 **(c)** .3980 **(d)** .0043

PROBLEM 8-33 A bag of jawbreakers contains 3 grape, 5 cherry, and 2 orange candies. You select a jawbreaker at random from the bag, note its flavor, and then put it back into the bag. You repeat this operation 8 times. (This is called a random sample of size 8 v.' en *sampling with replacement*.) Let X equal the number of grape jawbreakers that you observe. **(a)** Give the distribution of X. **(b)** Give the mean and variance of X. **(c)** Find the probability that you observe at least 3 grape jawbreakers. **(d)** Find the probability that you observe at most 2 grape jawbreakers.

Answer **(a)** $b(8, .3)$ **(b)** 2.4, 1.68 **(c)** .4482 **(d)** .5518

PROBLEM 8-34 On a 15-question multiple choice test, each question has 5 possible answers, exactly one of which is correct. Let X equal the number of questions answered correctly by guessing. Assume that X has a binomial distribution. **(a)** Give the values of n and p. **(b)** Find $P(X \leq 3)$. **(c)** Find $P(X \geq 7)$.

Answer **(a)** $n = 15, p = .20$ **(b)** .6481 **(c)** .0181

PROBLEM 8-35 In a certain city, 60% of the adults bank at more than one bank. Select a random sample of 19 adults and let X equal the number who use more than one bank. **(a)** How is X distributed? **(b)** Give the mean and variance of X. **(c)** Find $P(X = 10)$, $P(X \leq 10)$, $P(X > 10)$.

Answer **(a)** $b(19, .60)$ **(b)** 11.4, 4.56 **(c)** .1464, .3324, .6675

PROBLEM 8-36 For a pair of grackles, let p equal the probability that the male tarsus (a leg bone) is longer than the female tarsus. Suppose that $p = .75$. Among 12 pairs of grackles selected at random, let X equal the number of pairs for which the male tarsus is longer than the female tarsus. **(a)** Give the distribution of X. **(b)** Give the mean, variance, and standard deviation of X. **(c)** Find $P(X \geq 8)$. **(d)** Find $P(X \leq 5)$.

Answer **(a)** $b(12, .75)$ **(b)** 9, 2.25, 1.5 **(c)** .8424 **(d)** .0143

9 SOME OTHER DISCRETE DISTRIBUTIONS

THIS CHAPTER IS ABOUT

☑ **Hypergeometric Distribution**
☑ **Poisson Distribution**
☑ **Multinomial Distribution**

There are many different discrete distributions. In this chapter, we discuss two distributions of a single discrete random variable—the hypergeometric and Poisson—and the extension of the hypergeometric and binomial distributions to distributions of several discrete variables.

9-1. Hypergeometric Distribution

Suppose we have a collection of n objects that consists of n_1 objects of one kind and n_2 objects of another kind, so that $n = n_1 + n_2$. If we take a random sample of r objects from this collection without replacement and let X equal the number of objects of the first kind in the random sample, we say that X has a **hypergeometric distribution**. In this case, the probability function of X is defined by

HYPERGEOMETRIC PROBABILITY FUNCTION

$$f(x) = P(X = x) = \frac{\binom{n_1}{x}\binom{n_2}{r-x}}{\binom{n}{r}}, \quad x = 0, 1, \ldots, r \qquad \textbf{(9.1)}$$

EXAMPLE 9-1 From an urn that contains only $n_1 = 7$ orange balls and $n_2 = 3$ blue balls, take a random sample of $r = 2$ balls without replacement. Let the random variable X be the number of orange balls in the sample, and define the probability function of X.

Solution To find the number of samples of size 2 that you can select from 10 objects—i.e., the number of combinations of 10 things taken 2 at a time (see Section 6-3E), use formula (6.7):

$$\binom{n}{r} = \frac{n!}{r!(n-r)!} = \frac{10!}{2!8!} = \frac{10 \cdot 9}{2} = 45$$

You are choosing balls at random, so each of these 45 combinations has the same probability for being selected. Thus $\frac{10!}{2!8!} = 45$ is the number of equally likely points in the sample space and this becomes the denominator for our probabilities.

To find the probabilities that $X = 0, 1,$ or 2 (i.e., that there will be 0, 1, or 2 orange balls in your sample), first find the number of sets of size 2 that could consist of 0, 1, or 2 orange balls. You know that $X = 0$ if both the balls in your sample are blue, so find out how many ways there are to obtain an all-blue sample—i.e., 7 orange balls taken 0 at a time and 3 blue balls taken 2 at a time. Again, from (6.7) and using the multiplication principle (see Section 6-3A),

$$\binom{7}{0}\binom{3}{2} = \left(\frac{7!}{0!7!}\right)\left(\frac{3!}{2!1!}\right) = 3$$

So, there are 3 possible ways in which you can select 0 orange and 2 blue balls.

The number of ways in which you can select 1 orange ball and 1 blue ball, $X = 1$, is

$$\binom{7}{1}\binom{3}{1} = \left(\frac{7!}{1!6!}\right)\left(\frac{3!}{1!2!}\right) = 21$$

And the number of ways in which you can select 2 orange balls and 0 blue balls, $X = 2$, is

$$\binom{7}{2}\binom{3}{0} = \left(\frac{7!}{2!5!}\right)\left(\frac{3!}{0!3!}\right) = \frac{7 \cdot 6}{2} = 21$$

Now you can see that the probability function of X is

$$f(x) = \frac{\binom{7}{x}\binom{3}{2-x}}{\binom{10}{2}} = \begin{cases} \frac{3}{45}, & x = 0 \\ \frac{21}{45}, & x = 1 \\ \frac{21}{45}, & x = 2 \end{cases}$$

Notice that since X has a hypergeometric distribution, you could have used formula (9.1) to define this probability function of X:

$$f(x) = P(X = x) = \frac{\binom{n_1}{x}\binom{n_2}{r-x}}{\binom{n}{r}} = \frac{\binom{7}{x}\binom{3}{2-x}}{\binom{10}{2}}, \quad x = 0, 1, 2$$

EXAMPLE 9-2 If you select 5 cards at random out of a deck of 52 playing cards, what's the probability that 3 of those cards will be red?

Solution Let X be the number of red cards among the 5, and find $P(X = 3)$. The random variable X has a hypergeometric distribution with $n = 52$ cards, $n_1 = 26$ red cards, $n_2 = 26$ black cards, and $r = 5$ cards selected. You want to find the probability that there will be 3 red cards and 2 black cards in the hand, so you substitute into formula (9.1),

$$f(3) = P(X = 3) = \frac{\binom{26}{3}\binom{26}{5-3}}{\binom{52}{5}}$$

$$= \frac{\left(\frac{26!}{3!23!}\right)\left(\frac{26!}{2!24!}\right)}{\left(\frac{52!}{5!47!}\right)} = \frac{\left(\frac{26 \cdot 25 \cdot 24}{3 \cdot 2 \cdot 1}\right)\left(\frac{26 \cdot 25}{2 \cdot 1}\right)}{\left(\frac{52 \cdot 51 \cdot 50 \cdot 49 \cdot 48}{5 \cdot 4 \cdot 3 \cdot 2 \cdot 1}\right)}$$

$$= \frac{(2600)(325)}{2{,}598{,}960} = .325$$

A. Hypergeometric probabilities using a table

To help with the calculations of hypergeometric probabilities, you can use Table 1 in the Appendix. This table gives values of the binomial coefficients $\binom{n}{r}$ for n between 0 and 26 and r between 0 and 13, inclusive. Recall that

$$\binom{n}{r} = \frac{n!}{r!(n-r)!} = \frac{n!}{(n-r)!r!} = \binom{n}{n-r}$$

So when, for example, we see that $\binom{18}{14}$ isn't in the table, we apply the identity $\binom{n}{r} = \binom{n}{n-r}$, obtaining

$$\binom{18}{14} = \binom{18}{18-14} = \binom{18}{4}$$

$\binom{18}{4}$ is in the table, and equals 3060.

EXAMPLE 9-3 Twenty students in a statistics class decide to have a post-exam victory party. They want to form an organizing committee by selecting 4 students at random out of the 13 women and 7 men in the class, but they don't want the committee to be all women or all men. What are the probabilities that the committee will consist of 0, 1, 2, 3, or 4 women?

Solution You must calculate the probabilities for $X = 0, 1, 2, 3,$ and 4 (the number of women that could be on the committee). Using $r = 4$ (the total number of committee members), $n = 20$, $n_1 = 13$ [you make $n_1 = 13$ (instead of 7) because the question asks about the number of *women* on the committee, not the number of *men*], and $n_2 = 7$, define the probability function of X:

$$f(x) = \frac{\binom{13}{x}\binom{7}{4-x}}{\binom{20}{4}}, \quad x = 0, 1, 2, 3, 4$$

Then using Table 1 in the Appendix to evaluate the combinatorials, construct the tabulation shown to give the probabilities.

x	$f(x)$		
0	(1)(35)/4845 =	35/4845 =	.007
1	(13)(35)/4845 =	455/4845 =	.094
2	(78)(21)/4845 =	1638/4845 =	.338
3	(286)(7)/4845 =	2002/4845 =	.413
4	(715)(1)/4845 =	715/4845 =	.148
		4845/4845	1.000

B. Mean and variance of a hypergeometric random variable

We can also find the mean μ and variance σ^2 of a hypergeometric random variable.

MEAN OF HYPERGEOMETRIC DISTRIBUTION

$$\mu = \frac{rn_1}{n} = r\left(\frac{n_1}{n}\right) \tag{9.2}$$

VARIANCE OF HYPERGEOMETRIC DISTRIBUTION

$$\sigma^2 = \frac{rn_1n_2(n-r)}{n^2(n-1)} = r\left(\frac{n_1}{n}\right)\left(\frac{n_2}{n}\right)\left(\frac{n-r}{n-1}\right) \tag{9.3}$$

EXAMPLE 9-4 Annie's mother sets out a dish of 18 mints: 5 are white and the rest are either pink, yellow, or green. Annie grabs a handful and eats 4 mints before her mother can stop her. Let X equal the number of white mints that Annie eats. **(a)** Define the probability function of X and **(b)** find the probability for each value of X. **(c)** Find the mean μ of X. **(d)** Find the variance σ^2 of X.

Solution

(a) The random variable X has a hypergeometric distribution, so

$$f(x) = \frac{\binom{5}{x}\binom{13}{4-x}}{\binom{18}{4}}, \quad x = 0, 1, 2, 3, 4$$

(b) You can then evaluate $f(x)$ at $x = 0, 1, 2, 3, 4$ using Table 1 in the Appendix, and construct the tabulation as shown.

x	$f(x)$		
0	(1)(715)/3060 =	715/3060 =	.234
1	(5)(286)/3060 =	1430/3060 =	.467
2	(10)(78)/3060 =	780/3060 =	.255
3	(10)(13)/3060 =	130/3060 =	.042
4	(5)(1)/3060 =	5/3060 =	.002
		3060/3060	1.000

(c) By formula (9.2):

$$\mu = \frac{rn_1}{n} = \frac{(4)(5)}{18} = 1.11$$

(d) By formula (9.3):

$$\sigma^2 = \frac{rn_1n_2(n-r)}{n^2(n-1)} = \frac{(4)(5)(13)(14)}{(18)^2(17)} = \frac{3640}{5508} = .661$$

C. Extension of a hypergeometric distribution

We can extend the hypergeometric distribution to determine probabilities for more than just two kinds of objects in a collection.

Suppose that a collection of n objects consists of k different kinds of objects, n_1 of one kind, n_2 of another kind, and on up to n_k of a kth kind. If we take a random sample of r objects from this collection and let X_1 equal the number of objects of the first kind, X_2 equal the number of objects of the second kind, and so on up to X_k objects of the kth kind, the joint probability function of these k random variables is

$$f(x_1, x_2, \ldots, x_k) = \frac{\binom{n_1}{x_1}\binom{n_2}{x_2}\cdots\binom{n_k}{x_k}}{\binom{n}{r}}, \quad x_1 + x_2 + \cdots + x_k = r \tag{9.4}$$

EXAMPLE 9-5 Adam has a box of colored plastic bands that he uses to band birds. He selects 3 bands at random without replacement from the 10 red bands, 8 white bands, and 7 blue bands in the box. Let X_1 equal the number of red bands Adam selects, X_2 the number of white bands, and X_3 the number of blue bands. Use formula (9.4) to define the probability function of X_1, X_2, and X_3.

Solution Let $f(x_1, x_2, x_3) = P(X_1 = x_1, X_2 = x_2, X_3 = x_3)$. The number of combinations of 3 bands that can be selected out of 25 bands is $\binom{25}{3} = 2300$, and each combination is equally likely. The number of these 2300 combinations that contains x_1 red, x_2 white, and x_3 blue bands is the product of $\binom{10}{x_1}$, $\binom{8}{x_2}$, and $\binom{7}{x_3}$. So that, if x_1, x_2, and x_3 are nonnegative integers, the probability function of X_1, X_2, and X_3 is

$$f(x_1, x_2, x_3) = \frac{\binom{10}{x_1}\binom{8}{x_2}\binom{7}{x_3}}{\binom{25}{3}}, \quad x_1 + x_2 + x_3 = 3$$

If you want to know the probability for selecting one band of each color, substitute $x_1 = 1$, $x_2 = 1$, $x_3 = 1$ into the formula.

$$f(1, 1, 1) = \frac{\binom{10}{1}\binom{8}{1}\binom{7}{1}}{\binom{25}{3}} = \frac{10 \cdot 8 \cdot 7}{2300} = .243$$

EXAMPLE 9-6 Annie's mother has replenished the dish of mints and it now contains 3 pink, 5 yellow, 6 white, and 4 green mints. Once again Annie takes 4 mints at random and eats them. What is the probability that she'll get one mint of each color?

Solution First set up the variables: $f(x_1, x_2, x_3, x_4) = f(1, 1, 1, 1)$; $n = 18$ (the total number of mints in the bowl before Annie gets her hands on them); $r = 4$ (the number of mints Annie eats); and $n_1 = 3$ pink mints, $n_2 = 5$ yellow mints, $n_3 = 6$ white mints, and $n_4 = 4$ green mints. Using these values in formula (9.4), you have

$$f(1, 1, 1, 1) = \frac{\binom{3}{1}\binom{5}{1}\binom{6}{1}\binom{4}{1}}{\binom{18}{4}} = \frac{3 \cdot 5 \cdot 6 \cdot 4}{3060} = .118$$

9-2. Poisson Distribution

The binomial random variable counts the number of successes in n trials, where n is specified. There are experiments in which we're interested in counting the number of successes or number of occurrences of something per "unit"—e.g., number of phone calls arriving at a switchboard per minute, number of flaws on a computer tape per 100 feet, number of customers arriving at a bank teller's window per 15 minutes. The random variable that counts the number of successes per "unit" is called a **Poisson random variable**.

We use the Poisson random variable if:

1. The probability of more than one success in a very small interval is approximately 0,
2. The numbers of successes in non-overlapping intervals are independent,
3. The average number of successes per unit is λ, where λ is a constant determined by the particular application.

We define the probability function of X by

POISSON PROBABILITY FUNCTION

$$f(x) = \frac{\lambda^x e^{-\lambda}}{x!}, \quad x = 0, 1, 2, \ldots \tag{9.5}$$

where λ is a parameter specific to the problem and e is the base of the natural logarithm with an approximate value of 2.72, and we say that X has a **Poisson distribution.**

A. Poisson probabilities from a probability table

To find probabilities for a Poisson random variable X, given its associated parameter, you can use Table 3 in the Appendix.

EXAMPLE 9-7 Let X have a Poisson distribution with $\lambda = 3.4$. Find **(a)** $P(X = 2)$, **(b)** $P(X \leq 2)$, and **(c)** $P(X > 2)$.

Solution Reproduced here is a portion of Table 3 from the Appendix.

x	λ = 3.2	3.4	3.6
0	.0408	.0334	.0273
1	.1304	.1135	.0984
2	.2087	.1929	.1771
3	.2226	.2186	.2125
4	.1781	.1858	.1912
5	.1140	.1264	.1377
6	.0608	.0716	.0826
7	.0278	.0348	.0425
8	.0111	.0148	.0191
9	.0040	.0056	.0076
10	.0013	.0019	.1128
11	.0004	.0006	.0009
12	.0001	.0002	.0003
13	.0000	.0000	.0001
14	.0000	.0000	.0000

(a) Given that $\lambda = 3.4$, locate the column headed by 3.4 and the row $x = 2$. The number at the intersection is the probability for an event given $\lambda = 3.4$ and $x = 2$, which is .1929. So $P(X = 2) = .1929$. If we had used formula (9.5) and a calculator that gives the value of $e^{-3.4} = .03337$, then

$$P(X = 2) = \frac{(3.4)^2 e^{-3.4}}{2!} = .1929$$

(b) To find $P(X \leq 2)$, add together the first 3 entries in the column headed by 3.4:

$$P(X \leq 2) = P(X = 0) + P(X = 1) + P(X = 2)$$
$$= .0334 + .1135 + .1929 = .3398$$

(c) You can find $P(X > 2)$ by adding together all of the entries in the column headed by 3.4 from $P(X = 3)$ on, or by using the formula for the complement of an event (6.3):

$$P(A) = 1 - P(A')$$
$$P(X > 2) = 1 - P(X \leq 2) = 1 - .3398 = .6602$$

B. Mean and variance for a Poisson random variable

For a Poisson distribution, the parameter λ is equal to both the mean μ and the variance σ^2—that is,

MEAN AND VARIANCE FOR A POISSON DISTRIBUTION

$$\mu = \lambda, \qquad \sigma^2 = \lambda \tag{9.6}$$

EXAMPLE 9-8 If X has a Poisson distribution with mean $\lambda = 8.5$, **(a)** what's the variance σ^2 of X? Find **(b)** $P(X = 8)$ and **(c)** $P(X = 9 \text{ or } 10)$.

Solution

(a) In a Poisson distribution the mean and variance are both equal to λ, so from formula (9.6), $\sigma^2 = \lambda = 8.5$.

(b) Then from Table 3 in the Appendix, locate the intersection of column $\lambda = 8.5$ and row $x = 8$ to obtain $P(X = 8) = .1375$.

(c) Similarly, you can find the probability that $X = 9$ or $X = 10$—the union of the two events.

$$P(X = 9 \text{ or } 10) = P(X = 9) + P(X = 10)$$
$$= .1299 + .1104 = .2403$$

Customer arrival times often tend to have a Poisson distribution.

EXAMPLE 9-9 Let X equal the number of customers who arrive at a bank teller window during a 15-min time period. Assume that X has a Poisson distribution with mean equal to 4. **(a)** Define the probability function of X. **(b)** Find $P(X \le 3)$. **(c)** Find $P(X > 3)$.

Solution

(a) Because the mean is equal to 4, the parameter $\lambda = 4$. Thus, using formula (9.5),

$$f(x) = \frac{\lambda^x e^{-\lambda}}{x!} = \frac{4^x e^{-4}}{x!}, \quad x = 0, 1, 2, \ldots$$

(b) From Table 3 in the Appendix,

$$P(X \le 3) = .0183 + .0733 + .1465 + .1954 = .4335$$

(c) $P(X > 3) = 1 - P(X \le 3) = 1 - .4335 = .5665$

EXAMPLE 9-10 A bank manager who wanted to make the bank's service more efficient had the tellers keep track of the number of people who arrived at their windows every 15 minutes. After 15 of these intervals, one teller had compiled the following numbers of customer arrivals:

3 3 5 5 7 5 6 8 3 5 3 4 1 3 1

If these are observations of a Poisson random variable, μ and σ^2 will be equal. It should also be true that the sample mean $\bar{x}$ and the sample variance s^2 are approximately equal. Find **(a)** $\bar{x}$ and **(b)** s^2. **(c)** Compare the proportion of times that x is less than or equal to 3 in this example with $P(X \le 3)$ in Example 9-9.

Solution First, to find $\bar{x}$ and s^2, construct a tabulation as shown.

x	f	fx	fx^2
1	2	2	2
2	0	0	0
3	5	15	45
4	1	4	16
5	4	20	100
6	1	6	36
7	1	7	49
8	1	8	64
	15	62	312

(a) Then from the tabulation and the formula for the sample mean for grouped discrete data (2.2),

$$\bar{x} = \frac{\sum fx}{n} = \frac{62}{15} = 4.13$$

(b) From the tabulation and formula (3.5) for the sample variance for grouped discrete data,

$$s^2 = \frac{n\sum fx^2 - (\sum fx)^2}{n(n-1)} = \frac{(15)(312) - (62)^2}{(15)(14)} = \frac{836}{210} = 3.98$$

You can see that $\bar{x}$ and s^2 are approximately equal.

(c) The relative frequency or proportion of observations less than or equal to 3 is $\frac{7}{15} = .467$, which is just a little larger than $P(X \le 3) = .4335$.

EXAMPLE 9-11 Jane and Harry are both reading the same novel. Harry finds the book well-written and very readable, but Jane thinks it's choppy with too many short, simple words. To prove her point, Jane counts the number of letters per word for a couple of paragraphs and comes up with an average of 4.8 letters per word. Then she assumes that the number of letters per word in the novel has an approximate Poisson distribution with $\lambda = 4.8$. That's a lot of work to prove a point! **(a)** What's the probability that the number of letters per word in the novel is equal to or less than 3—$P(X \leq 3)$? **(b)** What's the probability that there are 6 or more letters per word—$P(X \geq 6)$?

Solution

(a) Refer to Table 3 in the Appendix, and locate the values for $x = 0$, $x = 1$, $x = 2$, and $x = 3$ in column $\lambda = 4.8$.

$$P(X \leq 3) = .0082 + .0395 + .0948 + .1517 = .2942$$

(b)

$$P(X \geq 6) = .1398 + .0959 + .0575 + .0307 + .0147 + .0064 + .0026 + .0009 + .0003 + .0001 = .3489$$

You could also find $P(X \geq 6)$ by finding $1 - P(X < 6)$.

EXAMPLE 9-12 The 60 words of length X that Jane actually counted in the novel (see Example 9-11) are as follows:

3	3	7	3	6	5	6	3	7	3	3	4	11	5	4
2	5	3	7	6	7	4	6	4	6	4	3	4	3	2
9	11	5	4	10	2	4	8	2	3	7	3	5	4	3
4	8	4	2	4	4	4	5	2	3	3	8	7	3	8

For these data, find **(a)** $\bar{x}$ and **(b)** s^2. **(c)** Compare the proportion of observations that were less than or equal to 3 with $P(X \leq 3)$ in Example 9-11. **(d)** Compare the proportion of observations that were greater than or equal to 6 with $P(X \geq 6)$ in Example 9-11.

Solution First construct a tabulation as shown.

x	f	fx	fx^2
2	6	12	24
3	15	45	135
4	14	56	224
5	6	30	150
6	5	30	180
7	6	42	294
8	4	32	256
9	1	9	81
10	1	10	100
11	2	22	242
	60	288	1686

(a) From the table and (2.2),

$$\bar{x} = \frac{288}{60} = 4.8$$

(b) From the table and (3.5),

$$s^2 = \frac{(60)(1686) - (288)^2}{(60)(59)} = \frac{18{,}216}{3540} = 5.15$$

You can see that $\bar{x}$ and s^2 are approximately equal.

(c) The proportion of observations less than or equal to 3 is $\frac{21}{60} = .350$, which is a little larger than $P(X \leq 3) = .294$, the value you found in Example 9-11 using $\lambda = 4.8$.

(d) The proportion of observations greater than or equal to 6 is $\frac{19}{60} = .317$, which is a little less than the $P(X \geq 6) = .349$ you found in Example 9-11.

C. Approximation of binomial probabilities using the Poisson distribution

Let X have a binomial distribution with parameters n and p so that X is $b(n, p)$. When the number of observations n is large and the probability for success p is small, we can use the Poisson distribution with mean $\lambda = np$ to approximate binomial probabilities. This is very useful because most binomial tables don't include large values for n. We have

$$P(X = x) = \binom{n}{x} p^x (1 - p)^{n-x} \approx \frac{(np)^x e^{-np}}{x!} = \binom{n}{x} p^x (1 - p)^{n-x}$$

This approximation is quite good if $n \geq 20$ and $p \leq .05$ and very good if $n \geq 100$ and $p \leq .1$.

EXAMPLE 9-13 Michael decides to take a chance in a lottery for which 2,000,000 tickets are being sold. He has heard that 2000 of the tickets will win a large prize, and he figures that if he buys 300 tickets he'll have a pretty good chance of winning one. Let X be the number of winning tickets out of the 300 that Michael purchases. (**a**) How is X distributed? (**b**) What's the probability that Michael won't hold even one winning ticket—$P(X = 0)$? (**c**) What's the probability that he'll have at least one winning ticket—$P(X \geq 1)$?

Solution

(**a**) You can use a binomial distribution for X because there's a chance to win (success) or lose (failure) with each ticket purchased. The probability of winning is (essentially) $p = 2000/2{,}000{,}000 = .001$ for each ticket purchased and the $n = 300$ trials are independent. So the distribution of X is $b(300, .001)$.

(**b**) With $n = 300 > 100$ and $p = .001 < .01$, you can use the Poisson approximation with $\lambda = 300(.001) = .3$, which gives us

$$P(X = 0) \approx \frac{(np)^x e^{-np}}{x!} = \frac{(.3)^0 e^{-.3}}{0!} = e^{-.3} = .7408$$

Notice that this same answer can be found in Table 3 with $\lambda = .3$ and $x = 0$.

(**c**) You can use the answer to part (**b**) to find

$$P(X \geq 1) = 1 - P(X = 0) \approx 1 - .7408 = .2592$$

EXAMPLE 9-14 Let X have a binomial distribution so that X is $b(200, .03)$. (**a**) Find $P(X \leq 3)$ using the binomial probability function formula (8.1) (see Section 8-2B). (**b**) Use the Poisson distribution to find, approximately, $P(X \leq 3)$.

Solution

(**a**) From formula (8.1),

$$f(x) = P(X = x) = \binom{n}{x} p^x (1 - p)^{n-x}$$

$$= \frac{n!}{x!(n - x)!} p^x (1 - p)^{n-x}$$

$$P(X \leq 3) = P(X = 0) + P(X = 1) + P(X = 2) + P(X = 3)$$

$$= (.97)^{200} + 200(.03)(.97)^{199} + \frac{200 \cdot 199}{2 \cdot 1}(.03)^2(.97)^{198} + \frac{200 \cdot 199 \cdot 198}{3 \cdot 2 \cdot 1}(.03)^3(.97)^{197}$$

$$= .0023 + .0140 + .0430 + .0879 = .1472$$

(**b**) Using the Poisson approximation with $\lambda = np = (200)(.03) = 6$, you have

$$P(X = x) = \frac{(np)^x e^{-np}}{x!}$$

$$P(X \leq 3) = \frac{6^0 e^{-6}}{0!} + \frac{6^1 e^{-6}}{1!} + \frac{6^2 e^{-6}}{2!} + \frac{6^3 e^{-6}}{3!}$$

$$= e^{-6} + 6e^{-6} + 18e^{-6} + 36e^{-6}$$

$$= .0025 + .0149 + .0446 + .0892 = .1512$$

Notice that with a calculator you can find that $e^{-6} = .002479$. An alternate way to find this probability is to use Table 3 with $\lambda = 6$.

$$P(X \leq 3) = .0025 + .0149 + .0446 + .0892$$

$$= .1512$$

You can see that the approximation is quite good and it's much easier to use the Poisson approximation than to calculate the binomial probabilities.

9-3. Multinomial Distribution

A **multinomial experiment** must satisfy the following properties:

1. An experiment has k possible outcomes, say $A_1, A_2, \ldots, A_k$.
2. The experiment is repeated n times.
3. The n trials are independent—i.e., the outcome of one trial has no effect on the outcome of any of the other trials.
4. On each trial, p_i is the probability that the outcome of the experiment is A_i, $i = 1, 2, \ldots, k$—that is, $p_i = P(A_i)$. The probabilities remain the same from trial to trial.
5. The sum of the probabilities is

$$p_1 + p_2 + \cdots + p_k = 1$$

6. The random variables $X_1, X_2, \ldots, X_k$ count the number of times that $A_1, A_2, \ldots, A_k$ are observed, respectively. We use the lower case x_i to give the actual number of times that A_i occurs.

The **multinomial probability function** is defined as

MULTINOMIAL PROBABILITY FUNCTION

$$f(x_1, x_2, \ldots, x_k) = P(X_1 = x_1, X_2 = x_2, \ldots, X_k = x_k)$$

$$= \frac{n!}{x_1!x_2!\cdots x_k!} p_1^{x_1} p_2^{x_2} \cdots p_k^{x_k} \quad \textbf{(9.7)}$$

where $x_1, x_2, \ldots, x_k$ are nonnegative integers with $x_1 + x_2 + \cdots + x_k = n$, and we can say that $X_1, X_2, \ldots, X_k$ have a **multinomial distribution.**

Notice that the multinomial distribution is an extension of the binomial distribution, for which there are only two possible outcomes: $A_1 = \{\text{success}\}$ and $A_2 = \{\text{failure}\}$, with $p_1 = p = P(\text{success})$ and $p_2 = 1 - p = P(\text{failure})$.

The respective expected values or means of $X_1, X_2, \ldots, X_k$ are

EXPECTED VALUES FOR MULTINOMIAL DISTRIBUTION

$$E(X_1) = \mu_1 = np_1,\ E(X_2) = \mu_2 = np_2, \ldots, E(X_k) = \mu_k = np_k \quad \textbf{(9.8)}$$

EXAMPLE 9-15 Roll a fair 4-sided die 8 times. Let $A_1 = \{1\}$ (the event that you'll roll a 1), $A_2 = \{2\}$, $A_3 = \{3\}$, and $A_4 = \{4\}$. **(a)** Give the expected values of X_1, X_2, X_3, and X_4, the expected number of times you'll roll a 1, 2, 3, or 4, respectively. **(b)** Find $P(X_1 = 2, X_2 = 2, X_3 = 2, X_4 = 2)$, the probability that you'll roll each number twice in the 8 rolls.

Solution Because the die is fair, $p_1 = p_2 = p_3 = p_4 = \frac{1}{4}$, and the number of trials is $n = 8$.

(a) From formula (9.8),

$$E(X_1) = \mu_1 = np_1, \quad E(X_2) = \mu_2 = np_2, \ldots, E(X_k) = \mu_k = np_k$$

$$E(X_1) = E(X_2) = E(X_3) = E(X_4) = 8\left(\frac{1}{4}\right) = 2$$

(b) From formula (9.7),

$$f(x_1, x_2, \ldots, x_k) = \frac{n!}{x_1!x_2!\cdots x_k!} p_1^{x_1} p_2^{x_2} \cdots p_k^{x_k}$$

$$f(2,2,2,2) = P(X_1 = 2, X_2 = 2, X_3 = 2, X_4 = 2)$$

$$= \frac{8!}{2!2!2!2!}\left(\frac{1}{4}\right)^2\left(\frac{1}{4}\right)^2\left(\frac{1}{4}\right)^2\left(\frac{1}{4}\right)^2 = 2520\left(\frac{1}{4}\right)^8 = .038$$

You can see that although you expect each number to come up twice in 8 rolls, the probability that each number will *actually* come up twice in 8 rolls is very small—.038.

EXAMPLE 9-16 A particular college classifies applicants according to their high-school grade point average (GPA). Let A_1 be those applicants with GPAs that are lower than 2.5, A_2 be those with GPAs higher than or equal to 2.5 and lower than 3.2, and A_3 be those with GPAs higher than or equal to 3.2.

Let $p_1 = .15$, $p_2 = .30$, and $p_3 = .55$. For 10 applications selected at random (**a**) define the probability function of X_1, X_2, X_3, (**b**) give the expected values of X_1, X_2, X_3, and (**c**) find $P(X_1 = 2, X_2 = 3, X_3 = 5)$.

Solution

(**a**) From (9.7),

$$f(x_1, x_2, x_3) = \frac{10!}{x_1!x_2!x_3!}(.15)^{x_1}(.30)^{x_2}(.55)^{x_3}, \quad x_1 + x_2 + x_3 = 10$$

(**b**) From (9.8),

$$E(X_1) = \mu_1 = 10(.15) = 1.5$$
$$E(X_2) = \mu_2 = 10(.30) = 3.0$$
$$E(X_3) = \mu_3 = 10(.55) = \underline{5.5}$$
$$10.0$$

(**c**) Again using (9.7), you have

$$\begin{aligned} P(X_1 = 2, X_2 = 3, X_3 = 5) &= f(2, 3, 5) \\ &= \frac{10!}{2!3!5!}(.15)^2(.30)^3(.55)^5 \\ &= 2520(.0225)(.0270)(.0503) \\ &= .0770 \end{aligned}$$

EXAMPLE 9-17 Roll a fair 8-sided die twice. Let event $A_1 = \{1\}$, event $A_2 = \{2, 3, 4\}$, and event $A_3 = \{5, 6, 7, 8\}$, so that $p_1 = P(A_1) = \frac{1}{8}$, $p_2 = P(A_2) = \frac{3}{8}$, $p_3 = P(A_3) = \frac{4}{8}$. Let X_1, X_2, X_3 denote the number of times that A_1, A_2, A_3 will be observed on the two rolls. Define the probabilities for all the possible combinations of outcomes.

Solution There are 6 possible outcomes. Either each roll results in a different outcome (A_1 and A_2, or A_1 and A_3, or A_2 and A_3) or both rolls result in 2 A_1's or 2 A_2's or 2 A_3's. The probabilities for these 6 outcomes are

$$f(1, 1, 0) = \frac{2!}{1!1!0!}\left(\frac{1}{8}\right)^1\left(\frac{3}{8}\right)^1\left(\frac{4}{8}\right)^0 = \frac{6}{64}$$

$$f(1, 0, 1) = \frac{2!}{1!0!1!}\left(\frac{1}{8}\right)^1\left(\frac{3}{8}\right)^0\left(\frac{4}{8}\right)^1 = \frac{8}{64}$$

$$f(0, 1, 1) = \frac{2!}{0!1!1!}\left(\frac{1}{8}\right)^0\left(\frac{3}{8}\right)^1\left(\frac{4}{8}\right)^1 = \frac{24}{64}$$

$$f(2, 0, 0) = \frac{2!}{2!0!0!}\left(\frac{1}{8}\right)^2\left(\frac{3}{8}\right)^0\left(\frac{4}{8}\right)^0 = \frac{1}{64}$$

$$f(0, 2, 0) = \frac{2!}{0!2!0!}\left(\frac{1}{8}\right)^0\left(\frac{3}{8}\right)^2\left(\frac{4}{8}\right)^0 = \frac{9}{64}$$

$$f(0, 0, 2) = \frac{2!}{0!0!2!}\left(\frac{1}{8}\right)^0\left(\frac{3}{8}\right)^0\left(\frac{4}{8}\right)^2 = \frac{16}{64}$$

$$\frac{64}{64}$$

EXAMPLE 9-18 A manufacturer puts out fifty-piece packages of candy, which contain an assortment of brown, yellow, orange, tan, and green candies. The company produces the different colors of candies in the following proportions, respectively: .4, .2, .2, .1, .1. Because a large number of pieces of candy are produced, we may assume that the respective probabilities of selecting these 5 colors are $p_1 = .4$, $p_2 = .2$, $p_3 = .2$, $p_4 = .1$, $p_5 = .1$. Let X_1, X_2, X_3, X_4, X_5 be the number of pieces of each color of candy in a package that's selected at random. Give the expected values of X_1, X_2, X_3, X_4, X_5.

Solution By formula (9.8),

$$E(X_1) = \mu_1 = 50(.4) = 20$$
$$E(X_2) = \mu_2 = 50(.2) = 10$$
$$E(X_3) = \mu_3 = 50(.2) = 10$$
$$E(X_4) = \mu_4 = 50(.1) = 5$$
$$E(X_5) = \mu_5 = 50(.1) = \underline{5}$$
$$50$$

SUMMARY

1. If a random sample of r objects is taken without replacement from a collection of n objects consisting of n_1 objects of one kind and n_2 objects of another kind so that $n = n_1 + n_2$, and the random variable X is equal to the number of objects of the first kind in the sample, then X has a *hypergeometric distribution.*
2. The probability function for a hypergeometric distribution is

$$f(x) = P(X = x) = \frac{\binom{n_1}{x}\binom{n_2}{r-x}}{\binom{n}{r}}, \quad x = 0, 1, \ldots, r$$

3. A *Poisson random variable* counts such occurrences as the number of customer arrivals per minute at a ticket window, or the number of alpha particle emissions per second from a radioactive source.
4. The probability function for a Poisson distribution is

$$f(x) = \frac{\lambda^x e^{-\lambda}}{x!}, \quad x = 0, 1, 2, \ldots$$

5. For a Poisson distribution, $\mu = \lambda$ and $\sigma^2 = \lambda$, where λ is the parameter for the particular distribution.
6. The Poisson distribution can be used to approximate binomial probabilities when n is large and p is small.
7. The *multinomial distribution* is an extension of the binomial distribution.
8. For a multinomial experiment, there are k possible outcomes, and n independent trials are observed.
9. For a multinomial experiment, the probabilities remain the same from trial to trial.
10. If $X_1, X_2, \ldots, X_k$ have a multinomial distribution, $E(X_i) = \mu_i = np_i$ for $i = 1, 2, \ldots, k$.
11. The probability function for a multinomial distribution is

$$f(x_1, x_2, \ldots, x_k) = P(X_1 = x_1, X_2 = x_2, \ldots, X_k = x_k) = \frac{n!}{x_1!x_2!\cdots x_k!} p_1^{x_1} p_2^{x_2} \cdots p_k^{x_k}$$

RAISE YOUR GRADES

Can you . . . ?

☑ describe a hypergeometric distribution
☑ define the probability function for the hypergeometric distribution
☑ calculate hypergeometric probabilities
☑ find Poisson probabilities using Table 3 in the Appendix
☑ give the relationship between the mean and variance for the Poisson distribution
☑ approximate binomial probabilities using the Poisson distribution
☑ define the probability function for the multinomial distribution
☑ find the expected values of multinomial random variables

RAPID REVIEW

1. If 13 cards are selected at random and without replacement out of 52 cards, the probability that 6 are red and 7 are black is ______. (Don't simplify.)
2. If 13 cards are selected at random and without replacement out of 52 cards, the probability of selecting 2 clubs, 5 diamonds, 4 hearts, and 2 spades is ______. (Don't simplify.)
3. If 5 cards are selected at random and without replacement out of 52 cards, the probability of selecting exactly 2 hearts is ______. (Don't simplify.)
4. Let X have a Poisson distribution with mean $\lambda = 1.37$. Using the probability function of X, $P(X = 2) =$ ______. (Don't simplify.)
5. If X has a Poisson distribution with mean $\lambda = 6.4$, using Table 3 in the Appendix, you find $P(X = 3) =$ ______.
6. If X has a Poisson distribution with mean λ equal to 8.7, then the variance of X is ______.
7. If X has a Poisson distribution with $\lambda = 2.3$, then $P(X \geq 1) =$ ______.
8. If X has a binomial distribution with $n = 5000$ and $p = .002$, the Poisson distribution that can be used to approximate probabilities for X has mean $\lambda =$ ______.
9. In a multinomial experiment, if $n = 18$, $p_1 = \frac{1}{6}$, $p_2 = \frac{2}{6}$, $p_3 = \frac{3}{6}$, then $P(X_1 = 3,\ X_2 = 6,\ X_3 = 9) =$ ______. (Don't simplify.)
10. In a multinomial experiment, if $n = 18$, $p_1 = \frac{1}{6}$, $p_2 = \frac{2}{6}$, $p_3 = \frac{3}{6}$, then $E(X_1) =$ **(a)** ______, $E(X_2) =$ **(b)** ______, and $E(X_3) =$ **(c)** ______.
11. For a multinomial experiment, the trials must be ______.

Answers **(1)** $\dfrac{\binom{26}{6}\binom{26}{7}}{\binom{52}{13}}$ **(2)** $\dfrac{\binom{13}{2}\binom{13}{5}\binom{13}{4}\binom{13}{2}}{\binom{52}{13}}$ **(3)** $\dfrac{\binom{13}{2}\binom{39}{3}}{\binom{52}{5}}$ **(4)** $\dfrac{(1.37)^2 e^{-1.37}}{2!}$

(5) .0726 **(6)** 8.7 **(7)** .8997 **(8)** 10 **(9)** $\dfrac{18!}{3!6!9!}\left(\dfrac{1}{6}\right)^3\left(\dfrac{2}{6}\right)^6\left(\dfrac{3}{6}\right)^9$

(10) **(a)** 3 **(b)** 6 **(c)** 9 **(11)** independent

SOLVED PROBLEMS

Hypergeometric Distribution

PROBLEM 9-1 The cook in a restaurant stashes away a tub containing 15 oysters because he knows that there are pearls in 9 of the oysters. A busboy who also knows about the pearls finds the tub, but can make off with only 4 of the oysters before someone sees him. If you let X be the number of oysters that contain a pearl out of those the busboy has, **(a)** what's the probability function of X? **(b)** What's the probability that he has 0, 1, 2, 3, or 4 oysters with a pearl?

Solution

(a) You have a collection of $n = 15$ oysters ($n_1 = 9$ oysters with pearls and $n_2 = 6$ oysters without) and you have a sample of $r = 4$ oysters selected at random without replacement. Use the hypergeometric probability function (formula 9.1) to define the probability function of X.

$$f(x) = P(X = x) = \frac{\binom{n_1}{x}\binom{n_2}{r - x}}{\binom{n}{r}}, \quad x = 0, 1, \ldots, r$$

$$= \frac{\binom{9}{x}\binom{6}{4 - x}}{\binom{15}{4}}, \quad x = 0, 1, 2, 3, 4$$

(b) From Table 1 in the Appendix you find that $\binom{15}{4} = 1365$, and you can then complete the probabilities listed in the tabulation as shown.

x	$f(x)$
0	$\binom{9}{0}\binom{6}{4}/1365 = 15/1365 = .011$
1	$\binom{9}{1}\binom{6}{3}/1365 = 180/1365 = .132$
2	$\binom{9}{2}\binom{6}{2}/1365 = 540/1365 = .396$
3	$\binom{9}{3}\binom{6}{1}/1365 = 504/1365 = .369$
4	$\binom{9}{4}\binom{6}{0}/1365 = 126/1365 = .092$
	$1365/1365 \quad 1.000$

PROBLEM 9-2 Find **(a)** the mean μ and **(b)** the variance σ^2 for the random variable X in Problem 9-1.

Solution

(a) Use formula (9.2) with $n_1 = 9$, $n_2 = 6$, $n = 15$, and $r = 4$.

$$\mu = \frac{rn_1}{n} = \frac{4 \cdot 9}{15} = \frac{36}{15} = 2.4$$

(b) Use formula (9.3) to find

$$\sigma^2 = \frac{rn_1n_2(n-r)}{n^2(n-1)} = \frac{4 \cdot 9 \cdot 6 \cdot 11}{15 \cdot 15 \cdot 14} = \frac{2376}{3150} = .754$$

PROBLEM 9-3 In his freezer Simon has 12 pizzas that he forgot to label. He only remembers that 5 are pepperoni and 7 are sausage. When a few of his friends drop over, Simon decides to serve them pizza. **(a)** If he selects 3 pizzas at random, what's the probability that they're all pepperoni? **(b)** What's the expected number (mean number) of pepperoni pizzas in a sample of size 3?

Solution If you let X equal the number of pepperoni pizzas in the sample, then X has a hypergeometric distribution with $n_1 = 5$, $n_2 = 7$, $n = 12$, and $r = 3$.

(a) From (9.1),

$$P(X = 3) = \frac{\binom{5}{3}\binom{7}{0}}{\binom{12}{3}} = \frac{10}{220} = .045$$

(b) Use (9.2) to find the mean of X.

$$\mu = \frac{3 \cdot 5}{12} = \frac{15}{12} = 1.25$$

PROBLEM 9-4 From the data in Problem 9-3, **(a)** find the probability that all 3 of the pizzas Simon selects are sausage. **(b)** What's the expected number of sausage pizzas in a sample of size 3?

Solution This time let X equal the number of sausage pizzas in the sample of size 3. Then $n_1 = 7$, $n_2 = 5$, $n = 12$, and $r = 3$.

(a) $$f(3) = \frac{\binom{5}{0}\binom{7}{3}}{\binom{12}{3}} = \frac{35}{220} = .159$$

(b) $$\mu = \frac{3 \cdot 7}{12} = 1.75$$

PROBLEM 9-5 Meg has 6 blue, 3 yellow, 4 red, and 7 green marbles in a bag. She wants to give her cousin Charlie 2 marbles of each color. If she reaches in the bag and takes out 8 marbles without looking and without replacing any, what's the probability that she'll select 2 marbles of each color?

Solution Extend the hypergeometric distribution formula (9.4) so that

$$f(2, 2, 2) = \frac{\binom{6}{2}\binom{3}{2}\binom{4}{2}\binom{7}{2}}{\binom{20}{8}} = \frac{15 \cdot 3 \cdot 6 \cdot 21}{125{,}970} = \frac{5670}{125{,}970} = .045$$

Poisson Distribution

PROBLEM 9-6 Let X have a Poisson distribution with $\lambda = 4.5$. Give **(a)** the mean μ and **(b)** the variance σ^2 of X. Find **(c)** $P(X \leq 2)$ and **(d)** $P(X > 2)$.

Solution

(a) From formula (9.6) you know that $\mu = \lambda$, so $\mu = \lambda = 4.5$.

(b) From formula (9.6) you also know that $\sigma^2 = \lambda$, so $\sigma^2 = \lambda = 4.5$.

(c) You can use Table 3 in the Appendix to figure $P(X \leq 2) = .0111 + .0500 + .1125 = .1736$.

(d) Use the answer to part **(c)** and the formula for the complement of an event (6.3) to find $P(X > 2) = 1 - P(X \leq 2) = 1 - .1736 = .8264$.

PROBLEM 9-7 Let X equal the number of alpha-particle emissions of carbon-15 in .5 seconds as recorded by a Geiger counter. Assume that X has a Poisson distribution with mean $\lambda = 8.2$, and find **(a)** $P(X \leq 8)$, **(b)** $P(X > 8)$.

Solution

(a) Use Table 3 in the Appendix to find

$$P(X \leq 8) = .0003 + .0023 + .0092 + .0252 + .0517 + .0849 + .1160 + .1358 + .1392 = .5646$$

(b) Then use the answer to part **(a)** and formula (6.3) to find

$$P(X > 8) = 1 - P(X \leq 8) = 1 - .5646 = .4354$$

PROBLEM 9-8 Tabulated here are fifty observations of the random variable X described in Problem 9-7. Find **(a)** the sample mean $\bar{x}$ and **(b)** the sample variance s^2. **(c)** Compare the relative frequency of the observations less than or equal to 8 with $P(X \leq 8)$ in Problem 9-7.

9	6	4	9	11	8	10	11	13	8
7	4	5	8	6	5	8	7	6	3
4	6	4	12	7	14	5	9	8	13
7	10	9	5	6	12	10	7	12	13
7	8	8	9	11	14	4	13	9	6

Solution To find $\bar{x}$ and s^2, construct a table as shown.

(a) Then from the table and formula (2.2),

$$\bar{x} = \frac{\sum fx}{n} = \frac{410}{50} = 8.2$$

(b) And from formula (3.5),

$$s^2 = \frac{n\sum fx^2 - (\sum fx)^2}{n(n-1)}$$

$$= \frac{50(3796) - (410)^2}{50(49)}$$

$$= \frac{21{,}700}{2450} = 8.857$$

x	f	fx	fx^2
3	1	3	9
4	5	20	80
5	4	20	100
6	6	36	216
7	6	42	294
8	7	56	448
9	6	54	486
10	3	30	300
11	3	33	363
12	3	36	432
13	4	52	676
14	2	28	392
	50	410	3796

(c) The proportion of observations less than or equal to 8 is $29/50 = .58$, which is very close to $P(X \leq 8) = .5646$.

PROBLEM 9-9 Let X equal the number of customers who arrive per minute at Lawrence's checkout lane. Assume that X has a Poisson distribution with $\lambda = .3$, which means that on the average 3 customers arrive every 10 minutes. **(a)** Define the probability function of X. Find **(b)** $P(X = 0)$, and **(c)** $P(X \geq 1)$.

Solution

(a) Use formula (9.5) to define the probability function:

$$f(x) = \frac{\lambda^x e^{-\lambda}}{x!} = \frac{.3^x e^{-.3}}{x!}, \quad x = 0, 1, 2, \ldots$$

(b) Substitute $X = 0$ into the Poisson probability function you've just defined.

$$P(X = 0) = \frac{(.3)^0 e^{-.3}}{0!} = e^{-.3} = .7408$$

You could also get this answer from Table 3 in the Appendix.

(c) Use the answer to part **(b)** and formula (6.3) to find

$$P(X \geq 1) = 1 - P(X = 0) = 1 - .7408 = .2592$$

PROBLEM 9-10 There is a rare type of heredity change in *E. coli* that causes this bacterium to become resistant to the drug streptomycin. This type of change, called a mutation, can be detected by plating the bacteria on Petri dishes containing streptomycin. Any colonies that grow on this medium result from a single mutant cell. One hundred fifty Petri dishes of streptomycin agar were each plated with 10^6 bacteria, and after 3 days the number of colonies on each dish was counted. The observed results were as follows: 98 plates had 0 colonies, 40 plates had 1 colony, 8 plates had 2 colonies, 3 plates had 3 colonies, and 1 plate had 4 colonies. Let X equal the number of colonies per plate, and assume that X has a Poisson distribution. Find **(a)** the sample mean $\bar{x}$ and **(b)** the sample variance s^2 of these 150 observations of X. **(c)** Let $\lambda = x$, and calculate $P(X = 0)$, $P(X = 1)$, $P(X = 2)$, $P(X = 3)$, $P(X \geq 4)$. **(d)** Compare the probabilities for the 5 events you've found with the relative frequencies of these 5 events. **(e)** Is the assumption that X has a Poisson distribution valid?

Solution To find $\bar{x}$ and s^2, construct a table as shown.

x	f	fx	fx^2
0	98	0	0
1	40	40	40
2	8	16	32
3	3	9	27
4	1	4	16
	150	69	115

(a) $\bar{x} = \dfrac{69}{150} = .46$

(b) $s^2 = \dfrac{150(115) - (69)^2}{150(149)} = \dfrac{12{,}489}{22{,}350} = .559$

(c) Because $\lambda = .46$ is not in the Poisson probability table in the Appendix, you must calculate these probabilities from formula (9.5).

$$P(X = 0) = \frac{(.46)^0 e^{-.46}}{0!} = e^{-.46} = .6313$$

$$P(X = 1) = \frac{(.46)^1 e^{-.46}}{1!} = .2904$$

$$P(X = 2) = \frac{(.46)^2 e^{-.46}}{2!} = .0668$$

$$P(X = 3) = \frac{(.46)^3 e^{-.46}}{3!} = .0102$$

$$P(X \geq 4) = 1 - [P(X = 0) + P(X = 1) + P(X = 2) + P(X = 3)] = 1 - .9987 = .0013$$

(d) You can construct a table as shown to compare the probabilities with the relative frequencies (see Section 1-1B).

x	$P(X = x)$	Relative frequency
0	.6313	98/150 = .6533
1	.2904	40/150 = .2667
2	.0668	8/150 = .0533
3	.0102	3/150 = .0200
≥4	.0013	1/150 = .0067

(e) Because $\bar{x}$ and s^2 are approximately equal, and because the probabilities and relative frequencies for the 5 given outcomes agree quite well, you can assume that X has a Poisson distribution. In Section 20-4 we'll give a statistical test for answering questions such as this one.

PROBLEM 9-11 In a small manufacturing firm, the proportion of defective parts produced by one particular machine is .005. Let X equal the number of defective parts among 1000 that an assembler selects at random from the production line. **(a)** How is X distributed? **(b)** Use the Poisson distribution to approximate $P(X \leq 3)$.

Solution

(a) Each part is either defective (success, for this problem) or good (failure), so that X has a binomial distribution with $n = 1000$ and $p = .005$—X is $b(1000, .005)$.

(b) Since n is large and p is small, you can use the Poisson distribution with $\lambda = np = 1000(.005) = 5$ to approximate the probabilities. It's easiest to look up the probabilities in Table 3 in the Appendix:

$$P(X \leq 3) \approx .0067 + .0337 + .0842 + .1404 = .2650$$

Multinomial Distribution

PROBLEM 9-12 Roll a fair 6-sided die 12 times. Let event $A_1 = \{1, 2, 3\}$, $A_2 = \{4\}$, and event $A_3 = \{5, 6\}$. Let X_1, X_2, X_3 denote, respectively, the number of times events A_1, A_2, A_3 are observed on the 12 rolls. **(a)** Define the probability function of X_1, X_2, X_3. **(b)** Give the values of μ_1, μ_2, and μ_3. **(c)** Find $P(X_1 = 6, X_2 = 2, X_3 = 4)$.

Solution

(a) Use formula (9.7) with $n = 12$, $p_1 = 3/6$, $p_2 = 1/6$, and $p_3 = 2/6$, to define

$$f(x_1, x_2, \ldots, x_k) = \frac{n!}{x_1! x_2! \cdots x_k!} p_1^{x_1} p_2^{x_2} \cdots p_k^{x_k}$$

$$= \frac{12!}{x_1! x_2! x_3!} \left(\frac{3}{6}\right)^{x_1} \left(\frac{1}{6}\right)^{x_2} \left(\frac{2}{6}\right)^{x_3}, \quad x_1 + x_2 + x_3 = 12$$

(b) You can use formula (9.8) to find the expected values or means:

$$E(X_1) = \mu_1 = np_1, E(X_2) = \mu_2 = np_2, \ldots, E(X_k) = \mu_k = np_k$$

$$\mu_1 = E(X_1) = 12\left(\frac{3}{6}\right) = 6$$

$$\mu_2 = E(X_2) = 12\left(\frac{1}{6}\right) = 2$$

$$\mu_3 = E(X_3) = 12\left(\frac{2}{6}\right) = 4$$

(c) $P(X_1 = 6, X_2 = 2, X_3 = 4) = f(6, 2, 4) = \dfrac{12!}{6!2!4!} \left(\dfrac{3}{6}\right)^6 \left(\dfrac{1}{6}\right)^2 \left(\dfrac{2}{6}\right)^4 = .074$

PROBLEM 9-13 A geneticist crossed two prolific red-eyed fruit flies that produced 3200 progeny. Let X_1, X_2, X_3, and X_4 equal the observed number of red-eyed, brown-eyed, scarlet-eyed, and white-eyed offspring, respectively, and let $p_1 = P(\text{red eyes})$, $p_2 = P(\text{brown eyes})$, $p_3 = P(\text{scarlet eyes})$, and $p_4 = P(\text{white eyes})$. If you assume that $p_1 = 9/16$, $p_2 = p_3 = 3/16$, and $p_4 = 1/16$, what are the expected values of X_1, X_2, X_3, and X_4?

Solution By formula (9.8),

$$\mu_1 = E(X_1) = 3200\left(\frac{9}{16}\right) = 1800$$

$$\mu_2 = E(X_2) = 3200\left(\frac{3}{16}\right) = 600$$

$$\mu_3 = E(X_3) = 3200\left(\frac{3}{16}\right) = 600$$

$$\mu_4 = E(X_4) = 3200\left(\frac{1}{16}\right) = 200$$

PROBLEM 9-14 At the end of a vigorous health/fitness program, the instructor measured the percentage of body fat (pbf) of each of the first-year students who participated. Let $A_1 = \{\text{pbf} \le 10\}$, $A_2 = \{10 < \text{pbf} \le 20\}$, $A_3 = \{20 < \text{pbf} \le 30\}$, and $A_4 = \{30 < \text{pbf}\}$. Select 10 students at random, and let X_1, X_2, X_3, and X_4 equal the number of students who belong to A_1, A_2, A_3, and A_4, respectively. Assume that $p_1 = .12$, $p_2 = .44$, $p_3 = .38$, and $p_4 = .06$. **(a)** Define the probability function of X_1, X_2, X_3, and X_4. **(b)** Give the values of μ_1, μ_2, μ_3, and μ_4. **(c)** Find $P(X_1 = 2, X_2 = 4, X_3 = 3, X_4 = 1)$.

Solution

(a) $$f(x_1, x_2, x_3, x_4) = \frac{10!}{x_1!x_2!x_3!x_4!}(.12)^{x_1}(.44)^{x_2}(.38)^{x_3}(.06)^{x_4}, \quad x_1 + x_2 + x_3 + x_4 = 10$$

(b) $\mu_1 = E(X_1) = 10(.12) = 1.2 \qquad \mu_3 = E(X_3) = 10(.38) = 3.8$

$\mu_2 = E(X_2) = 10(.44) = 4.4 \qquad \mu_4 = E(X_4) = 10(.06) = .6$

(c) $f(2, 4, 3, 1) = P(X_1 = 2, X_2 = 4, X_3 = 3, X_4 = 1)$

$$= \frac{10!}{2!4!3!1!}(.12)^2(.44)^4(.38)^3(.06)^1 = .0224$$

PROBLEM 9-15 In a city we won't name, there are 5 banks, let's call them B_1, B_2, B_3, B_4, and B_5. Suppose that the proportions of people who claim that bank B_i is their bank, $i = 1, 2, 3, 4, 5$, are $p_1 = .25$, $p_2 = .31$, $p_3 = .06$, $p_4 = .27$, and $p_5 = .11$. We take a random sample of 200 people. For $i = 1, 2, 3, 4, 5$, let X_i equal the number of people who claim bank B_i as their bank. **(a)** Find $P(X_1 = 50, X_2 = 60, X_3 = 10, X_4 = 55, X_5 = 25)$. (Don't simplify.) **(b)** Give the means of X_1, X_2, X_3, X_4, and X_5.

Solution

(a) $$f(50, 60, 10, 55, 25) = \frac{200!}{50!60!10!55!25!}(.25)^{50}(.31)^{60}(.06)^{10}(.27)^{55}(.11)^{25}$$

(b) $\mu_1 = 200(.25) = 50$

$\mu_2 = 200(.31) = 62$

$\mu_3 = 200(.06) = 12$

$\mu_4 = 200(.27) = 54$

$\mu_5 = 200(.11) = 22$

Supplementary Problems

PROBLEM 9-16 Amy's golf bag contains 11 yellow golf balls and 7 orange ones. She reaches into the bag and pulls out 5 balls at random. Let X equal the number of yellow balls that she selects. Find **(a)** $P(X = 3)$, **(b)** the mean of X, and **(c)** the variance of X.

Answer **(a)** .4044 **(b)** $\mu = 3.056$ **(c)** $\sigma^2 = .9087$

PROBLEM 9-17 Herb buys 6 red and 9 pink rose bushes from a nursery. He selects 5 bushes at random to plant in his front yard. Let X equal the number of red rose bushes out of the 5 he selects. Find **(a)** $P(X = 2)$, **(b)** the mean of X, and **(c)** the variance of X.

Answer **(a)** .4196 **(b)** 2 **(c)** .8571

PROBLEM 9-18 A bag in the locker room of a country club contains 20 golf balls: 9 white, 6 yellow, and 5 orange. If a golfer in a hurry reaches into the bag and grabs 6 balls at random, what's the probability that he'll get 2 balls of each color?

Answer .1393

PROBLEM 9-19 If X has a Poisson distribution with mean $\lambda = 4.2$, find $P(X \leq 3)$.

Answer .3955

PROBLEM 9-20 Let X equal the number of telephone calls per minute that enter a particular switchboard. Assume that X has a Poisson distribution with mean 2.8. **(a)** Give the variance of X. **(b)** Find $P(X > 1)$.

Answer **(a)** 2.8 **(b)** .7689

PROBLEM 9-21 The probability for winning in the Michigan Daily Lottery with a 3-way boxed bet is .003. Let X equal the number of winning tickets if 500 tickets are purchased on 500 different days. **(a)** How is X distributed? Use the Poisson distribution to approximate **(b)** $P(X = 0)$ and **(c)** $P(X \geq 1)$.

Answer **(a)** $b(500, .003)$ **(b)** .2231 **(c)** .7769

PROBLEM 9-22 A random sample of size 9 is selected from an urn that contains 10 orange, 20 white, and 30 blue balls. Give the probability that 3 balls of each color will be selected if the sampling is done **(a)** without replacement and **(b)** with replacement. (Don't simplify.)

Answer **(a)** $\dfrac{\binom{10}{3}\binom{20}{3}\binom{30}{3}}{\binom{60}{9}}$ **(b)** $\dfrac{9!}{3!3!3!}\left(\dfrac{1}{6}\right)^3\left(\dfrac{2}{6}\right)^3\left(\dfrac{3}{6}\right)^3$

PROBLEM 9-23 Roll a 12-sided die 6 times and let $A_1 = \{1, 2\}$, $A_2 = \{3, 4, 5, 6\}$, $A_3 = \{7, 8, 9, 10, 11, 12\}$. Let X_1, X_2, X_3 equal the number of times that A_1, A_2, A_3 is observed, respectively. Find **(a)** μ_1, **(b)** μ_2, **(c)** μ_3, **(d)** $P(X_1 = 1, X_2 = 2, X_3 = 3)$.

Answer **(a)** 1 **(b)** 2 **(c)** 3 **(d)** .1389

PROBLEM 9-24 The Crackly Crunch Cereal Company randomly puts one of 4 different prizes into each box of cereal. If you purchase 8 boxes of this cereal, what's the probability that you will get 2 of each of the 4 prizes?

Answer .0385

10 THE NORMAL DISTRIBUTION

THIS CHAPTER IS ABOUT

- ☑ **Continuous Distributions**
- ☑ **The Standard Normal Distribution**
- ☑ **Other Normal Distributions**

10-1. Continuous Distributions

In this chapter we'll consider experiments for which the set of possible outcomes is a continuous set of points in an interval or a union of intervals.

A. Continuous random variables

If an experiment measures some quantity and its sample space is an interval or a union of intervals, we call the experiment a **continuous-type experiment**. When we let a random variable equal the outcome of such an experiment, it's called a **continuous random variable** or a **random variable of the continuous type**. If, for example, we measure the heights of college students, the pounds of milk produced by a cow per day, or the lengths of time between customer arrivals at a supermarket checkout lane, we're performing a continuous-type experiment.

EXAMPLE 10-1 Let t denote the time a female runner, selected at random, needs to complete a 25-km race. If the times from a random sample of 73 women ranged from 92.30 to 204.10 minutes (see Problem 4-21), what is the sample space?

Solution You want to set the outer limits of the sample space so that all the possible outcomes are included and yet the interval isn't too wide. To include all the possible times, you could let the sample space be the set of times between 80 and 250 minutes, but suppose you choose the set between 50 and 300 minutes. The times near the ends of this interval would not be very likely, and you must be ready to change the sample space if the experiment results in times that are shorter or longer than the range you've selected. Let the sample space for this experiment be

$$S = \{t: 50 < t < 300\}$$

and read it as the set of times t that are between 50 and 300 minutes. You would denote the associated random variable by T, which will take a specific value t in this interval for each runner you select. (Remember that you denote the random variable itself by a capital letter and any particular value that you observe by a lowercase letter.)

EXAMPLE 10-2 Let w equal the weight of apples in a "three-pound" bag of apples selected at random. A random sample of 24 bags weighed from 3.02 to 3.62 pounds (see Example 1-4). Describe the sample space.

Solution To include all possible values, you might define S by

$$S = \{w: 2.9 < w < 3.9\}$$

so that S equals the set of weights w that are between 2.9 and 3.9. You would denote the associated random variable by W.

note: If you used a different produce distributor, you might have to define the sample space differently, because some bags might weigh less than 2.9 pounds or more than 3.9 pounds.)

• ***remember:*** There is not just one way to define S. The only restriction on the sample space is that all known observations must fall within the intervals.

note: Continuous data are often rounded off so that they look like discrete data. But we still consider an experiment to be a continuous-type experiment if the associated random variables can assume any number in an interval of numbers—such as variables representing height, weight, or length—even though the data produced *appear* to be integers.

B. Probability density function

A **probability density function**, abbreviated **pdf**, is a function $f(x)$ that can be used to find probabilities for any interval of events associated with a continuous random variable X. In order to find the probability that X lies within a certain interval, we can graph the pdf of all possible outcomes for a particular experiment and then find the area under the pdf that lies above that interval. It's the area under the pdf that gives us the probability value. (Finding areas is often difficult, so there are tables in most statistics books that give the probabilities for many types of continuous random variables.)

The pdf defines the probabilities for all possible outcomes of a continuous random variable, so the "sum" of these probabilities must be 1. Since it's the area that gives the probability values, this means that the area under the entire pdf must also be 1.

EXAMPLE 10-3 Let the random variable X equal a number selected at random from the interval $(0, 1)$. (Table 9 in the Appendix is a table of random numbers that you can use to simulate just such a random selection from the interval $(0, 1)$, or from other intervals.) The pdf of X is defined by

$$f(x) = 1, \quad 0 \le x \le 1$$

We say that X comes from a **uniform distribution** since the probability is uniformly distributed over the interval from 0 to 1. Notice that for a uniform distribution, the value of $f(x)$ is the constant 1 for all values of x between 0 and 1. Find **(a)** $P(0 \le X \le .35)$ and **(b)** $P(.6 \le X \le .9)$, and show the corresponding area under the pdf for each.

Solution On a graph, mark both the x- and y-axes with appropriate intervals between 0 and 1. The pdf $f(x) = 1$ is drawn as a line segment between the points $(0, 1)$ and $(1, 1)$.

(a) The probability that X is between 0 and .35 is given by the area of the rectangle whose base is the interval $(0, .35)$ and height is 1 (see Fig. 10-1a). The area enclosed by this rectangle equals the probability that X lies between 0 and .35: $P(0 \le X \le .35)$. The area of a rectangle equals bh, so

$$P(0 \le X \le .35) = bh = (.35)(1) = .35$$

(b) The probability that X is between .6 and .9 is given by the area of the rectangle whose base is the interval $(.6, .9)$ and whose height is 1 (see Fig. 10-1b).

$$P(.6 \le X \le .9) = (.9 - .6)(1) = .3$$

note: You can see that the area under the entire pdf is

$$P(0 \le X \le 1) = bh = (1)(1) = 1$$

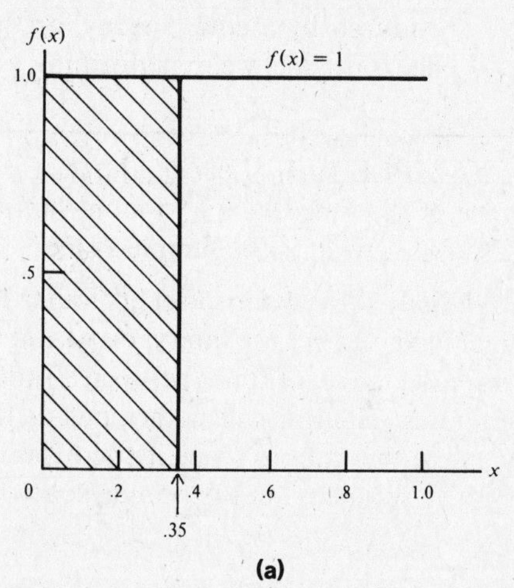

(a)

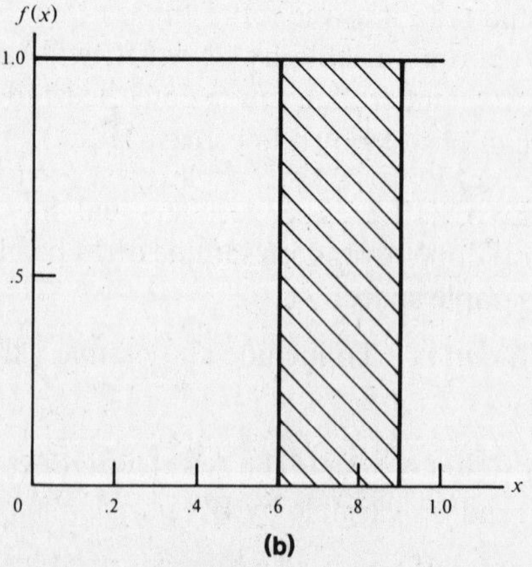

(b)

Figure 10-1

EXAMPLE 10-4 Let the pdf of X be defined by

$$f(x) = \frac{1}{2}x, \quad 0 \le x \le 2$$

Notice that this pdf is not uniform—the value of the probability function $f(x)$ depends on the value of x in the interval $(0, 2)$. Find **(a)** $P(0 \le X \le 1.4)$, **(b)** $P(1.4 < X \le 2)$, and **(c)** $P(1.0 \le X \le 1.4)$, and show for each of these the corresponding area under the pdf that relates to the probability.

Solution The pdf $f(x) = \frac{1}{2}x$ is a straight line with a slope of $m = \frac{1}{2}$. (Recall that the slope-intercept equation for a line is $y = mx + b$, with $m =$ the slope). Graph this line as x goes from 0 to 2.

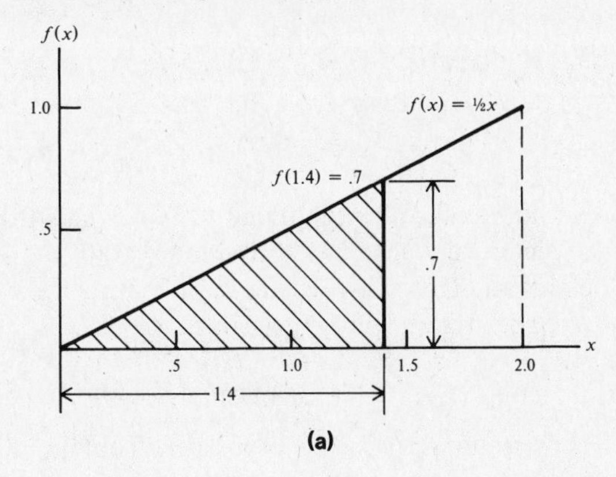

(a) To find $P(0 \le X \le 1.4)$, you need to find the area of the shaded triangle graphed in Fig. 10-2a, whose base is $(0, 1.4)$ and whose height is $f(1.4) = \frac{1}{2}(1.4) = .7$. Thus, since the area of a triangle is $\frac{1}{2}bh$,

$$P(0 \le X \le 1.4) = \frac{1}{2}(1.4)(.7) = .49$$

Notice that the area under the entire pdf curve is $\frac{1}{2}bh = \frac{1}{2}(2)(1) = 1$, as it must be to be a pdf.

(b) $P(1.4 < X \le 2)$ is the same as $P(X > 1.4)$, since X has probability between 0 and 2, only. Knowing that the area under the entire pdf is 1, you find that

$$P(X > 1.4) = 1 - P(0 \le X \le 1.4) = 1 - .49$$
$$= .51$$

This is the area that falls under the pdf between 1.4 and 2 and isn't shaded in Fig. 10-2a.

(c) To find $P(1.0 \le X \le 1.4)$, you need to find the area of the trapezoid shaded in Fig. 10-2b. The area of a trapezoid is given by $\frac{1}{2}$(sum of the parallel sides)(height) $= \frac{1}{2}(b + B)(h)$. Thus

$$P(1.0 \le X \le 1.4) = \tfrac{1}{2}[f(1.0) + f(1.4)] \times [1.4 - 1.0]$$
$$= \tfrac{1}{2}[.5 + .7][.4] = .24$$

You can also find this probability by noticing that this area is the difference between the area of the triangle with base $(0, 1.4)$ and the area of the triangle with base $(0, 1)$. From part **(a)**, you know that the area of the first triangle is .49. Similarly, the area of the second triangle is

$$\left(\frac{1}{2}\right)bh = \left(\frac{1}{2}\right)(1.0)[f(1.0)] = \left(\frac{1}{2}\right)(1.0)(.5) = .25$$

Thus $P(1.0 \le X \le 1.4) = .49 - .25 = .24$

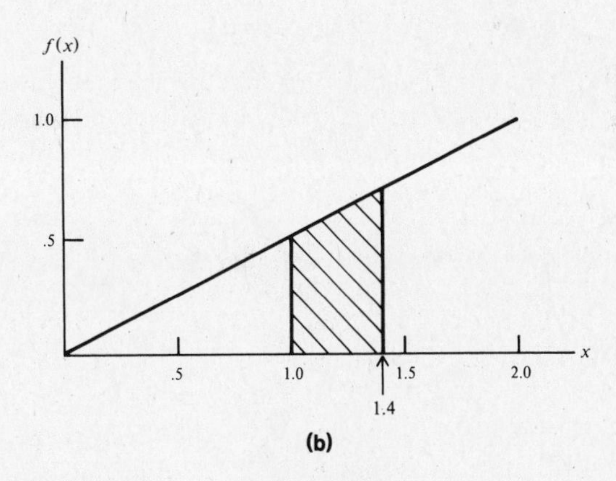

Figure 10-2

C. The distribution function

The distribution function for continuous random variables is defined the same way as it is for discrete random variables. The **distribution function of X** gives the cumulative probability up to and including x:

$$F(x) = P(X \le x)$$

Values of the distribution function for several types of continuous random variables at selected values of x are given in most statistics books. Some tables give "right tail" probabilities—$P(X > x)$—which are values of $1 - F(x)$ rather than $F(x)$. And one table gives values of $P(0 \le X \le x)$. We will illustrate the use of probability tables for continuous random variables in later sections.

EXAMPLE 10-5 Define and graph the distribution function for the uniform distribution given in Example 10-3.

Solution All of the probability lies between 0 and 1. Thus

$$F(x) = P(X \le x) = 0 \qquad \text{for } x < 0$$

For values of x between 0 and 1, $F(x)$ equals the area of the rectangle whose base is the interval $(0, x)$ and whose height is 1. Thus

$$F(x) = P(X \le x) = P(0 \le X \le x)$$
$$= (x)(1) = x \qquad \text{for } 0 \le x < 1$$

There is no probability beyond 1. That is, all of the probability is less than x if $x > 1$. Thus

$$F(x) = P(X \le x) = 1 \qquad \text{for } 1 \le x$$

The graph of this distribution function is given in Fig. 10-3.

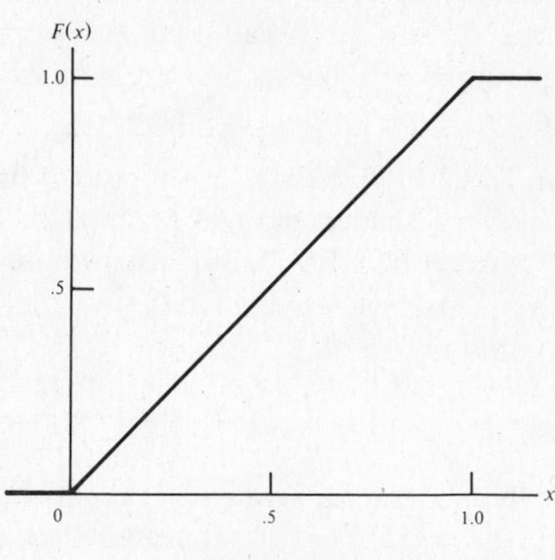

Figure 10-3

EXAMPLE 10-6 Define and graph the distribution function for the random variable defined in Example 10-4.

Solution All of the probability lies between 0 and 2. Thus

$$F(x) = P(X \le x) = 0 \qquad \text{for } x < 0$$

For values of x between 0 and 2, $F(x)$ is the area of the triangle whose base is the interval $(0, x)$ and whose height is $f(x) = \frac{1}{2}x$. Because the area of a triangle is $\frac{1}{2}bh$, you have

$$F(x) = P(X \le x) = \left(\frac{1}{2}\right)(x)f(x) = \left(\frac{1}{2}\right)(x)\left(\frac{x}{2}\right)$$
$$= \frac{x^2}{4} \qquad \text{for } 0 \le x < 2$$

Because all of the probability lies in the interval $(0, 2)$, the cumulative probability curve no longer increases after $x = 2$. Thus

$$F(x) = P(X \le x) = 1 \qquad \text{for } 2 \le x$$

The graph of this distribution function is given in Fig. 10-4.

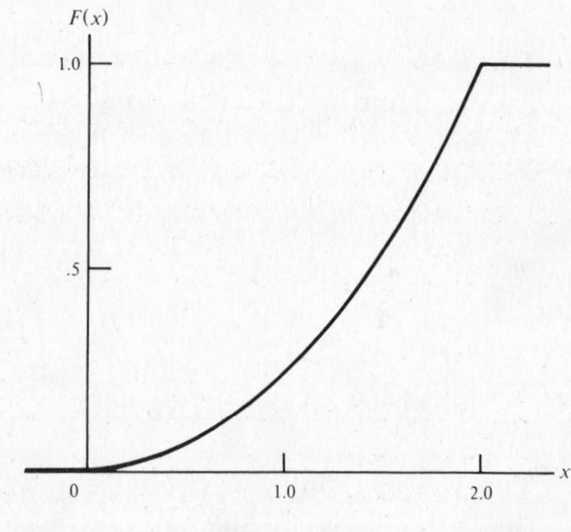

Figure 10-4

D. The mean, variance, and standard deviation of a continuous distribution

The mean and variance (standard deviation) give measures of the center (or balancing point) and the spread of the distribution of probability. For symmetric distributions, the mean is the center point. For continuous distributions, the mean and variance are calculated using integration—i.e., we can't do them without calculus. For completeness we will give the definitions but we won't use them for calculating means and variances in this book.

- The **mean of** X is defined by

$$\mu = E(X) = \int x f(x)\,dx$$

where $f(x)$ is the pdf of X.

- The **variance of** X is defined by

$$\sigma^2 = E[(X - \mu)^2] = \int (x - \mu)^2 f(x)\,dx$$

- The **standard deviation of** X is defined by

$$\sigma = \sqrt{\sigma^2}$$

EXAMPLE 10-7 For the uniform distribution in Example 10-3, the mean is $\mu = \frac{1}{2}$ and the variance is $\sigma^2 = \frac{1}{12}$. Illustrate this empirically by taking a sample of 20 random numbers from Table 9 in the Appendix and showing that $\bar{x}$ is close to .5 and s^2 is close to $\frac{1}{12} = .08333$.

Solution Use the last four blocks of 5 numbers in the first column:

.87098	.51870	.14247	.91298	.67642
.41632	.15764	.55692	.41541	.64886
.31103	.98353	.75329	.07829	.62589
.53808	.65320	.93711	.16041	.04206

For these 20 numbers,

$$\sum x = 10.39959$$

and

$$\sum x^2 = 7.09221$$

Thus $\bar{x} = \dfrac{10.39959}{20} = .51998$ and $s^2 = \dfrac{(20)(7.09221) - (10.39959)^2}{20(19)} = \dfrac{33.69273}{380} = .08867$

We see that $\bar{x}$ is close to .5 and s^2 is close to $\frac{1}{12} = .08333$.

note: In later chapters we will use the sample mean $\bar{x}$ to estimate an unknown distribution mean μ and the sample variance s^2 to estimate an unknown distribution variance σ^2. At that time we'll say more about the closeness of the sample mean to the distribution mean and the closeness of the sample variance to the distribution variance.

10-2. The Standard Normal Distribution

Probably the most important continuous distribution is the **normal distribution**, which is characterized by its "*bell-shaped*" *curve* (see Fig. 10-5). The mean is the middle value of this symmetrical distribution. (We'll see that approximately 68% of the probability is within one standard deviation of the mean.) Heights, weights, and SAT scores are typically normal random variables that exhibit this bell-shaped probability density function. A normal random variable is described by two parameters—its mean and its variance.

A particularly important normal random variable is the **standard normal random variable**, which is designated by the letter Z. This random variable is normally distributed with a mean μ of 0 and variance σ^2 of 1. We say that Z is $N(0, 1)$.

The pdf of Z is defined for all real numbers z $(-\infty < z < \infty)$ by

$$f(z) = \frac{1}{\sqrt{2\pi}} e^{-z^2/2}$$

A graph of this pdf is given in Fig. 10-5. Notice that the graph is symmetric about its mean, $z = 0$. The maximum value of this function is

$$f(0) = \frac{1}{\sqrt{2\pi}} = .3989$$

Because the variance is equal to 1, it's also true that the standard deviation is equal to 1: $\sigma = \sqrt{1} = 1$. From the graph you can see that most of the probability lies between -3 and $+3$. That is, most of the probability is within 3 standard deviations of the mean, 0.

When we're finding probabilities for the normal distribution, it's a good idea first to sketch a bell-shaped curve like the one in Fig. 10-5. Next, we shade in the region for which we're finding the area, i.e., the probability. (Recall that areas and probabilities are equivalent.) Then use Table 4 in the Appendix, which gives areas under the standard normal curve. The following examples explain how to use this table.

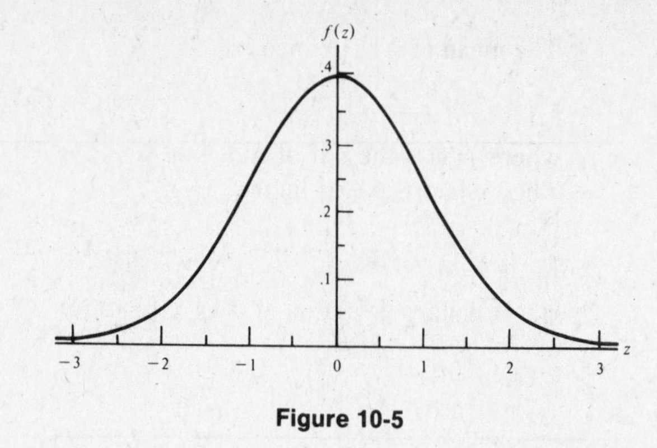

Figure 10-5

EXAMPLE 10-8 Let Z have a standard normal distribution $N(0, 1)$. Find the following probabilities: **(a)** $P(0 \leq Z \leq 1.43)$, **(b)** $P(0 \leq Z)$, **(c)** $P(Z > 1.61)$, **(d)** $P(-1.52 \leq Z \leq 0)$, **(e)** $P(-1.52 \leq Z \leq 1.43)$. For each of these, sketch a bell-shaped curve and shade the area under the curve that equals the probability.

Solution Following is a part of Table 4 from the Appendix:

	Second decimal place in z			
z	.00	.01	.02	.03
1.2	.3849	.3869	.3888	.3907
1.3	.4032	.4049	.4066	.4082
1.4	.4192	.4207	.4222	.4236
1.5	.4332	.4345	.4357	.4370
1.6	.4452	.4463	.4474	.4484

(a) To find $P(0 \leq Z \leq 1.43)$, first sketch a normal curve and shade the area under the curve for $0 \leq z \leq 1.43$ (see Fig. 10-6a). The values in Table 4 give the probabilities for Z between 0 and a positive number, z. You can see from your sketch that you want to find the probability that Z falls between 0 and 1.43. In that part of Table 4 reproduced above, read down the z column to $z = 1.4$ and then read across to the column headed by .03 (the second decimal place in the number 1.43) to find

$$P(0 \leq Z \leq 1.43) = .4236$$

(b) Because the normal pdf is symmetric about $z = 0$ (see Fig. 10-5), one-half of the probability lies to the right of $z=0$ and one-half of the probability lies to the left of $z = 0$. Thus

$$P(0 \leq Z) = .5000$$

(c) To find $P(Z > 1.61)$, sketch the curve as in Fig. 10-6b, and you'll see that you're looking for a right-tail probability. First find $P(0 \leq Z \leq 1.61)$ by reading down the z column in Table 4 to $z = 1.6$, and then read across to the column headed by .01 to find

$$P(0 \leq Z \leq 1.61) = .4463$$

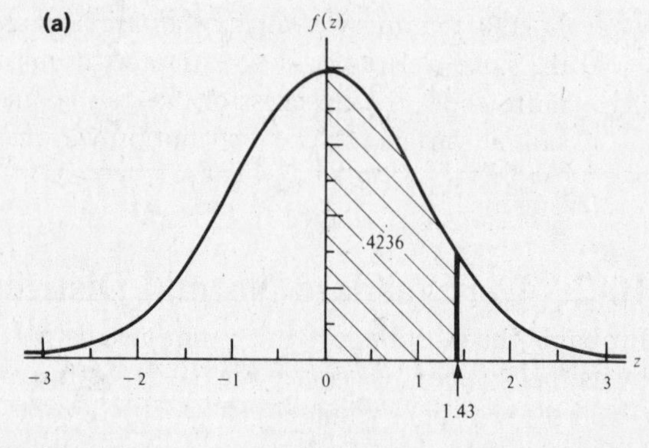

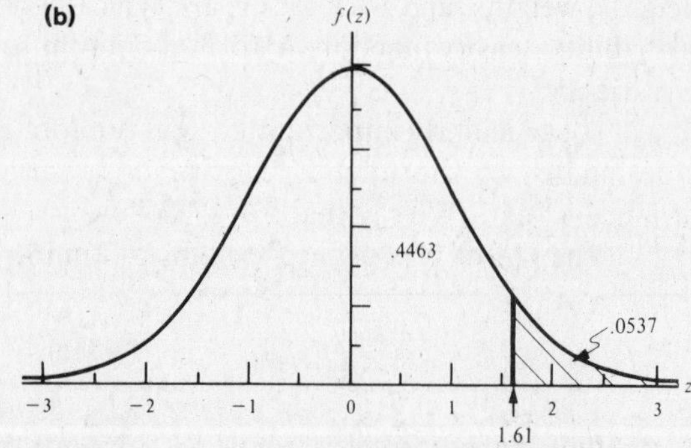

Figure 10-6

Because the probability to the right of $z = 0$ is .5000,

$$P(Z > 1.61) = .5000 - P(0 \le Z \le 1.61)$$
$$= .5000 - .4463 = .0537$$

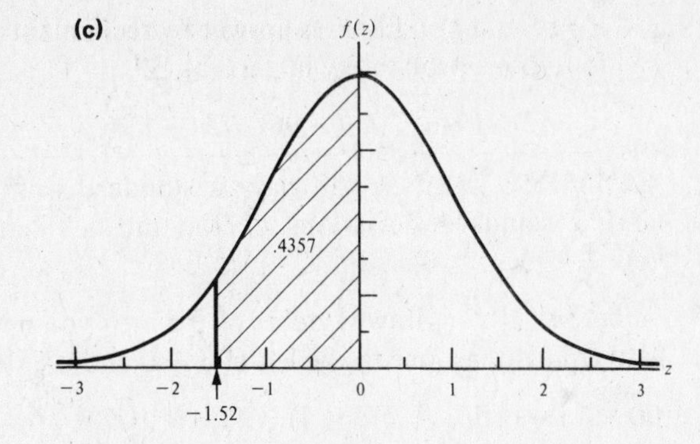

(d) To find $P(-1.52 \le Z \le 0)$, sketch a normal curve and shade the area under the curve for z between -1.52 and 0 (see Fig. 10-6c). Because of the symmetry of the normal pdf, this area is equal to the area under the normal pdf and that between $z = 0$ and $z = 1.52$. This latter probability can be read directly from Table 4.

$$P(-1.52 \le Z \le 0) = P(0 \le Z \le 1.52)$$
$$= .4357$$

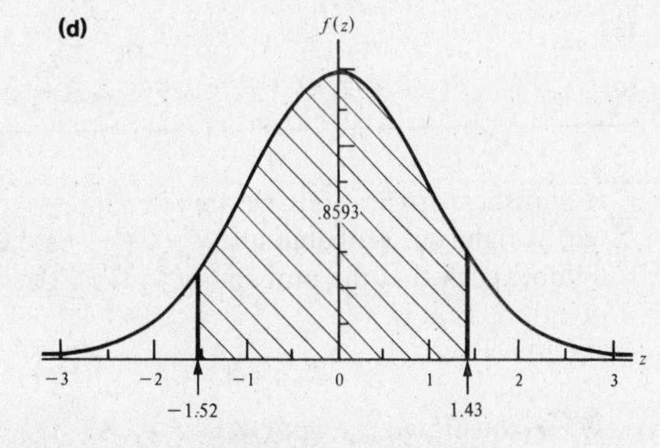

Figure 10-6 (*continued*)

(e) To find $P(-1.52 \le Z \le 1.43)$, again sketch a normal curve and shade in the region under the curve for z between -1.52 and 1.43 (see Fig. 10-6d). Notice that this area is a combination of the shaded areas in Figs. 10-6c and 10-6a. So,

$$P(-1.52 \le Z \le 1.43) = P(-1.52 \le Z \le 0) + P(0 \le Z \le 1.43)$$
$$= .4357 + .4236$$
$$= .8593$$

note: $P(Z = 0) = 0$, so you can include 0 in both of the probabilities on the right. In fact, $P(Z = c) = 0$ for any constant c. That is, the area of a line segment has zero area. We sometimes use $<$ and other times $\le$, so that

$$P(0 < Z < 1.61) = P(0 \le Z \le 1.61) = .4463$$

EXAMPLE 10-9 Let the distribution of Z be $N(0, 1)$. Find **(a)** $P(0 \le Z \le 1.96)$, **(b)** $P(Z > 1.96)$, **(c)** $P(Z < -1.96)$, and **(d)** $P(-1.96 \le Z \le 1.96)$.

Solution Sketch your bell-shaped curve and label -1.96 and 1.96 on the z-axis, shading the areas to the right of $z = 1.96$ and to the left of $z = -1.96$ (see Fig. 10-7).

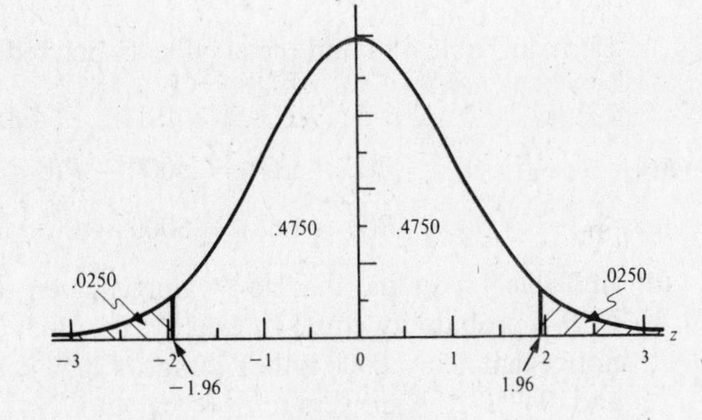

Figure 10-7

(a) From Table 4:

$$P(0 \le Z \le 1.96) = .4750$$

(b) Because $P(Z \ge 0) = .5000$ and $P(0 \le Z \le 1.96) + P(Z > 1.96) = P(Z \ge 0)$, you can use your answer to part **(a)** and write

$$P(Z > 1.96) = P(Z \ge 0) - P(0 \le Z \le 1.96)$$
$$= .5000 - .4750 = .0250$$

(c) Because of the symmetry of the normal curve,

$$P(Z < -1.96) = P(Z > 1.96) = .0250$$

(d) For $P(-1.96 \le Z \le 1.96)$, you can see that the area you wish to find includes the entire area under the curve except for that to the right of $z = 1.96$ and to the left of $z = -1.96$. Since you found these probabilities in **(b)** and **(c)**, you need only to subtract them from 1:

$$P(-1.96 \le Z \le 1.96) = 1 - P(Z < -1.96) - P(Z > 1.96)$$
$$= 1 - .0250 - .0250 = .9500$$

You could also find this answer by recognizing that $P(-1.96 \le Z \le 0) = P(0 \le Z \le 1.96)$ and you found this probability in part (**a**). So,

$$P(-1.96 \le Z \le 1.96) = P(-1.96 \le Z \le 0) + P(0 \le Z \le 1.96) = .4750 + .4750 = .9500$$

EXAMPLE 10-10 Let Z have a standard normal distribution. Find the probabilities within 1, 2, and 3 standard deviations of the mean. That is, find (**a**) $P(-1 \le Z \le 1)$, (**b**) $P(-2 \le Z \le 2)$, (**c**) $P(-3 \le Z \le 3)$.

Solution If you draw sketches of the area you need to find in each case, you can see that it consists of two equal areas: one to the left of $z = 0$ and one to the right of $z = 0$, similar to Fig. 10-7.

(**a**) $$P(-1 \le Z \le 1) = P(-1 \le Z \le 0) + P(0 \le Z \le 1) = .3413 + .3413 = .6826$$

(**b**) $$P(-2 \le Z \le 2) = P(-2 \le Z \le 0) + P(0 \le Z \le 2) = .4772 + .4772 = .9544$$

(**c**) $$P(-3 \le Z \le 3) = P(-3 \le Z \le 0) + P(0 \le Z \le 3) = .4987 + .4987 = .9974$$

- In statistical applications, we are often interested in right-tail probabilities. We let z_α be a number such that the probability to the right of z_α is α. That is,

$$P(Z > z_\alpha) = \alpha \qquad \textbf{(10.1)}$$

This is illustrated in Fig. 10-8.

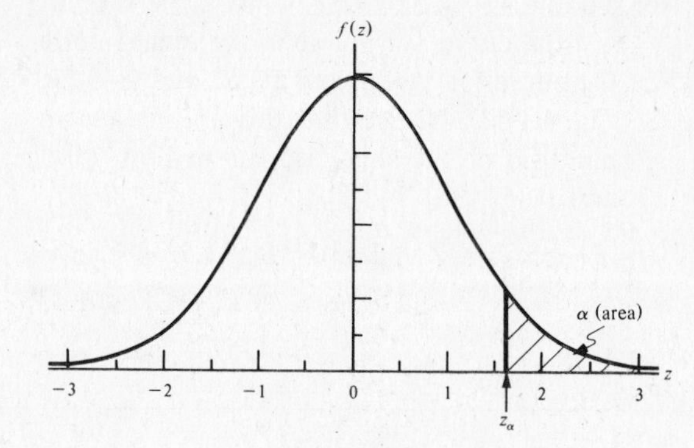

Figure 10-8

EXAMPLE 10-11 Find (**a**) $z_{.025}$, (**b**) $P(Z > 1.64)$, (**c**) $P(Z > 1.65)$, and (**d**) $z_{.05}$.

Solution

(**a**) From formula (10.1), $P(Z > z_{.025}) = .0250$. So

$$P(0 \le Z \le z_{.025}) = .5000 - .0250 = .4750$$

Look in Table 4 to find the z-value associated with a probability of .4750 (see Fig. 10-7):

$$z_{.025} = 1.96$$

(**b**) $$P(Z > 1.64) = .5000 - P(0 \le Z \le 1.64) = .5000 - .4495 = .0505$$

(**c**) $$P(Z > 1.65) = .5000 - P(0 \le Z \le 1.65) = .5000 - .4505 = .0495$$

(**d**) In Table 4, you find that no z-value is given with a probability equal to exactly .05. But notice that $.05 = .0500$ is the average of .0505 and .0495—the answers to parts (**b**) and (**c**):

$$\frac{.0505 + .0495}{2} = \frac{.1000}{2} = .05$$

Thus, $z_{.05}$ is equal to the average of $z_{.0505}$ and $z_{.0495}$—1.64 and 1.65, respectively.

$$z_{.05} = \frac{1.64 + 1.65}{2} = \frac{3.29}{2} = 1.645$$

This is illustrated in Fig. 10-9.

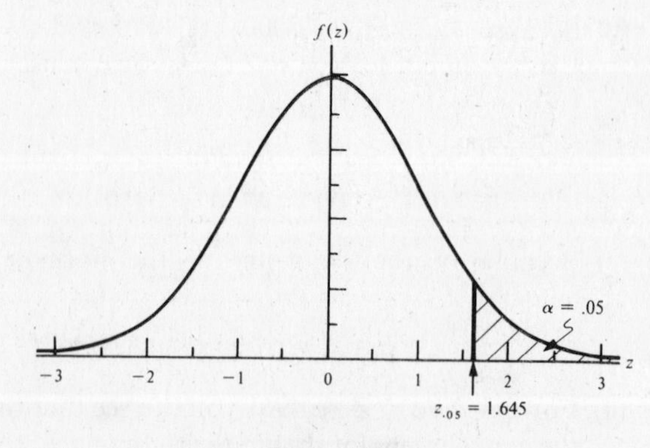

Figure 10-9

note: Since the probability $z_{.05}$ doesn't appear in Table 4, you should try to memorize it. It's one you'll often need to know.

Because of the symmetry of the normal distribution, $-z_\alpha$ is a number such that the probability to the left of $-z_\alpha$ is α. That is,

$$P(Z < -z_\alpha) = \alpha \qquad \textbf{(10.2)}$$

This is illustrated in Fig. 10-10.

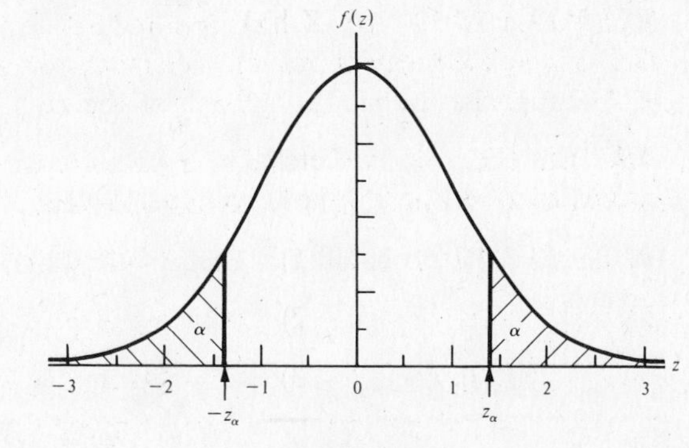

Figure 10-10

EXAMPLE 10-12 Find (**a**) $-z_{.025}$ and (**b**) $-z_{.05}$.

Solution

(**a**) From formula (10.2), $P(Z < -z_{.025}) = .0250$. Because of the symmetry of the normal probability curve (see Fig. 10-11a), $P(Z > z_{.025}) = .0250$, and you found in Example 10-11(a) that $z_{.025} = 1.96$. It follows that

$$-z_{.025} = -1.96$$

(**b**) You should remember that $z_{.05} = 1.645$ [see Example 10-11(d)]. Thus by symmetry (see Fig. 10-11b),

$$-z_{.05} = -1.645$$

(a)

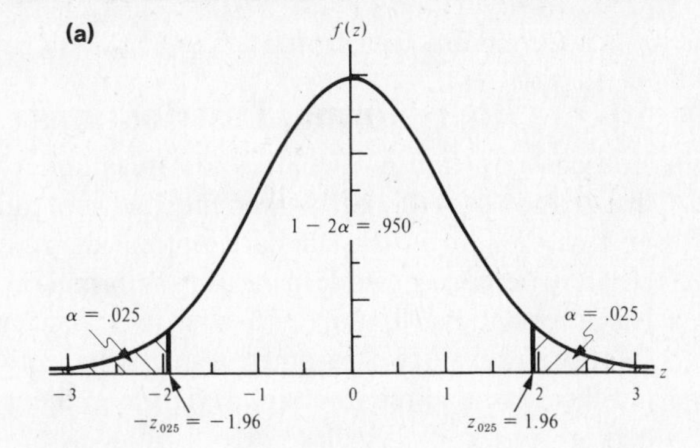

(b)

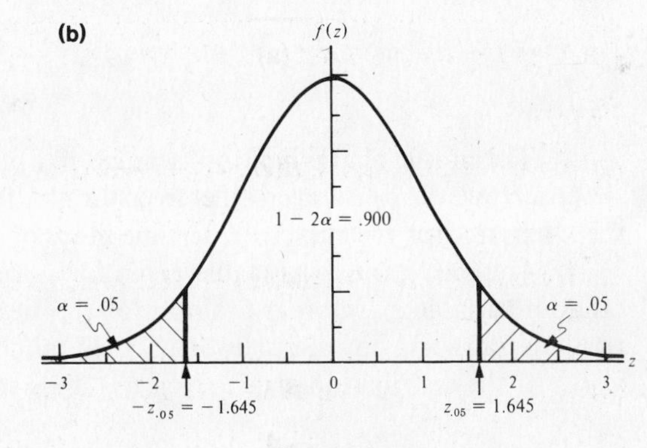

Figure 10-11

EXAMPLE 10-13 Find (**a**) $P(-z_{.025} \le Z \le z_{.025})$, (**b**) $P(-z_{.05} \le Z \le z_{.05})$, (**c**) $P(-z_\alpha \le Z \le z_\alpha)$ and (**d**) $P(-z_{\alpha/2} \le Z \le z_{\alpha/2})$.

Solution

(**a**) From Fig. 10-11a you can see that you're being asked to find the area under the curve excluding that to the left of $-z_{.025} = -1.96$ and that to the right of $z_{.025} = 1.96$. Since $\alpha = .025$ in both cases, the total excluded area equals 2(.025), or .050. Therefore,

$$\begin{aligned} &P(-z_{.025} \le Z \le z_{.025}) \\ &= P(-1.96 \le Z \le 1.96) \\ &= 1 - .050 = .950 \end{aligned}$$

(a)

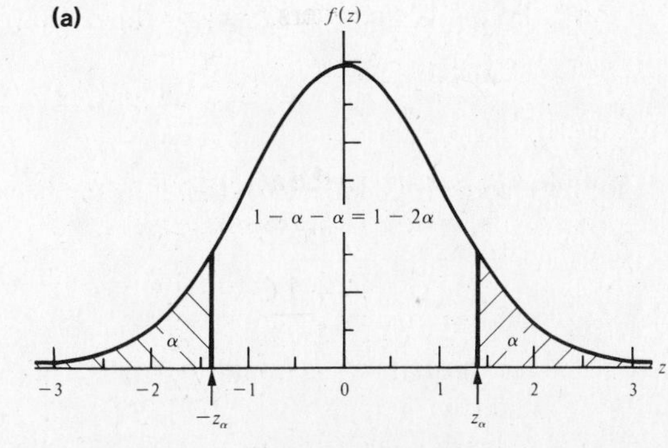

Figure 10-12(a)

(b) Similarly, you find (see Fig. 10-11b),

$$P(-z_{.05} \le Z \le z_{.05})$$
$$= P(-1.645 \le z \le 1.645)$$
$$= 1 - 2(.050) = .900$$

(c) Because the probability in each of the tails is α,

$$P(-z_\alpha \le Z \le z_\alpha) = 1 - \alpha - \alpha = 1 - 2\alpha$$

(d) Because the probability in each of the tails is $\alpha/2$,

$$P(-z_{\alpha/2} \le Z \le z_{\alpha/2}) = 1 - \frac{\alpha}{2} - \frac{\alpha}{2} = 1 - \alpha$$

Figure 10-12a and b illustrates these probabilities.

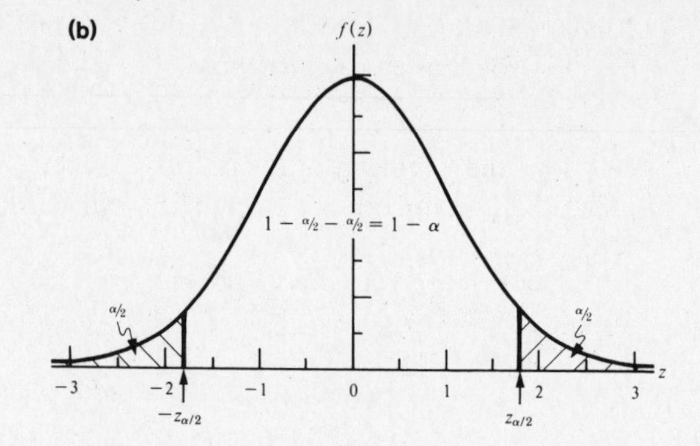

Figure 10-12(b)

10-3. Other Normal Distributions

You can now find probabilities for areas under the standard normal curve—that is, a curve with a distribution of $N(0, 1)$. However, most normal random variables don't have a mean of 0 and a variance of 1 as they do in the standard normal distribution. In fact, an infinite number of mean–variance combinations are possible in defining a normal random variable. For a continuous random variable that has a normal distribution with mean μ and variance σ^2, we say that the distribution of X is $N(\mu, \sigma^2)$. [Notice that the two parameters in the parentheses are the mean and the variance. Some authors put the mean and the standard deviation in the parentheses.] The pdf of X is defined for all real numbers $(-\infty < x < \infty)$ by

$$f(x) = \frac{1}{\sigma\sqrt{2\pi}} e^{-(x-\mu)^2/2\sigma^2}$$

The graph of this pdf looks much like the standard normal curve (Fig. 10-5)—it's still bell-shaped. However, as μ and σ^2 differ from 0 and 1, the center of the curve will shift to the right or the left on the x-axis and/or the curve will become more or less spread out.

The only probability table given for normal distributions is the one for the standard normal random variable Z. In order to find probabilities for other normal distributions, we use the following result to transform any normally distributed random variable X into its standard normal counterpart Z.

If the distribution of X is $N(\mu, \sigma^2)$, then the distribution of

$$Z = \frac{X - \mu}{\sigma} \tag{10.3}$$

is $N(0, 1)$.

To use this result, suppose that the distribution of X is $N(\mu, \sigma^2)$ and we're interested in finding $P(X \le c)$. We first find the value of

$$z = \frac{c - \mu}{\sigma} \tag{10.4}$$

It follows from (10.3) and (10.4) that

$$P(X \le c) = P\left(\frac{X - \mu}{\sigma} \le \frac{c - \mu}{\sigma}\right) = P(Z \le z)$$

and this last probability is found in Table 4 in the Appendix.

Similarly,

$$P(c \le X) = P\left(\frac{c - \mu}{\sigma} \le \frac{X - \mu}{\sigma}\right) = P(z \le Z)$$

EXAMPLE 10-14 Let X have a normal distribution with mean $\mu = 72$ and variance $\sigma^2 = 100$. That is, the distribution of X is $N(72, 100)$. Find and graph (**a**) $P(X \leq 86.3)$ and (**b**) $P(61.8 \leq X \leq 94.7)$.

Solution

(**a**) Your sketch should look like Fig. 10-13a. Using formula (10.4) with $c = 86.3$, $\mu = 72$, and $\sigma = \sqrt{100} = 10$, you find

$$z = \frac{c - \mu}{\sigma} = \frac{86.3 - 72}{10} = 1.43$$

Therefore, from Table 4,

$$\begin{aligned} P(X \leq 86.3) &= P(Z \leq 1.43) = P(Z \leq 0) \\ &\quad + P(0 \leq Z \leq 1.43) \\ &= .5000 + .4236 = .9236 \end{aligned}$$

(**b**) From (10.4),

$$z = \frac{61.8 - 72}{10} = -1.02 \qquad z = \frac{94.7 - 72}{10} = 2.27$$

Thus (see Fig. 10-13b),

$$\begin{aligned} P(61.8 \leq X \leq 94.7) &= P(-1.02 \leq Z \leq 2.27) \\ &= P(-1.02 \leq Z \leq 0) \\ &\quad + P(0 \leq Z \leq 2.27) \\ &= .3461 + .4884 = .8345 \end{aligned}$$

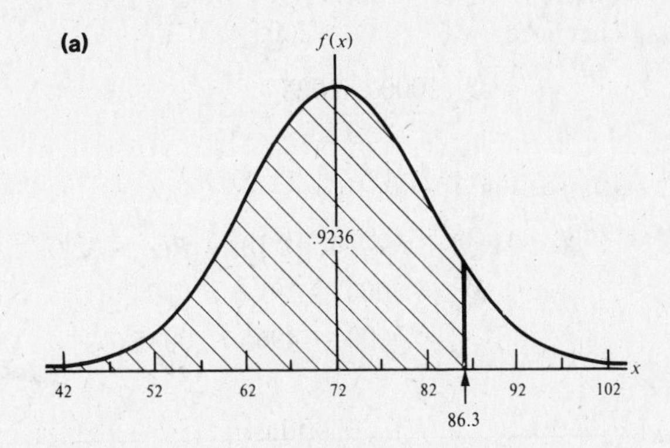

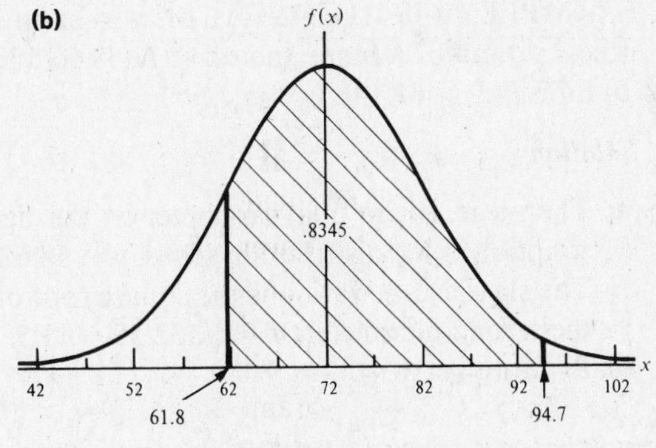

Figure 10-13

EXAMPLE 10-15 For a large dairy herd, let X equal the amount of milk (in lb) produced per day by a cow selected at random. Assume that X is $N(41.3, 225)$. That is, X has a normal distribution with mean $\mu = 41.3$ and standard deviation $\sigma = \sqrt{225} = 15$. Find (**a**) $P(X \leq 69.2)$, (**b**) $P(X > 69.2)$, and (**c**) $P(22.1 \leq X \leq 60.5)$.

Solution

(**a**) Using formula (10.4),

$$z = \frac{69.2 - 41.3}{15} = 1.86$$

Thus (see Fig. 10-14a),

$$\begin{aligned} P(X \leq 69.2) &= P(Z \leq 1.86) \\ &= .5000 + .4686 = .9686 \end{aligned}$$

(**b**) Using your answer to (**a**), you find

$$\begin{aligned} P(X > 69.2) &= 1 - P(X \leq 69.2) \\ &= 1 - P(Z \leq 1.86) \\ &= 1 - .9686 = .0314 \end{aligned}$$

(**c**)

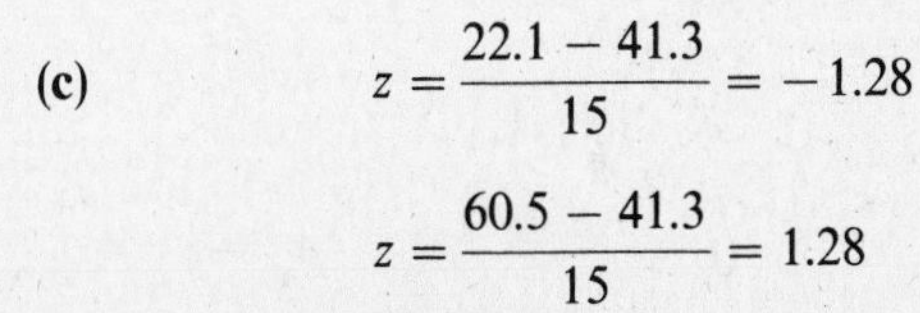

$$z = \frac{22.1 - 41.3}{15} = -1.28$$

$$z = \frac{60.5 - 41.3}{15} = 1.28$$

Therefore (see Fig. 10-14b),

$$\begin{aligned} P(22.1 \leq X \leq 60.5) &= P(-1.28 \leq Z \leq 1.28) \\ &= .3997 + .3997 = .7994 \end{aligned}$$

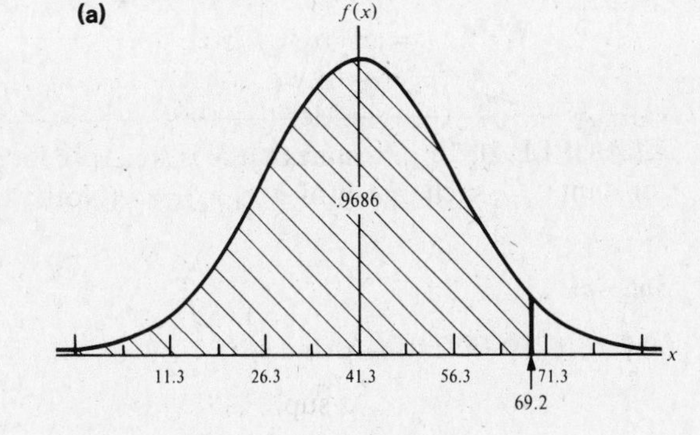

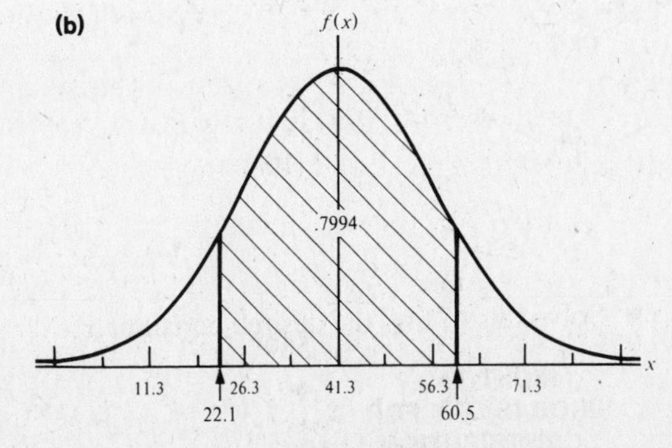

Figure 10-14

EXAMPLE 10-16 Let X equal the weight of liquid soap in a "1000-g" bottle. Assume that X is $N(1027, 100)$. Find $P(X < 1000)$.

Solution Since $\sigma^2 = 100$, $\sigma = \sqrt{100} = 10$.
Therefore,

$$z = \frac{1000 - 1027}{10} = -2.70$$

and (see Fig. 10-15),

$$\begin{aligned} P(X < 1000) &= P(Z < -2.70) = P(Z > 2.70) \\ &= .5000 - P(0 \le Z \le 2.70) \\ &= .5000 - .4965 = .0035 \end{aligned}$$

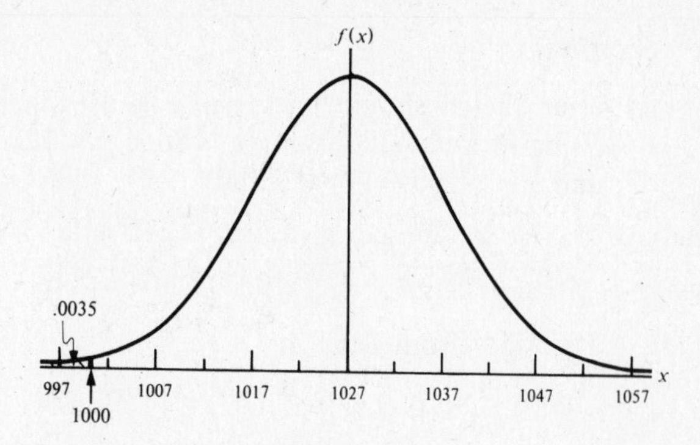

Figure 10-15

EXAMPLE 10-17 Let X equal the weight (in kg) of a female student selected at random from a health fitness program. Assume that X is $N(58.60, 132.25)$. **(a)** Give the mean and standard deviation of X. **(b)** Find $P(X > 81.14)$.

Solution

(a) The mean is the first parameter in the description of the distribution. Thus, $\mu = 58.60$. The standard deviation is the square root of the second parameter: $\sigma = \sqrt{132.25} = 11.5$.

(b) By formula (10.4),

$$\begin{aligned} z &= \frac{81.14 - 58.60}{11.5} \\ &= 1.96 \end{aligned}$$

Thus (see Fig. 10-16),

$$\begin{aligned} P(X > 81.14) &= P(Z > 1.96) \\ &= .5000 - P(0 \le Z \le 1.96) \\ &= .5000 - .4750 \\ &= .0250 \end{aligned}$$

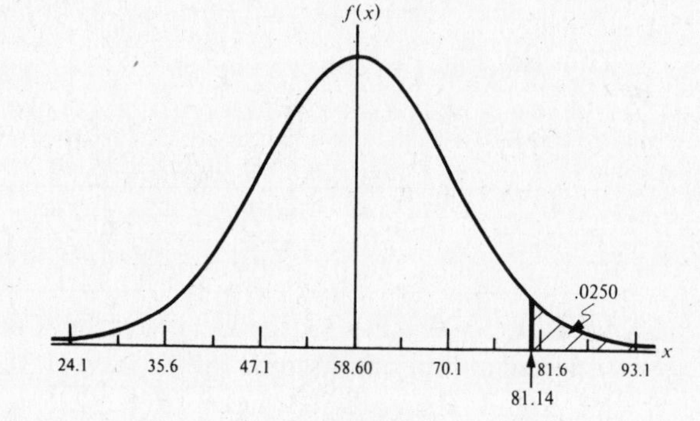

Figure 10-16

EXAMPLE 10-18 Assume that X is $N(\mu, \sigma^2)$. Find **(a)** a constant x_α such that $P(X > x_\alpha) = \alpha$, and **(b)** a constant $x_{1-\alpha}$ such that $P(X < x_{1-\alpha}) = \alpha$. (Notice that the subscript gives the probability to the right of $x_{1-\alpha}$.)

Solution

(a) From (10.4), $z = \frac{x_\alpha - \mu}{\sigma}$, and so

$$\alpha = P(X > x_\alpha) = P\left(Z > \frac{x_\alpha - \mu}{\sigma}\right)$$

You know from (10.1) that $\alpha = P(Z > z_\alpha)$. Thus you can equate the following:

$$\frac{x_\alpha - \mu}{\sigma} = z_\alpha$$

Solve this to find the desired constant x_α.

RIGHT-TAIL PROBABILITY FOR X DISTRIBUTION

$$x_\alpha = \mu + z_\alpha \sigma \tag{10.5}$$

Just as for z_α, x_α is a number such that the probability to the *right* of x_α is α. This formula is used generally when $\alpha < .50$ (see Fig. 10-17).

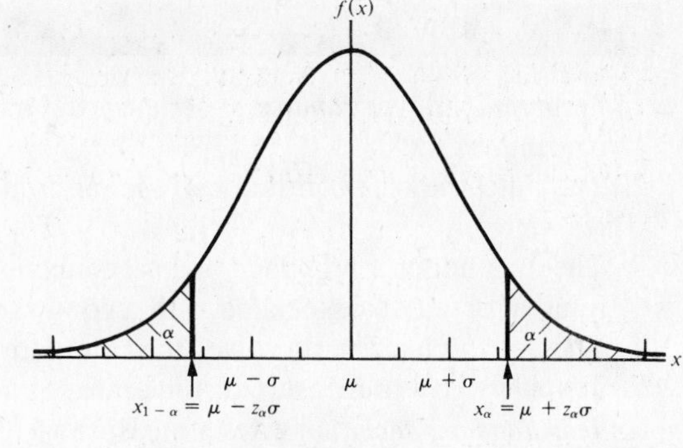

Figure 10-17

(b) From (10.4), $z = \frac{x_{1-\alpha}-\mu}{\sigma}$, and so

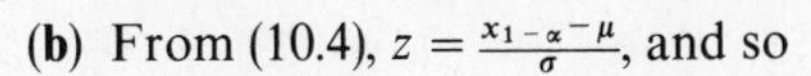

$$\alpha = P(X < x_{1-\alpha}) = P\left(Z < \frac{x_{1-\alpha} - \mu}{\sigma}\right)$$

From (10.2), $\alpha = P(Z < -z_\alpha)$. Thus,

$$\frac{x_{1-\alpha} - \mu}{\sigma} = -z_\alpha$$

Solve this to find the desired constant $x_{1-\alpha}$.

LEFT-TAIL PROBABILITY FOR X DISTRIBUTION

$$x_{1-\alpha} = \mu - z_\alpha \sigma \qquad \textbf{(10.6)}$$

You can see that by subtracting the right-tail probability, $1 - \alpha$, from 1, you can find the left-tail probability for x: $1 - (1 - \alpha) = \alpha$ is the probability to the *left* of $x_{1-\alpha}$. This formula is used generally when $\alpha < .50$, so that $1 - \alpha > .50$.

EXAMPLE 10-19 If X is $N(1027, 100)$, find **(a)** $x_{.05}$ and **(b)** $x_{.975}$. **(c)** Draw a graph of this normal pdf and label both $x_{.05}$ and $x_{.975}$ on the x-axis.

Solution

(a) Remember that $z_{.05} = 1.645$. (see Example 10-11). Using formula (10.5), you find

$$x_{.05} = x_\alpha = \mu + (z_{.05})\sigma$$
$$= 1027 + (1.645)(10) = 1043.45$$

(b) Here, .975 is the right-tail probability for x. If we let $1 - \alpha = .975$, then $\alpha = .025$. The z_α corresponding to a probability of .025 is $z_{.025} = 1.96$. Since $.975 > .500$, use formula (10.6):

$$x_{.975} = x_{1-\alpha} = \mu - (z_{.025})\sigma$$
$$= 1027 - 1.96(10) = 1007.4$$

(c) See Fig. 10-18.

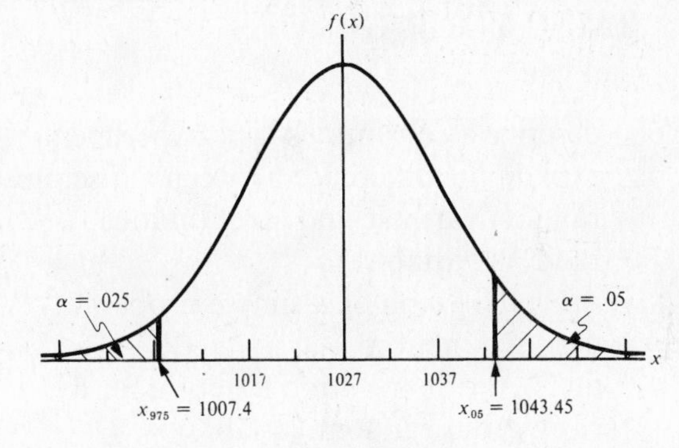

Figure 10-18

EXAMPLE 10-20 If X is $N(41.3, 225)$, find **(a)** $x_{.025}$ and **(b)** $x_{.95}$. **(c)** Draw a graph of this normal pdf, and label both $x_{.025}$ and $x_{.95}$ on the x-axis.

Solution

(a) Use formula (10.5):

$$x_{.025} = 41.3 + (1.96)(15) = 70.7$$

(b) Use formula (10.6):

$$x_{.95} = 41.3 - (1.645)(15) = 16.625$$

(c) See Fig. 10-19.

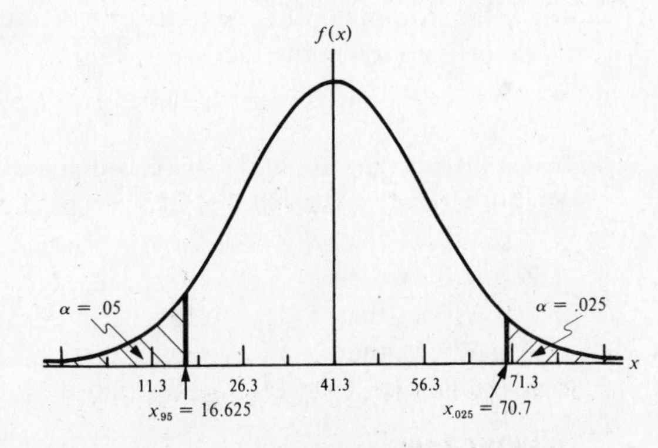

Figure 10-19

SUMMARY

1. A *continuous-type random experiment* has as its set of possible outcomes any numerical value on a continuous scale.
2. The outcome of a continuous-type random experiment is an observation of a *continuous random variable.*
3. The area under a *probability density function* (pdf) curve can be used to find probabilities for any interval of events associated with a continuous random variable.
4. The *distribution function* gives the cumulative probability up to and including x.
5. The *mean* gives a measure of the center of a probability distribution.
6. The *standard deviation* gives a measure of the spread of a probability distribution.
7. The standard normal random variable, designated by the letter Z, has a mean of 0 and a variance of 1, and thus is denoted by $N(0, 1)$.
8. Table 4 in the Appendix gives probability values for a standard normal random variable for intervals between 0 and a value z.
9. The probability to the right of z_α is α.
10. The probability to the left of $-z_\alpha$ is α.
11. A normal random variable X that has mean μ and variance σ^2 is denoted by $N(\mu, \sigma^2)$.
12. If the random variable X is $N(\mu, \sigma^2)$, then $Z = (X - \mu)/\sigma$ has a standard normal distribution, $N(0, 1)$.
13. If the distribution of X is $N(\mu, \sigma^2)$, then the probability to the right of x_α is α, and $x_\alpha = \mu + z_\alpha\sigma$.
14. If the distribution of X is $N(\mu, \sigma^2)$, then the probability to the left of $x_{1-\alpha}$ is α, and $x_{1-\alpha} = \mu - z_\alpha\sigma$.

RAISE YOUR GRADES

Can you . . . ?

☑ describe a continuous-type experiment
☑ explain the difference between a discrete random variable and a continuous random variable
☑ draw the graph of a simple probability density function
☑ define the distribution function for a continuous random variable
☑ use tables to find probabilities for the standard normal distribution
☑ define and find the value of z_α
☑ define and find the value of $-z_\alpha$
☑ find probabilities for any normal distribution
☑ define and find the value of x_α
☑ define and find the value of $x_{1-\alpha}$

RAPID REVIEW

1. An obstetrician used an ultrasound examination on 15 patients to measure the femur length x of each fetus. This is an example of a ______________ experiment.
2. If x is the maximum lung capacity of a member of the volleyball team, then $S = \{x: 2.8 < x < 6.1\}$ could be used to denote the ______________.
3. If $f(x) = 1,\ 0 \le x \le 1$, is the pdf of the random variable X, then $P(.25 \le X \le .75) =$ ______________.
4. If the distribution function of X is defined by $F(x) = 0$ if $x \le 0$, $F(x) = x^2/4$ if $0 < x \le 2$, and $F(x) = 1$ if $2 < x$, then $P(X \le 2/3) =$ ______________.
5. If Z is $N(0, 1)$, then $P(Z \le -2.34) =$ ______________.
6. If Z is $N(0, 1)$, then $P(Z \le 2.13) =$ ______________.
7. If Z is $N(0, 1)$, then $z_{.2005} =$ ______________.
8. If Z is $N(0, 1)$, then $-z_{.0099} =$ ______________.
9. If X is $N(90, 16)$, then $P(82 \le X \le 98) =$ ______________.
10. If X is $N(75, 100)$, then $x_{.025} =$ ______________.

Answers **(1)** continuous-type **(2)** sample space **(3)** .5 **(4)** $\frac{1}{9}$ **(5)** .0096 **(6)** .9834 **(7)** .84 **(8)** -2.33 **(9)** .9544 **(10)** 94.6

SOLVED PROBLEMS

Continuous Distributions

PROBLEM 10-1 Select a 2.2K (2200)-ohm resistor at random. Let x equal its resistance. Using the data in Problem 1-13, define the sample space.

Solution The sample space should include all possible values of the data, as suggested by the random selection of $n = 100$ resistors in Problem 1-13. For example, you could let

$$S = \{x: 2.05 \leq x \leq 2.35\}$$

This answer, however, isn't the only one possible—any interval containing all observed values would be equally acceptable.

PROBLEM 10-2 Let *pbf* equal the percentage of body fat of a person selected at random. Describe the sample space.

Solution To be certain that no one is excluded, you could let

$$S = \{pbf: 0 \leq pbf \leq 100\}$$

For most collections of people, there could be a "little less fat" in the limits of this sample space—that is, the limits could be closer together.

PROBLEM 10-3 Let the pdf of X be defined by

$$f(x) = 1 - \left(\frac{1}{2}\right)x, \qquad 0 \leq x \leq 2$$

(a) Sketch a graph of this pdf. Find **(b)** $P(X > 4/3)$ and **(c)** $P(X \leq 4/3)$.

Solution

(a) See Fig. 10-20a. This pdf gives the equation of a straight line. At $x = 0$, $f(0) = 1$; at $x = 2$, $f(2) = 1 - 1 = 0$.

(b) To find $P(X > 4/3)$, find the area of the triangle whose base b is the interval $(4/3, 2)$ and whose height h is the function $f(x)$ evaluated at $4/3$ (see Fig. 10-20b):

$$f\left(\frac{4}{3}\right) = 1 - \left(\frac{1}{2}\right)\left(\frac{4}{3}\right) = \frac{1}{3}$$

Since the area of a triangle is $(1/2)bh$,

$$P\left(X > \frac{4}{3}\right) = \frac{1}{2}\left(2 - \frac{4}{3}\right)\left(\frac{1}{3}\right) = \frac{1}{9}$$

(c) As the total area under the pdf must equal 1,

$$P\left(X \leq \frac{4}{3}\right) = 1 - P\left(X > \frac{4}{3}\right) = 1 - \frac{1}{9} = \frac{8}{9}$$

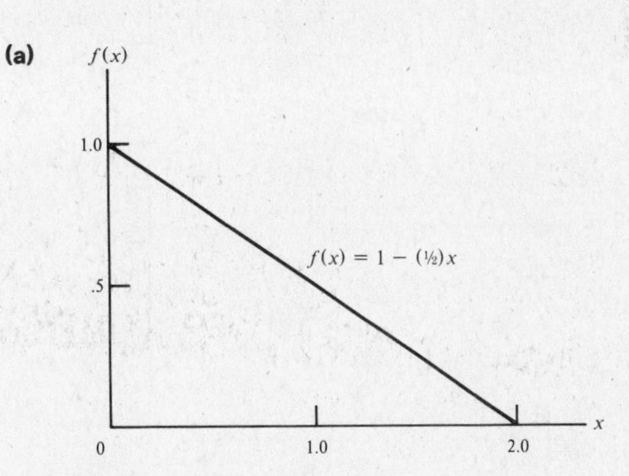

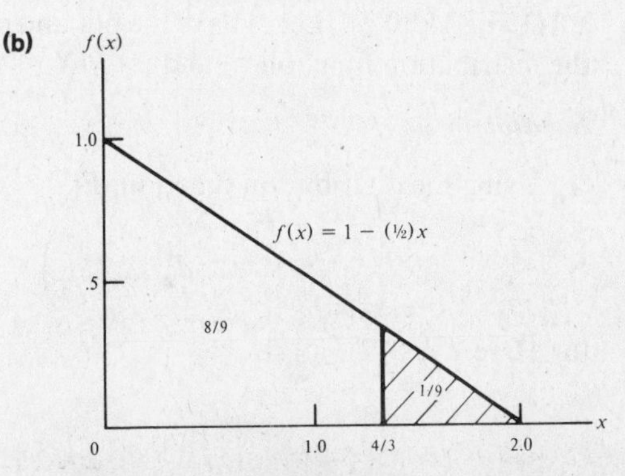

Figure 10-20

PROBLEM 10-4 (**a**) Define the distribution function for the pdf defined in Problem 10-3. (**b**) Sketch a graph of this distribution function.

Solution

(**a**) All of the probability is contained in the interval $[0, 2]$. Thus the cumulative probability up to the lower end of the interval is 0, or

$$F(x) = P(X \leq x) = 0 \qquad \text{for } x < 0$$

For values of x between 0 and 2, find $P(X > x)$ and subtract this answer from 1. $P(X > x)$ is given by the area of the triangle whose base is the interval $[x, 2]$ and whose height is $f(x) = 1 - (1/2)x$. Thus,

$$P(X > x) = \frac{1}{2}(2 - x)\left(1 - \frac{x}{2}\right) = \frac{1}{2}\left(2 - x - x + \frac{x^2}{2}\right)$$
$$= 1 - x + \frac{x^2}{4}$$

So for $0 \leq x \leq 2$,

$$F(x) = P(X \leq x) = 1 - P(X > x) = 1 - \left(1 - x + \frac{x^2}{4}\right) = x - \frac{x^2}{4}$$

For $x > 2$, $F(x) = P(X \leq x) = 1$.

(**b**) See Fig. 10-21.

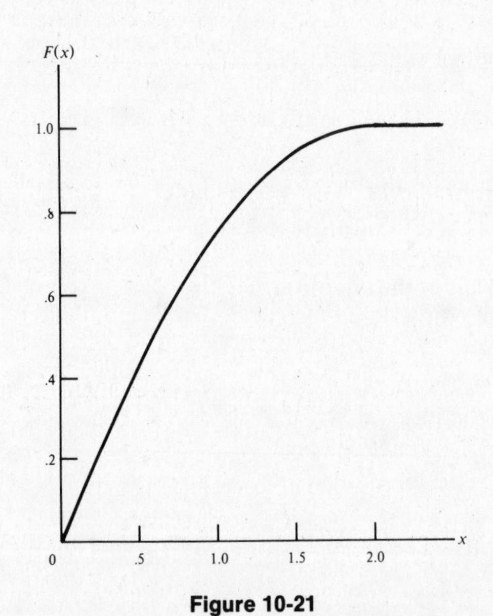

Figure 10-21

PROBLEM 10-5 Let X have the pdf and distribution function defined in Problems 10-3 and 10-4. Use the distribution function to find (**a**) $P(X \leq \frac{2}{3})$ and (**b**) $P(X \leq \frac{3}{2})$.

Solution

(**a**) Using the distribution function, $F(x) = x - (x^2/4)$, where $x = 2/3$, you find

$$P\left(X \leq \frac{2}{3}\right) = F\left(\frac{2}{3}\right) = \frac{2}{3} - \frac{(2/3)^2}{4} = \frac{6}{9} - \frac{1}{9} = \frac{5}{9}$$

(**b**) Here $x = 3/2$, and so

$$P\left(X \leq \frac{3}{2}\right) = F\left(\frac{3}{2}\right) = \frac{3}{2} - \frac{(3/2)^2}{4} = \frac{24}{16} - \frac{9}{16} = \frac{15}{16}$$

The Standard Normal Distribution

PROBLEM 10-6 Let the distribution of Z be $N(0, 1)$. Find **(a)** $P(0 \le Z \le 1.29)$, **(b)** $P(Z > 1.72)$, and **(c)** $P(.36 \le Z \le 2.07)$.

Solution

(a) In Table 4 in the Appendix, read down the z column to 1.2 and across to the column headed by .09 to find (see Fig. 10-22a):

$$P(0 \le Z \le 1.29) = .4015$$

(b) Since half of the probability is to the right of 0,

$$P(Z > 1.72) = P(Z \ge 0) - P(0 \le Z \le 1.72)$$
$$= .5000 - .4573 = .0427$$

See Fig. 10-22b.

(c) See Fig. 10-22c:

$$P(.36 \le Z \le 2.07) = P(0 \le Z \le 2.07) - P(0 \le Z \le .36)$$
$$= .4808 - .1406$$
$$= .3402$$

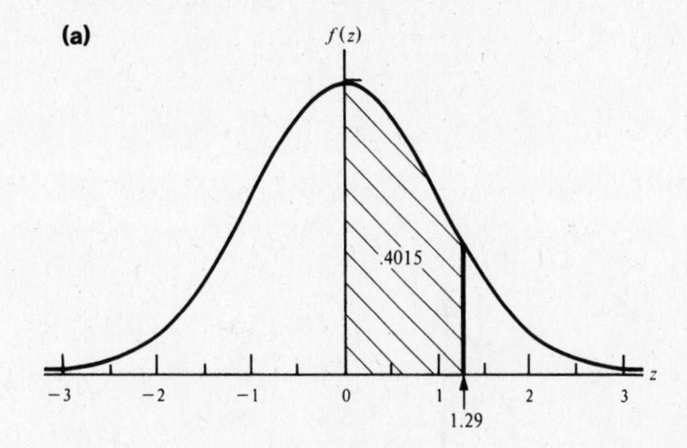

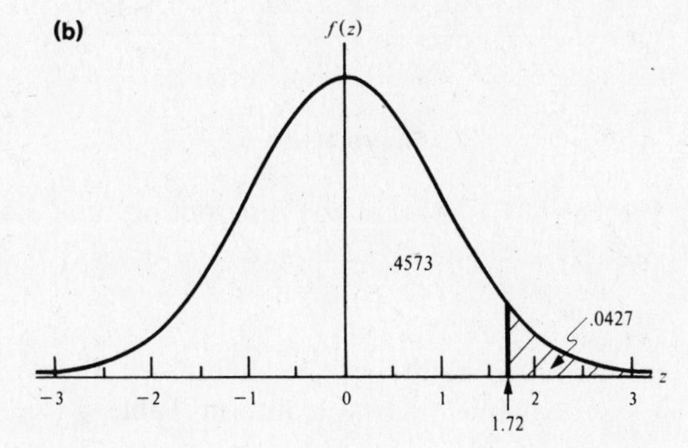

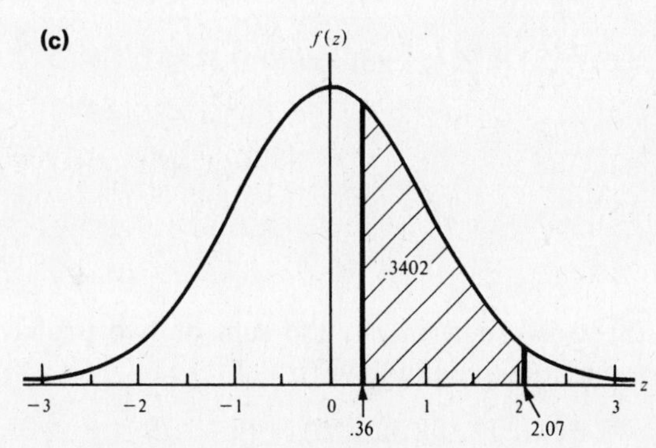

Figure 10-22

PROBLEM 10-7 Let the distribution of Z be $N(0, 1)$. Find **(a)** $P(-.72 \le Z \le 0)$, **(b)** $P(Z < -.91)$, and **(c)** $P(-2.93 \le Z \le -.76)$.

Solution

(a) Because of the symmetry of the normal curve (see Fig. 10-23),

$$P(-.72 \le Z \le 0) = P(0 \le Z \le .72) = .2642$$

You'll find the answer in Table 4.

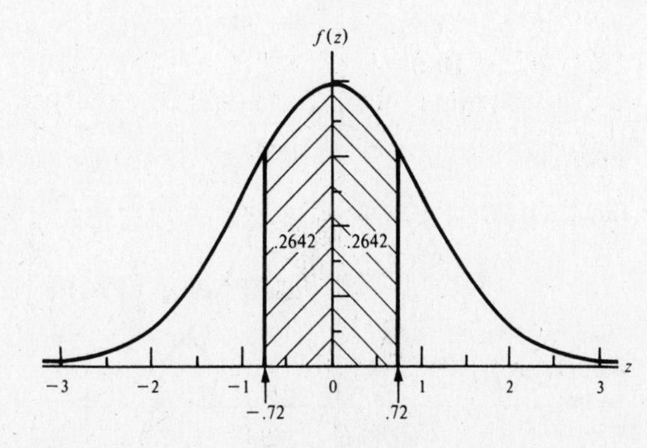

Figure 10-23

(b) See Fig. 10-24:

$$P(Z < -.91) = P(Z > .91)$$
$$= P(Z \geq 0) - P(0 \leq Z \leq .91)$$
$$= .5000 - .3186 = .1814$$

(c) See Fig. 10-25:

$$P(-2.93 \leq Z \leq -.76) = P(.76 \leq Z \leq 2.93)$$
$$= P(0 \leq Z \leq 2.93)$$
$$- P(0 \leq Z \leq .76)$$
$$= .4983 - .2764$$
$$= .2219$$

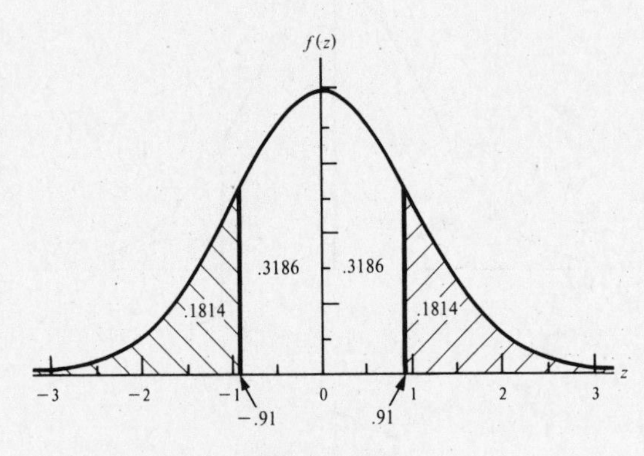

Figure 10-24

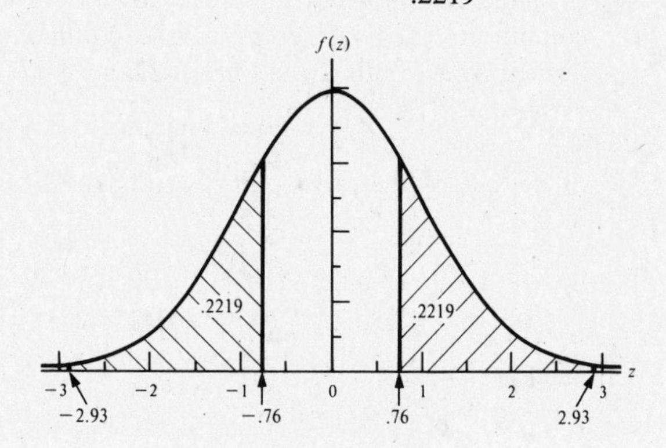

Figure 10-25

PROBLEM 10-8 Let the distribution of Z be $N(0, 1)$. Find **(a)** $P(.84 \leq Z \leq 2.33)$, **(b)** $P(-1.79 \leq Z \leq 2.03)$, and **(c)** $P(-2.89 \leq Z \leq -1.67)$.

Solution

(a) This probability is the difference of two probabilities that you find in Table 4 (see Fig. 10-26):

$$P(.84 \leq Z \leq 2.33) = P(0 \leq Z \leq 2.33)$$
$$- P(0 \leq Z \leq .84)$$
$$= .4901 - .2995 = .1906$$

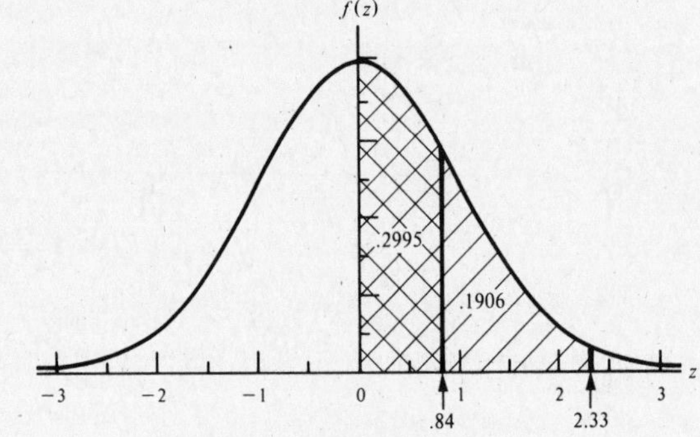

Figure 10-26

(b) This probability is the sum of two probabilities (see Fig. 10-27):

$$P(-1.79 \leq Z \leq 2.03) = P(-1.79 \leq Z \leq 0)$$
$$+ P(0 \leq Z \leq 2.03)$$
$$= P(0 \leq Z \leq 1.79)$$
$$+ P(0 \leq Z \leq 2.03)$$
$$= .4633 + .4788$$
$$= .9421$$

(c) Use the symmetry of the normal curve and then find the difference of two probabilities (see Fig. 10-28):

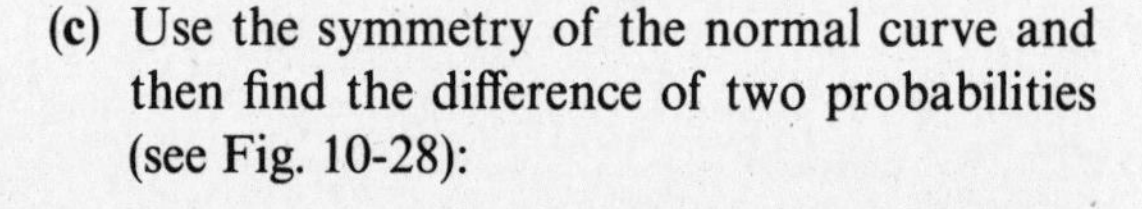

$$P(-2.89 \leq Z \leq -1.67) = P(1.67 \leq Z \leq 2.89)$$
$$= P(0 \leq Z \leq 2.89)$$
$$- P(0 \leq Z \leq 1.67)$$
$$= .4981 - .4525$$
$$= .0456$$

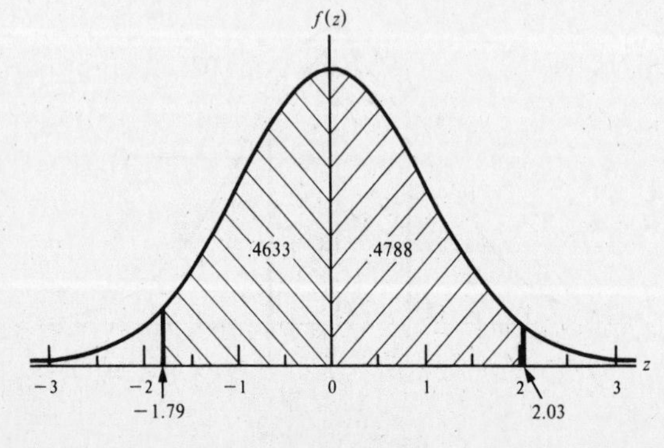

Figure 10-27

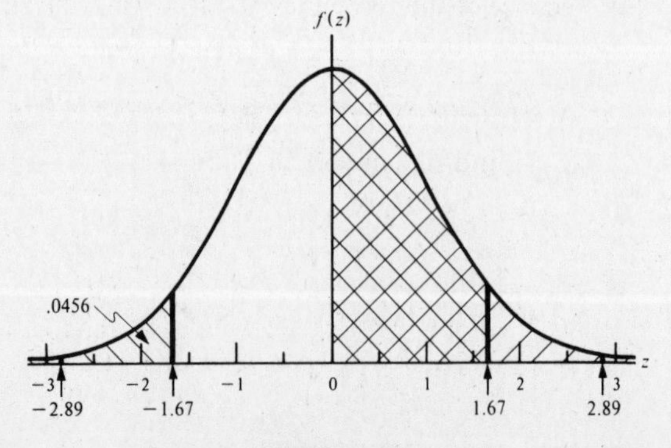

Figure 10-28

PROBLEM 10-9 Let the distribution of Z be $N(0, 1)$. Find the values of **(a)** $z_{.0125}$, **(b)** $z_{.1230}$, **(c)** $z_{.33}$.

Solution Each of these values is of the form z_α, where α is the right-tail probability. First find the probability between 0 and z_α, and then look up the desired z-values in Table 4.

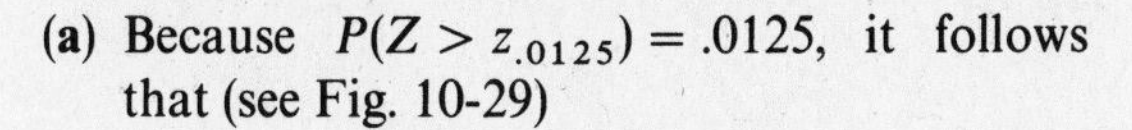

(a) Because $P(Z > z_{.0125}) = .0125$, it follows that (see Fig. 10-29)

$$P(0 \le Z \le z_{.0125}) = .5000 - .0125$$
$$= .4875$$

From Table 4, $z_{.0125} = 2.24$.

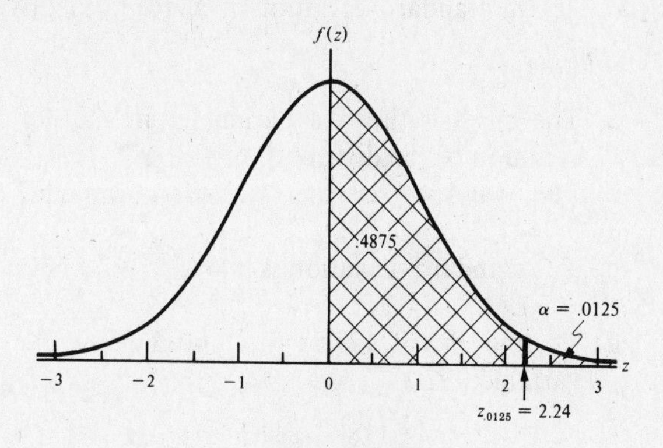

Figure 10-29

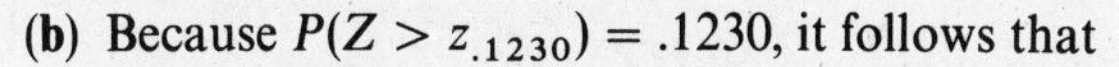

(b) Because $P(Z > z_{.1230}) = .1230$, it follows that

$$P(0 \le Z \le z_{.1230}) = .5000 - .1230$$
$$= .3770$$

Thus, from Table 4, $z_{.1230} = 1.16$.

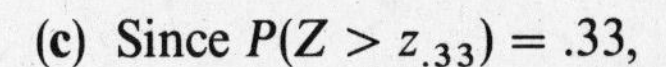

(c) Since $P(Z > z_{.33}) = .33$,

$$P(0 \le Z \le z_{.33}) = .5000 - .3300$$
$$= .1700$$

Then from Table 4, $z_{.33} = .44$.

PROBLEM 10-10 Let the distribution of Z be $N(0, 1)$. Find the values of the following common right-tail probabilities: **(a)** $z_{.005}$, **(b)** $z_{.01}$, **(c)** $z_{.10}$.

Solution

(a)

$$P(Z > z_{.005}) = .005$$
$$P(0 \le Z \le z_{.005}) = .5000 - .0050$$
$$= .4950$$

So, $z_{.005} = 2.58$.

(b)

$$P(Z > z_{.01}) = .01$$
$$P(0 \le Z \le z_{.01}) = .5000 - .0100$$
$$= .4900$$

From Table 4, $z_{.01} = 2.33$.

(c)

$$P(Z > z_{.10}) = .10$$
$$P(0 \le Z \le z_{.10}) = .5000 - .1000$$
$$= .4000$$

From Table 4, $z_{.10} = 1.28$.

PROBLEM 10-11 Construct a table summarizing the values of $z_{.10}, z_{.05}, z_{.025}, z_{.01}$, and $z_{.005}$. (You'll want to refer back to Example 10-11 and Problem 10-10.)

Solution You should construct a tabulation as shown. These right-tail probabilities are used so often in statistics that you'd find it worth your time to memorize them.

α:	.10	.05	.025	.01	.005
z_α:	1.28	1.645	1.96	2.33	2.58

PROBLEM 10-12 Let the distribution of Z be $N(0, 1)$. Find the value of **(a)** $-z_{.005}$, **(b)** $-z_{.01}$, **(c)** $-z_{.10}$.

Solution Each of these values of $-z$ has the probability α to its left. Due to the symmetry of a normal curve, you can look up the values in the table in Problem 10-11. The only difference is that the values of z are now negative.

(a) $-z_{.005} = -2.58$ **(b)** $-z_{.01} = -2.33$ **(c)** $-z_{.10} = -1.28$

Other Normal Distributions

PROBLEM 10-13 Let X have a normal distribution $N(112, 121)$. Give (**a**) the mean, (**b**) the variance, and (**c**) the standard deviation of X. (**d**) Find $P(X \leq 128.5)$.

Solution

(**a**) The mean is the first parameter in the description of the distribution, $\mu = 112$.

(**b**) The variance is the second parameter, $\sigma^2 = 121$.

(**c**) The standard deviation is $\sigma = \sqrt{\sigma^2} = \sqrt{121} = 11$.

(**d**) To find $P(X \leq 128.5)$, first standardize the variable:

$$z = \frac{128.5 - 112}{11} = 1.5$$

Now read the desired area from Table 4 (see Fig. 10-30),

$$\begin{aligned} P(X \leq 128.5) &= P(Z \leq 1.5) \\ &= P(Z < 0) + P(0 \leq Z \leq 1.5) \\ &= .5000 + .4332 \\ &= .9332 \end{aligned}$$

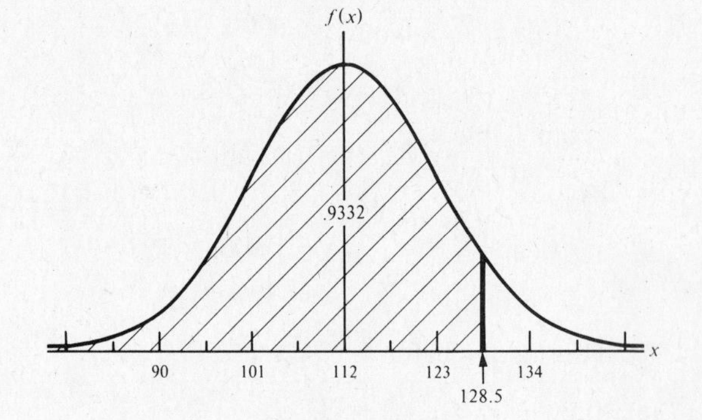

Figure 10-30

PROBLEM 10-14 Let X equal the resistance of a 2.2K (2200)-ohm resistor. Assume that X is $N(2.200, .03^2)$. Find (**a**) $P(X \leq 2.164)$ and (**b**) $P(X \leq 2.245)$.

Solution

(**a**) See Fig. 10-31.

$$\begin{aligned} z &= \frac{2.164 - 2.200}{.03} \\ &= -1.20 \end{aligned}$$

$$\begin{aligned} P(X \leq 2.164) &= P(Z \leq -1.20) \\ &= P(Z \geq 1.20) \\ &= P(0 \leq Z) \\ &\quad - P(0 \leq Z \leq 1.20) \\ &= .5000 - .3849 \\ &= .1151 \end{aligned}$$

(**b**) See Fig. 10-32.

$$\begin{aligned} z &= \frac{2.245 - 2.200}{.03} \\ &= 1.50 \end{aligned}$$

$$\begin{aligned} P(X \leq 2.245) &= P(Z \leq 1.50) \\ &= P(Z < 0) \\ &\quad + P(0 \leq Z \leq 1.50) \\ &= .5000 + .4332 \\ &= .9332 \end{aligned}$$

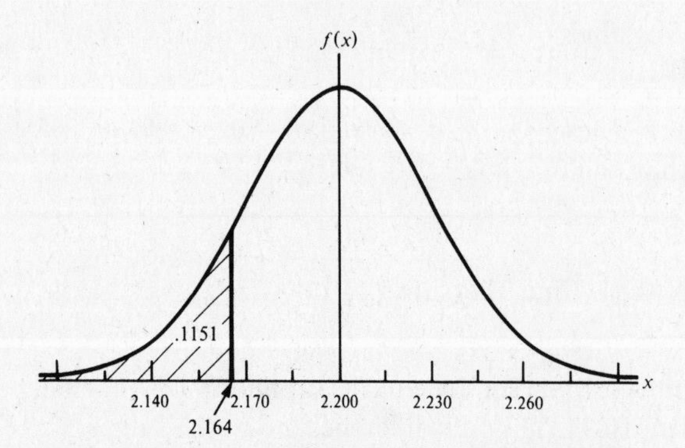

Figure 10-31

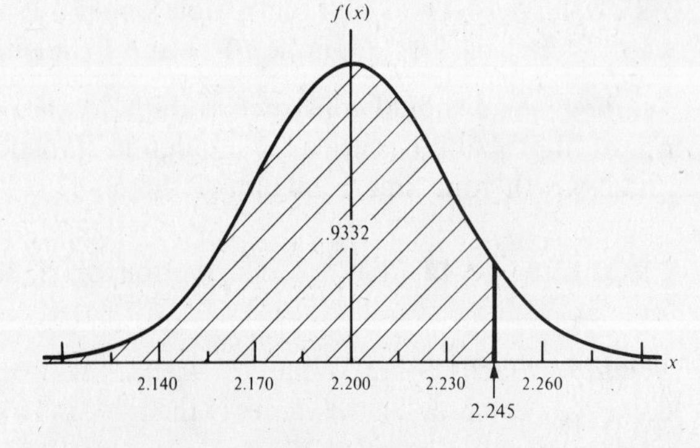

Figure 10-32

PROBLEM 10-15 Let Y equal the weight change (in kg) for a freshman male in a health-fitness program. Assume that Y is $N(2.8, 2.89)$. Find **(a)** $P(Y \le 6.2)$, **(b)** $P(Y \le 1.1)$, and **(c)** $P(Y \le 0)$.

Solution

(a) See Fig. 10-33.

$$z = \frac{6.2 - 2.8}{1.7} = 2.0$$

$$\begin{aligned} P(Y \le 6.2) &= P(Z \le 2.0) \\ &= P(Z < 0) + P(0 \le Z \le 2.0) \\ &= .5000 + .4772 = .9772 \end{aligned}$$

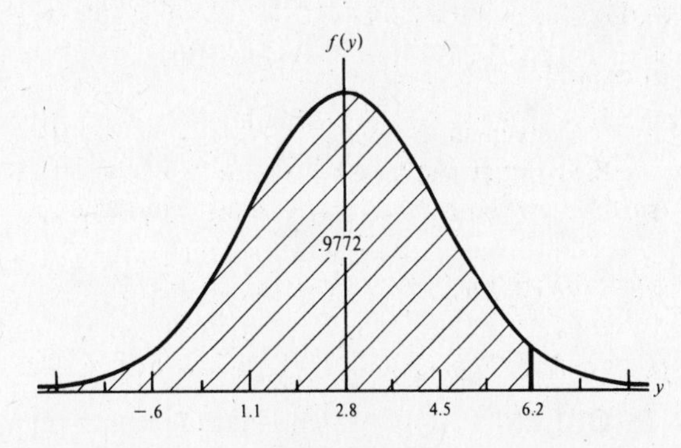

Figure 10-33

(b) See Fig. 10-34.

$$z = \frac{1.1 - 2.8}{1.7} = -1.0$$

$$\begin{aligned} P(Y \le 1.1) &= P(Z \le -1.0) = P(Z \ge 1.0) \\ &= P(0 \le Z) - P(0 \le Z \le 1.0) \\ &= .5000 - .3413 = .1587 \end{aligned}$$

(c) See Fig. 10-35.

$$z = \frac{0 - 2.8}{1.7} = -1.647$$

$$\begin{aligned} P(Y \le 0) &= P(Z \le -1.647) = P(Z \ge 1.647) \\ &= P(0 \le Z) - P(0 \le Z \le 1.647) \\ &= .5000 - .4502 = .0498 \end{aligned}$$

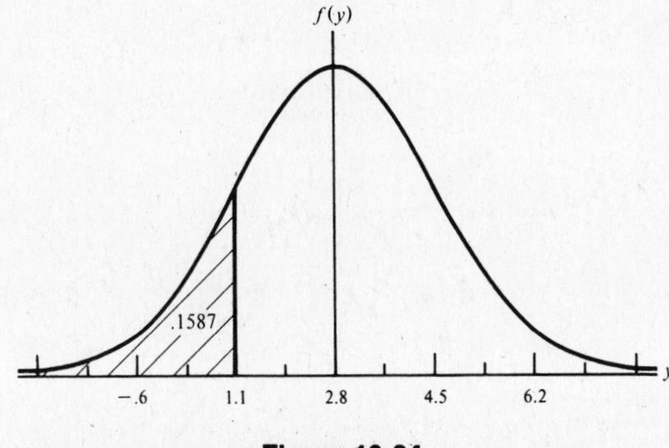

Figure 10-34

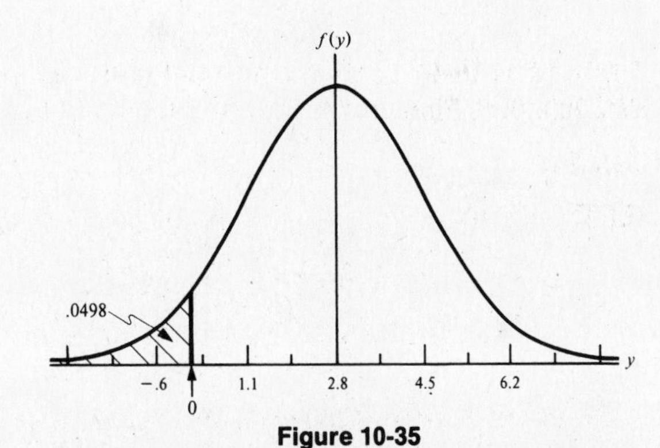

Figure 10-35

PROBLEM 10-16 A chemistry student uses an atomic absorption spectrophotometer to read the measurement of zinc in a sample taken from a standard zinc solution in which the concentration of zinc in parts per million (ppm) is .2000. Let X equal this measurement, and assume that X is $N(.2000, .005^2)$. Find **(a)** $P(X \le .2091)$ and **(b)** $P(X \le .1897)$.

Solution

(a) See Fig. 10-36.

$$z = \frac{.2091 - .2000}{.005} = 1.82$$

$$\begin{aligned} P(X \le .2091) &= P(Z \le 1.82) \\ &= P(Z \le 0) \\ &\quad + P(0 \le Z \le 1.82) \\ &= .5000 + .4656 = .9656 \end{aligned}$$

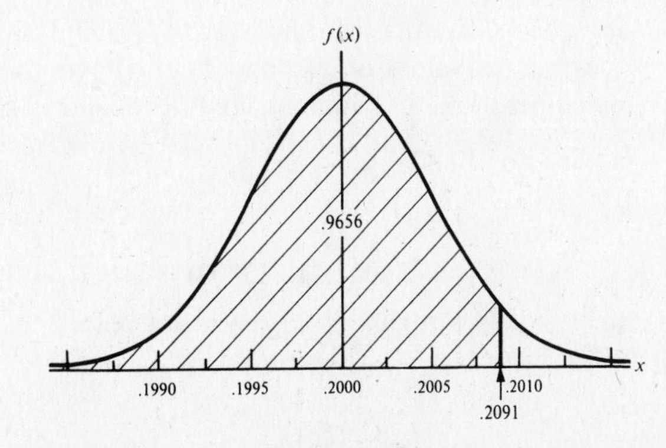

Figure 10-36

(b) See Fig. 10-37.

$$z = \frac{.1897 - .2000}{.005} = -2.06$$

$$\begin{aligned} P(X \leq .1897) &= P(Z \leq -2.06) \\ &= P(Z \geq 2.06) \\ &= P(0 \leq Z) \\ &\quad - P(0 \leq Z \leq 2.06) \\ &= .5000 - .4803 = .0197 \end{aligned}$$

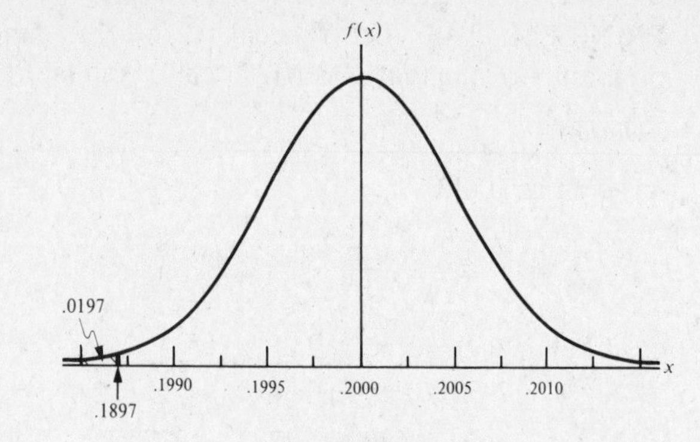

Figure 10-37

PROBLEM 10-17 Assume that the length (in mm) of the culmen (frontal shield) of a female common moorhen (gallinule) is $N(40.9, 2.25)$. Find **(a)** $P(X \leq 37.18)$ and **(b)** $P(X > 43.51)$.

Solution

(a) See Fig. 10-38.

$$z = \frac{37.18 - 40.9}{1.5} = -2.48$$

$$\begin{aligned} P(X \leq 37.18) &= P(Z \leq -2.48) \\ &= P(Z > 2.48) \\ &= P(0 \leq Z) \\ &\quad - P(0 \leq Z \leq 2.48) \\ &= .5000 - .4934 = .0066 \end{aligned}$$

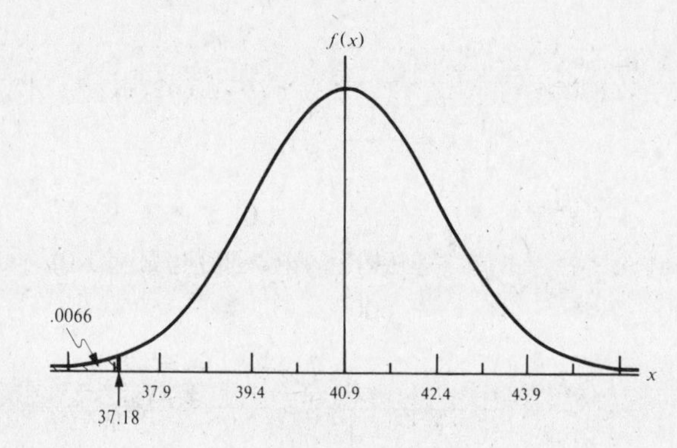

Figure 10-38

(b) See Fig. 10-39.

$$z = \frac{43.51 - 40.9}{1.5} = 1.74$$

$$\begin{aligned} P(X > 43.51) &= P(Z > 1.74) \\ &= .5000 - P(0 \leq Z \leq 1.74) \\ &= .5000 - .4591 = .0409 \end{aligned}$$

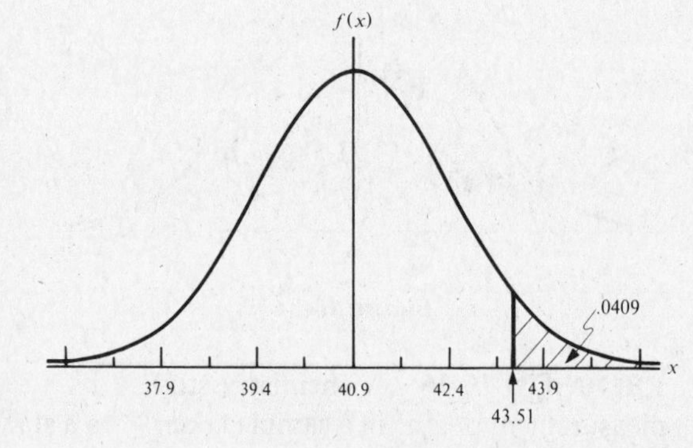

Figure 10-39

PROBLEM 10-18 For the resistors in Problem 10-14, find **(a)** $x_{.05}$ and **(b)** $x_{.025}$.

Solution You can use formula (10.5). You only need to know the values of $z_{.05}$ and $z_{.025}$. (If you can't remember these, look in the table in Problem 10-11.)

(a) See Fig. 10-40.

$$x_\alpha = \mu + z_\alpha \sigma$$

$$x_{.05} = 2.200 + 1.645(.03) = 2.249$$

(b) $$x_{.025} = 2.200 + 1.96(.03) = 2.259$$

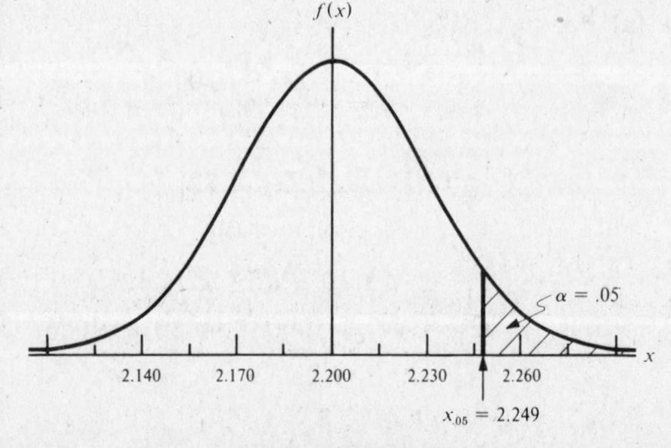

Figure 10-40

PROBLEM 10-19 For the weight changes in Problem 10-15, find (**a**) $x_{.95}$ and (**b**) $x_{.975}$.

Solution Use formula (10.6):

(**a**) Since $(1 - \alpha) = .95$, $\alpha = .05$ and you know that $z_{.05} = 1.645$, (see Fig. 10-41)

$$x_{1-\alpha} = \mu - z_\alpha \sigma$$

$$x_{.95} = 2.8 - 1.645(1.7) = .0035$$

(**b**) Since $(1 - \alpha) = .975$, $\alpha = .025$, and $z_{.025} = 1.96$,

$$x_{.975} = 2.8 - 1.96(1.7) = -.532$$

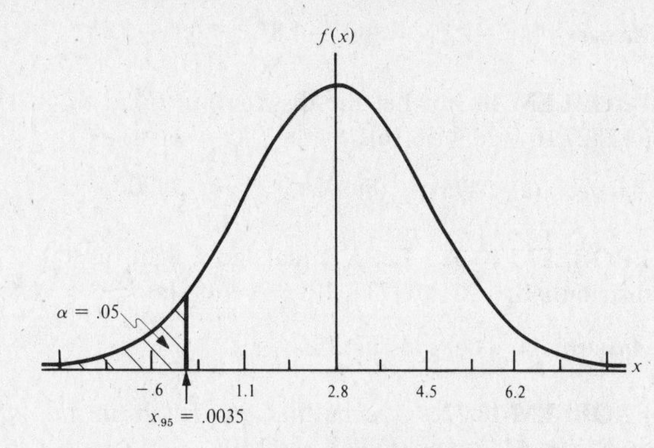

Figure 10-41

Supplementary Problems

PROBLEM 10-20 Select a student at random, and measure the student's height h (in inches). Define any appropriate sample space.

Answer $S = \{h: 48 \le h \le 86\}$ is one.

PROBLEM 10-21 Let the pdf of X be defined by

$$f(x) = 2 - 2x, \qquad 0 \le x \le 1$$

Find (**a**) $P(X > 1/3)$ and (**b**) $P(X \le 1/3)$.

Answer (**a**) 4/9 (**b**) 5/9

PROBLEM 10-22 For the pdf in Problem 10-21, give the definition of the distribution function for $0 \le x \le 1$.

Answer $F(x) = 2x - x^2, \qquad 0 \le x \le 1$

PROBLEM 10-23 Use the distribution function in Problem 10-22 to find $P(X \le 1/4)$.

Answer $F(1/4) = 7/16$

PROBLEM 10-24 Let the distribution of Z be $N(0, 1)$. Find (**a**) $P(Z \le -2.65)$, (**b**) $P(Z \le -1.27)$, and (**c**) $P(Z \le -.28)$.

Answer (**a**) .0040 (**b**) .1020 (**c**) .3897

PROBLEM 10-25 Let the distribution of Z be $N(0, 1)$. Find (**a**) $P(Z \le .36)$, (**b**) $P(Z \le 1.82)$, and (**c**) $P(Z \le 2.61)$.

Answer (**a**) .6406 (**b**) .9656 (**c**) .9955

PROBLEM 10-26 Let the distribution of Z be $N(0, 1)$. Find (**a**) $P(-2.48 \le Z \le -.84)$, (**b**) $P(-1.64 \le Z \le 1.64)$, and (**c**) $P(0 \le Z \le 2.56)$.

Answer (**a**) .1939 (**b**) .8990 (**c**) .4948

PROBLEM 10-27 Let the distribution of Z be $N(0, 1)$. Find (**a**) $P(Z > 2.76)$, (**b**) $P(Z > 1.55)$, and (**c**) $P(Z > -1.39)$.

Answer (**a**) .0029 (**b**) .0606 (**c**) .9177

PROBLEM 10-28 Find (**a**) $z_{.015}$, (**b**) $z_{.0351}$, and (**c**) $z_{.0495}$.

Answer (**a**) 2.17 (**b**) 1.81 (**c**) 1.65

PROBLEM 10-29 Find (a) $-z_{.015}$, (b) $-z_{.0351}$, and (c) $-z_{.0495}$.

Answer (a) -2.17 (b) -1.81 (c) -1.65

PROBLEM 10-30 Let the distribution of X be $N(43.66, 100)$. Find (a) $P(X \leq 27.16)$, (b) $P(X \leq 58.16)$, and (c) $P(27.16 \leq X \leq 58.16)$.

Answer (a) .0495 (b) .9265 (c) .8770

PROBLEM 10-31 Let X equal the weight (in kg) of a freshman male selected at random. Assume that the distribution of X is $N(73.8, 129.96)$. Find (a) $P(X > 90.9)$ and (b) $P(X < 59.55)$.

Answer (a) .0668 (b) .1056

PROBLEM 10-32 Assume that the length (in mm) of the culmen of a male gallinule is $N(44.63, 2.25)$. Find (a) $P(X \leq 47.15)$ and (b) $P(X > 42.50)$.

Answer (a) .9535 (b) .9222

PROBLEM 10-33 A product has a label weight of 450 g. Let X equal the weight of a package that is selected at random from the production line. Assume that X is $N(456.2, 5.76)$. (a) Give the value of the standard deviation σ. Find (b) $P(X < 450)$ and (c) $P(451.52 \leq X \leq 460.88)$.

Answer (a) 2.4 (b) .0049 (c) .9488

PROBLEM 10-34 For the weights in Problem 10-33, find (a) $x_{.025}$ and (b) $x_{.05}$.

Answer (a) 460.904 (b) 460.148

PROBLEM 10-35 For the weights in Problem 10-33, find (a) $x_{.975}$ and (b) $x_{.99}$.

Answer (a) 451.496 (b) 450.608

SAMPLING DISTRIBUTIONS

THIS CHAPTER IS ABOUT

- ☑ Random Samples: The Mean, Variance, and Standard Deviation of $\bar{X}$
- ☑ The Distribution of the Sample Mean
- ☑ Normal Approximation of Binomial Probabilities
- ☑ The Chi-Square Distribution
- ☑ Student's t Distribution
- ☑ The F Distribution

In this chapter we take repeated independent observations of a random variable or independent observations from some population and investigate the distribution of functions of these observations, called **statistics.** For example, what can we say about the probability distribution of the sample mean $\bar{X}$ or of the sample variance S^2? Each such distribution is called a **sampling distribution of the statistic.**

11-1. Random Samples: The Mean, Variance, and Standard Deviation of $\bar{X}$

In this section we investigate the distribution of the sample mean, which is called the **sampling distribution of $\bar{X}$.** It is important to understand the difference between the population distribution and the sampling distribution of a statistic. The *population distribution* is the underlying distribution of a random variable, often unknown. It indicates the probabilities for the occurrence of various outcomes or sets of outcomes in an entire population—for example, sets of heights taken from *all* statistics students. The *sampling distribution of $\bar{X}$* indicates the probabilities for obtaining various outcomes or sets of outcomes of the mean of a random sample of size n taken from the population. We could, for example, take a random sample of four statistics students and determine their average height. If we perform this experiment many times, we begin to get a feel for the distribution of mean heights for samples of size four. We may think of this mean height $\bar{X}$ as a random variable for which we can find its mean, variance, and other characteristics.

EXAMPLE 11-1 Roll a fair 4-sided die 5 times and calculate the sample mean $\bar{x}$. Repeat this process 50 times (i.e., roll the die 5 times for each of the 50 trials), recording the 5 outcomes and the sample mean for each trial. Find **(a)** the sample mean and **(b)** the sample variance of the entire set of 250 rolls of the die. Find **(c)** the sample mean, **(d)** the sample variance, and **(e)** the sample standard deviation of the 50 sample means. **(f)** Draw a histogram of the 250 rolls of the die. **(g)** Draw a histogram of the 50 sample means.

Solution The following results were simulated on a computer:

Trial	Outcomes	$\bar{x}$	Trial	Outcomes	$\bar{x}$
1	3 3 3 2 1	2.4	9	1 3 4 2 4	2.8
2	1 3 1 4 3	2.4	10	4 2 4 3 1	2.8
3	1 2 2 3 4	2.4	11	4 1 2 1 1	1.8
4	2 1 3 3 1	2.0	12	4 4 1 4 1	2.8
5	2 2 3 1 1	1.8	13	2 2 4 1 4	2.6
6	3 3 4 3 4	3.4	14	1 1 2 3 3	2.0
7	3 2 2 1 4	2.4	15	1 1 3 4 1	2.0
8	3 2 2 4 3	2.8	16	2 2 2 1 1	1.6

(continued)

Trial	Outcomes	$\bar{x}$	Trial	Outcomes	$\bar{x}$
17	2 3 2 1 4	2.4	34	1 1 2 4 2	2.0
18	2 2 1 1 2	1.6	35	2 1 2 1 4	2.0
19	1 2 3 3 1	2.0	36	1 4 4 4 1	2.8
20	4 4 1 3 1	2.6	37	2 3 2 1 4	2.4
21	3 2 4 4 3	3.2	38	1 3 4 4 4	3.2
22	3 2 1 1 1	1.6	39	4 2 3 1 3	2.6
23	4 3 1 3 4	3.0	40	2 1 1 4 2	2.0
24	1 2 1 3 2	1.8	41	1 4 2 4 2	2.6
25	4 2 4 3 3	3.2	42	1 4 3 2 1	2.2
26	4 1 4 4 2	3.0	43	2 1 2 2 4	2.2
27	4 3 1 3 2	2.6	44	1 2 1 2 2	1.6
28	4 4 3 1 3	3.0	45	3 1 2 2 2	2.0
29	3 4 4 2 4	3.4	46	3 1 2 1 4	2.2
30	2 2 3 3 4	2.8	47	3 3 1 2 4	2.6
31	2 2 4 1 3	2.4	48	3 1 3 4 2	2.6
32	2 1 3 1 4	2.2	49	3 4 4 2 2	3.0
33	4 2 2 1 3	2.4	50	3 2 1 2 3	2.2

(a) For these 250 rolls of the die, $\sum x = 607$. Thus the sample mean of the entire set of 250 observations is $\bar{x} = 607/250 = 2.428$.

(b) For the 250 rolls of the die, $\sum x^2 = 1787$. Thus the sample variance of the 250 rolls of the die is

$$s^2 = \frac{(250)(1787) - (607)^2}{(250)(249)} = \frac{78{,}301}{62{,}250} = 1.258$$

(c) For the 50 observations of $\bar{x}$, $\sum \bar{x} = 121.4$. Thus the sample mean of the 50 sample means is $121.4/50 = 2.428$.

(d) For the 50 observations of $\bar{x}$, $\sum \bar{x}^2 = 306.44$. Thus the sample variance of the 50 sample means is

$$s_{\bar{x}}^2 = \frac{(50)(306.44) - (121.4)^2}{(50)(49)} = \frac{584.04}{2450} = .238$$

(e) The sample standard deviation is $s = \sqrt{.238} = .488$.

(f) See Fig. 11-1A.

(g) See Fig. 11-1B.

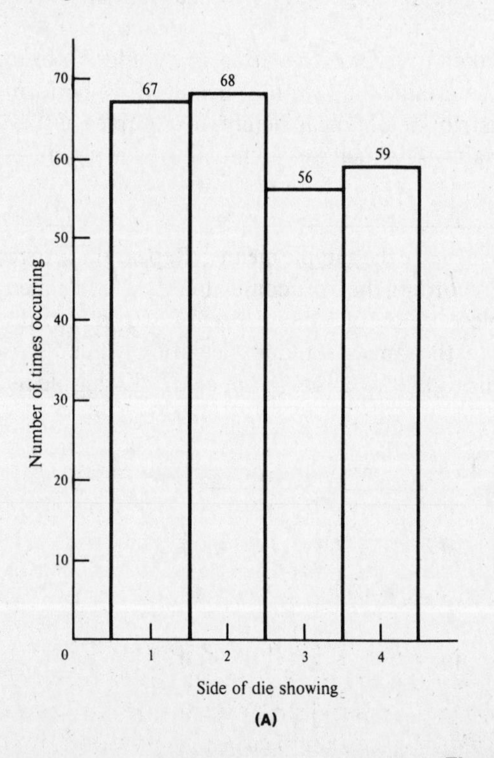

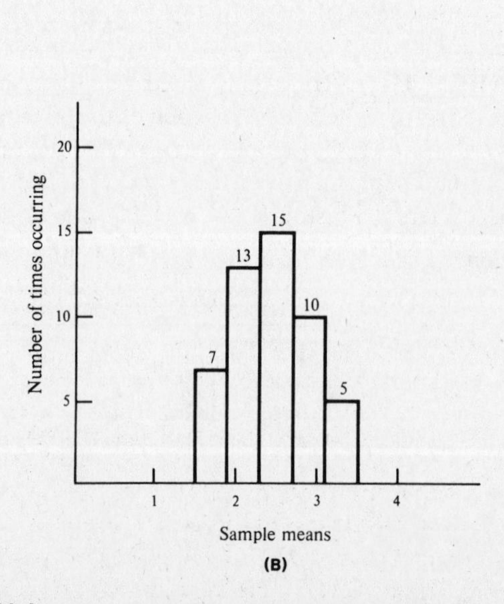

Figure 11-1

There are four observations to make about the results in Example 11-1:

1. The sample mean of the 50 sample means is equal to the sample mean of the 250 observations.
2. The sample variance of the 50 sample means is smaller than the sample variance of the 250 observations. In fact, it's approximately 1/5th as large.
3. The histogram in Fig. 11-1A shows that the original distribution is uniform (the numbers of ones, twos, threes, and fours rolled are approximately equal), while the histogram in Fig. 11-1B shows that the distribution of the sample means is more triangular or bell-shaped, with the probability being the largest in the center of the distribution.
4. The sample mean can be thought of as a random variable, and as such, it has its own probability distribution. In other words, we may think of each of the 50 sample means as an observation of a random variable $\bar{X}$.

A. Random samples

A random sample can be thought of as n independent observations from a distribution or population.

First, if we observe n independent observations of a random variable X that has some probability distribution, these n observations are called a **random sample** of size n from this *distribution.* We let $X_1, X_2, \ldots, X_n$ denote the items of a random sample and $x_1, x_2, \ldots, x_n$ denote the observed values of the random sample. To illustrate, let X in Example 11-1 be the random variable that equals the outcome of a roll of a 4-sided die. Each set of $n = 5$ observations of X is a random sample of size 5 from this uniform distribution on the integers 1, 2, 3, and 4. The random variables associated with these 5 rolls of the die would be X_1, X_2, X_3, X_4, and X_5. The observed values would be, e.g., $x_1 = 3$, $x_2 = 2$, $x_3 = 2$, $x_4 = 1$, $x_5 = 4$.

Second, if we select n independent observations from a population in such a way that each member of the population has the same chance of being selected, these n observations are called a **random sample** of size n from this population. To illustrate, we might take a random sample of 5 bottles to estimate the mean weight of soap in "1000-g" bottles. Such a random sample could yield, for example, $x_1 = 1048$, $x_2 = 1042$, $x_3 = 1036$, $x_4 = 1039$, and $x_5 = 1038$. These 5 observations would be the values of a random sample of size $n = 5$ from the population of "1000-g" bottles of soap.

We speak of random samples from distributions or from populations, depending on the context of the situation.

B. The mean, variance, and standard deviation of $\bar{X}$

Let $X_1, X_2, \ldots, X_n$ be a random sample of size n from either a distribution or a population that has mean μ and variance σ^2. Let $\bar{X} = (1/n)\sum X_i$.

The probability distribution of $\bar{X}$ is called the **sampling distribution of the mean.** We can determine the following characteristics of the random variable $\bar{X}$:

- The mean of the distribution of $\bar{X}$ is

MEAN OF SAMPLING DISTRIBUTION OF $\bar{X}$

$$\mu_{\bar{x}} = \mu \tag{11.1}$$

- The variance of the distribution of $\bar{X}$ is

VARIANCE OF SAMPLING DISTRIBUTION OF $\bar{X}$

$$\sigma_{\bar{x}}^2 = \frac{\sigma^2}{n} \tag{11.2}$$

- The standard deviation of the distribution of $\bar{X}$ is

STANDARD DEVIATION OF SAMPLING DISTRIBUTION OF $\bar{X}$

$$\sigma_{\bar{x}} = \sqrt{\sigma_{\bar{x}}^2} = \frac{\sigma}{\sqrt{n}} \tag{11.3}$$

We often call $\sigma_{\bar{x}} = \sigma/\sqrt{n}$ the **standard error of the mean.**

This means that given a random sample of size n from a population that has mean μ and variance σ^2, the mean of the sampling distribution of $\bar{X}$ is equal to the population mean; the variance of the sampling distribution of $\bar{X}$ is equal to the population variance divided by the sample size n; and the standard deviation of the sampling distribution for $\bar{X}$ (or the standard error of the mean) is equal to the standard deviation of the population divided by the square root of n.

EXAMPLE 11-2 Let X equal the outcome when you roll a 4-sided die. The probability function of X is $f(x) = 1/4$, $x = 1, 2, 3, 4$, and so $\mu = 2.5$, $\sigma^2 = 1.25$, and $\sigma = 1.118$. Let $\bar{X}$ be the sample mean of 5 observations of X. That is, $\bar{X}$ will equal the sample mean of 5 rolls of this 4-sided die. Give the values of **(a)** $\mu_{\bar{x}}$, **(b)** $\sigma_{\bar{x}}^2$, and **(c)** $\sigma_{\bar{x}}$.

Solution

(a) By formula (11.1), the mean of $\bar{X}$ is the same as the mean of X:

$$\mu_{\bar{x}} = \mu = 2.5$$

(b) By formula (11.2), the variance of the sample mean is the variance of X divided by the number of observations in the sample:

$$\sigma_{\bar{x}}^2 = \frac{\sigma^2}{n} = \frac{1.25}{5} = .25$$

(c) By formula (11.3),

$$\sigma_{\bar{x}} = \frac{\sigma}{\sqrt{n}}$$

$$= \frac{\sqrt{1.25}}{\sqrt{5}} = .5$$

Notice how close these theoretical results are to the respective empirical results found in Example 11-1(c), (d), and (e).

EXAMPLE 11-3 Let X equal the weight of liquid soap in a "1000-g" bottle. Assume that the distribution of X is known to be $N(1027, 100)$. Let $\bar{X}$ be the sample mean of $n = 25$ bottles that are selected at random from the production line. Give the values of **(a)** $\mu_{\bar{x}}$, **(b)** $\sigma_{\bar{x}}^2$, and **(c)** $\sigma_{\bar{x}}$.

Solution

(a) Using formula (11.1), you find $\mu_{\bar{x}} = 1027$.
(b) By formula (11.2), $\sigma_{\bar{x}}^2 = 100/25 = 4$.
(c) By formula (11.3), $\sigma_{\bar{x}} = 10/\sqrt{25} = 2$.

EXAMPLE 11-4 Let X equal the weight (in grams) of a bag of potato chips that has a label weight of 450 g. Assume that the distribution of X is $N(456.2, 5.76)$. Let $\bar{X}$ denote the sample mean of 16 bags of chips that are selected at random from the production line. Give the values of **(a)** $\mu_{\bar{x}}$, **(b)** $\sigma_{\bar{x}}^2$, and **(c)** $\sigma_{\bar{x}}$.

Solution

(a) By (11.1), $\mu_{\bar{x}} = 456.2$.
(b) By (11.2), $\sigma_{\bar{x}}^2 = 5.76/16 = .36$.
(c) By (11.3), $\sigma_{\bar{x}} = \sqrt{5.76}/\sqrt{16} = 2.4/4 = .6$.

EXAMPLE 11-5 Let $X_1, X_2, \ldots, X_{36}$ denote a random sample of size 36 from a distribution that has a mean of $\mu = 24.7$ and a standard deviation of $\sigma = 3.9$. Let $\bar{X}$ equal the mean of this random sample. Give the values of **(a)** $\mu_{\bar{x}}$ and **(b)** the standard error of the mean.

Solution

(a) By (11.1), $\mu_{\bar{x}} = 24.7$.
(b) The standard error of the mean is the same as the standard deviation of $\bar{X}$, or $\sigma_{\bar{x}} = 3.9/\sqrt{36} = .65$.

11-2. The Distribution of the Sample Mean

Although in Section 11-1 we gave the mean, variance, and standard deviation for the sampling distribution of the sample mean $\bar{X}$, we said nothing about the *distribution* of $\bar{X}$. This section gives the distribution of $\bar{X}$ when sampling from normal populations and the approximate distribution of $\bar{X}$ when sampling from any population.

A. Distribution of the sample mean X for normal populations

Let $X_1, X_2, \ldots, X_n$ be the items of a random sample of size n from a normal distribution, $N(\mu, \sigma^2)$. Then the distribution of $\bar{X}$ is $N(\mu, \sigma^2/n)$. That is, the distribution of $\bar{X}$ is also normal with the same mean as the distribution from which the sample was taken and with a variance of σ^2/n. The standard deviation of the distribution of $\bar{X}$ is $\sigma/\sqrt{n}$. Recall that any normal random variable minus its mean and divided by its standard deviation is $N(0, 1)$. We can thus say that the distribution of

$$Z = \frac{\bar{X} - \mu}{\sigma/\sqrt{n}}$$

is $N(0, 1)$.

EXAMPLE 11-6 In Example 11-3, let $X_1, X_2, \ldots, X_{25}$ denote the items of a random sample of size $n = 25$ from the normal distribution $N(1027, 100)$. **(a)** Find $P(1022.5 \leq X_1 \leq 1031.5)$. **(b)** How is $\bar{X}$ distributed? **(c)** Give the standard error of the mean. **(d)** Find $P(1022.5 \leq \bar{X} \leq 1031.5)$.

Solution

(a) The distribution of X_1, the first observation, is $N(1027, 100)$. Subtract the mean and divide by the standard deviation to obtain the z values.

$$z = \frac{1022.5 - 1027}{10} = -.45 \qquad z = \frac{1031.5 - 1027}{10} = .45$$

Then use Table 4 in the Appendix to find this probability (see Fig. 11-2A).

$$\begin{aligned} P(1022.5 \leq X_1 \leq 1031.5) &= P(-.45 \leq Z \leq .45) \\ &= P(-.45 \leq Z \leq 0) + P(0 \leq Z \leq .45) \\ &= 2P(0 \leq Z \leq .45) = 2(.1736) = .3472 \end{aligned}$$

(b) Because $\bar{X}$ is the sample mean of a random sample taken from a normal distribution, the distribution of $\bar{X}$ is $N(1027, 100/25)$. That is, $\bar{X}$ has a normal distribution with the same mean $\mu = 1027$ as the distribution from which the sample is taken and a variance of $\sigma^2/n = 100/25$.

(c) The standard error of the mean is

$$\sigma_{\bar{x}} = \sigma/\sqrt{n} = 10/\sqrt{25} = 2$$

(d) The distribution of $Z = (\bar{X} - 1027)/(10/5)$ is $N(0, 1)$. Therefore,

$$z = \frac{1022.5 - 1027}{2} = -2.25 \qquad z = \frac{1031.5 - 1027}{2} = 2.25$$

Thus you find (Fig. 11-2B)

$$\begin{aligned} P(1022.5 \leq \bar{X} \leq 1031.5) &= P(-2.25 \leq Z \leq 2.25) \\ &= 2P(0 \leq Z \leq 2.25) = 2(.4878) = .9756 \end{aligned}$$

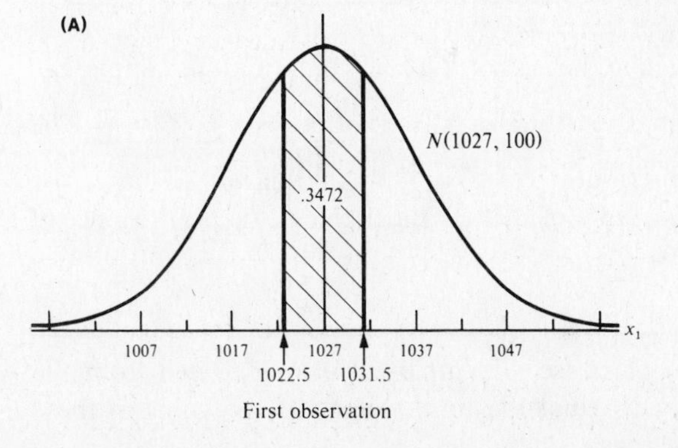

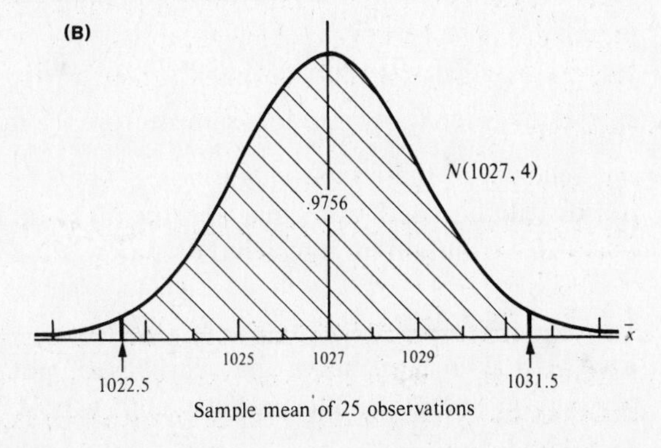

Figure 11-2

EXAMPLE 11-7 In Example 11-4, let $X_1, X_2, \ldots, X_{16}$ denote the weights of a random sample of size $n = 16$ bags of chips from the normal distribution $N(456.2, 5.76)$. **(a)** Find $P(X_7 \le 450.2)$. **(b)** Give the distribution of the sample mean $\bar{X}$. **(c)** Give the value of the standard error of the mean. **(d)** Find $P(\bar{X} \le 454.7)$.

Solution

(a) X_7 is the seventh bag in the random sample, and the distribution of X_7 (as well as the distribution of each individual item of the random sample) is $N(456.2, 5.76)$. First standardize to find the z value,

$$z = \frac{450.2 - 456.2}{2.4} = -2.50$$

and then find (see Fig. 11-3A)

$$\begin{aligned} P(X_7 \le 450.2) &= P(Z \le -2.50) \\ &= P(Z \ge 2.50) \\ &= .5000 - P(0 \le Z \le 2.50) \\ &= .5000 - .4938 = .0062 \end{aligned}$$

(b) Because $\bar{X}$ is the sample mean of a random sample of size $n = 16$ from a normal distribution $N(456.2, 5.76)$, the distribution of $\bar{X}$ is $N(456.2, 5.76/16)$.

(c) The standard error of the mean is

$$\sigma_{\bar{x}} = \sqrt{5.76}/\sqrt{16} = 2.4/4 = .6$$

(d) You find that

$$z = \frac{454.7 - 456.2}{2.4/4} = -2.50$$

and therefore (Fig. 11-3B)

$$P(\bar{X} \le 454.7) = P(Z \le -2.50) = .0062$$

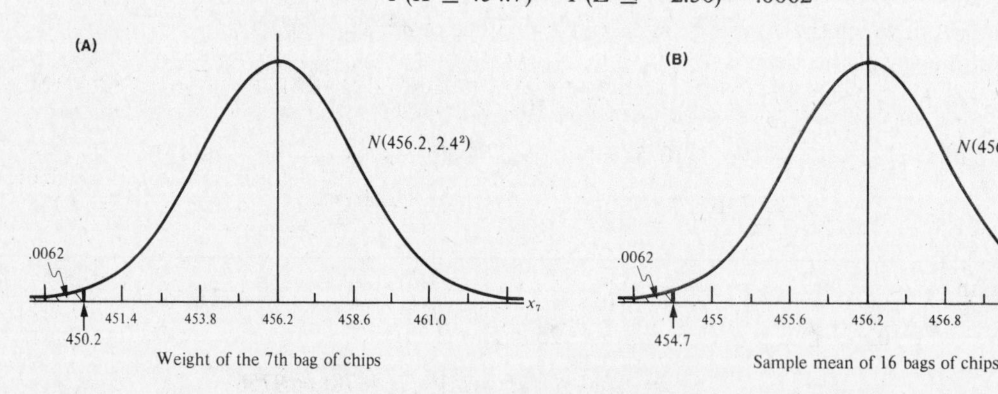

Figure 11-3

There are times when you may be interested in the sum of the items of a random sample. Let $X_1, X_2, \ldots, X_n$ be a random sample from the normal distribution $N(\mu, \sigma^2)$. Let $Y = \sum_{i=1}^{n} X_i = \sum X$ be the sum of these random variables. Then the distribution of $Y = \sum X$ is $N(n\mu, n\sigma^2)$. That is, the distribution of Y is normal, and the mean and variance of Y are n times the mean and variance of the distribution from which the sample is taken.

EXAMPLE 11-8 Suppose that 16 bags of potato chips (see Example 11-4) are selected from the production line at random and put into a box. Let Y equal the total weight of the chips in these 16 packages. Thus $Y = \sum_{i=1}^{16} X_i = \sum X$ where each X_i is $N(456.2, 5.76)$. **(a)** How is Y distributed? Find **(b)** $P(Y \le 7270.4)$, and **(c)** $P(Y > 7313.6)$.

Solution

(a) The distribution of $Y = \sum X$ is

$$N(n\mu, n\sigma^2) = N[(16)(456.2), (16)(5.76)] = N(7299.2, 92.16)$$

(b) Since Y has a distribution of $N(7299.2, 92.16)$,

$$z = \frac{7270.4 - 7299.2}{\sqrt{92.16}} = -3.00$$

and (see Fig. 11-4)

$$\begin{aligned} P(Y \le 7270.4) &= P(Z \le -3.00) \\ &= P(Z \ge 3.00) \\ &= .5000 - P(0 \le Z \le 3.00) \\ &= .5000 - .4987 = .0013 \end{aligned}$$

(c)

$$z = \frac{7313.6 - 7299.2}{9.6} = 1.50$$

and (see Fig. 11-4)

$$\begin{aligned} P(Y > 7313.6) &= P(Z > 1.50) \\ &= .5000 - P(0 \le Z \le 1.50) \\ &= .5000 - .4332 = .0668 \end{aligned}$$

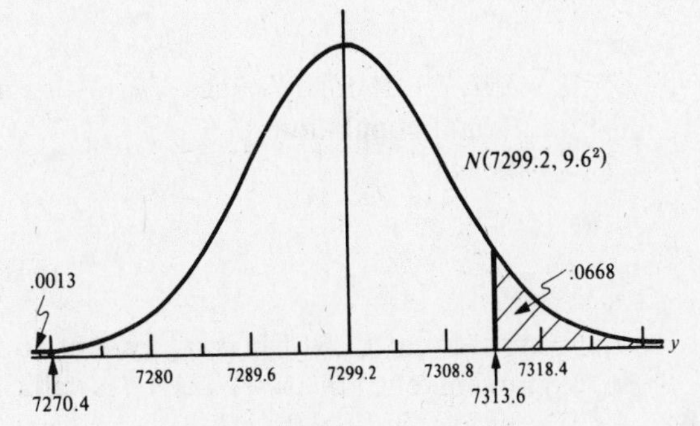

Figure 11-4

B. The central limit theorem

Once you have learned about the distribution of the sample mean for normal populations, you can use the following theorem to generalize those concepts to distributions that are not necessarily normal.

Central limit theorem

Let $X_1, X_2, \ldots, X_n$ be the items of a random sample of size n from any population (or distribution) that has mean μ and variance σ^2. Then the distribution of

$$Z = \frac{\bar{X} - \mu}{\sigma/\sqrt{n}} \tag{11.4}$$

or (multiplying numerator and denominator by n)

$$Z = \frac{\sum X - n\mu}{\sqrt{n}\,\sigma} = \frac{Y - n\mu}{\sqrt{n}\,\sigma} = \frac{Y - n\mu}{\sqrt{n\sigma^2}} \tag{11.5}$$

has a distribution that is approximately $N(0, 1)$, provided that n is sufficiently large.

To use the Central Limit Theorem to find approximate probabilities, suppose that $\bar{X}$ is the sample mean of a random sample of size n from a population that has a mean of μ and a variance of σ^2. To find $P(\bar{X} \le c)$, you first find

$$z = \frac{c - \mu}{\sigma/\sqrt{n}} \tag{11.6}$$

Then

$$P(\bar{X} \le c) = P\left(\frac{\bar{X} - \mu}{\sigma/\sqrt{n}} \le \frac{c - \mu}{\sigma/\sqrt{n}}\right) \approx P(Z \le z)$$

where z is given by formula (11.6). You'll find this latter probability in Table 4 in the Appendix.

If $Y = \sum X$, to find $P(Y \le c)$, you first find

$$z = \frac{c - n\mu}{\sqrt{n}\,\sigma} = \frac{c - n\mu}{\sqrt{n\sigma^2}} \tag{11.7}$$

Then

$$P(Y \le c) = P\left(\frac{Y - n\mu}{\sqrt{n}\,\sigma} \le \frac{c - n\mu}{\sqrt{n}\,\sigma}\right) \approx P(Z \le z)$$

where z is given by formula (11.7). You find this latter probability in Table 4 in the Appendix.

Several points must be made in connection with the Central Limit Theorem:

1. If the sample is taken from a population that is *normally distributed* with a mean of μ and a variance of σ^2, $N(\mu, \sigma^2)$, then

$$Z = \frac{\bar{X} - \mu}{\sigma/\sqrt{n}}$$

is *exactly* $N(0, 1)$ for all values of n. (This agrees with what you learned about the distribution of $\bar{X}$ for normal populations.)

2. To say that

$$Z = \frac{\bar{X} - \mu}{\sigma/\sqrt{n}} = \frac{Y - n\mu}{\sqrt{n}\,\sigma}$$

is approximately $N(0, 1)$ is equivalent to saying that $\bar{X}$ is approximately $N(\mu, \sigma^2/n)$ and Y is approximately $N(n\mu, n\sigma^2)$ when n is sufficiently large.

3. It is always true that the mean of the distribution of $\bar{X}$ equals the mean of the population, and the variance of the distribution of $\bar{X}$ equals the population variance divided by n. But the conclusion that the sampling distribution is *normally distributed* with a mean of 0 and a variance of 1 is only an approximation.

4. Many authors suggest that in order for n to be sufficiently large, n should be at least 30. This is a very conservative rule. The size of n that is needed for the Central Limit Theorem to hold depends on the underlying distribution. If the underlying distribution is almost normal or is symmetric about its mean, a value of $n < 30$ is sufficiently large.

EXAMPLE 11-9 Example 11-5 specified a random sample of size 36 from a distribution with a mean of $\mu = 24.7$ and a standard deviation of $\sigma = 3.9$. (Notice that this distribution is not specified as normal.) Let $\bar{X}$ equal the sample mean of this random sample. **(a)** How is $\bar{X}$ distributed, approximately? **(b)** Give a function of $\bar{X}$ that has an approximate $N(0, 1)$ distribution. **(c)** Find $P(\bar{X} \le 23.79)$, approximately.

Solution

(a) According to point 2, the approximate distribution of $\bar{X}$ is $N(24.7, (3.9)^2/36)$. You know nothing about the underlying distribution, but $n = 36$ is sufficiently large for this approximation to hold.

(b) By formula (11.4), the distribution of

$$Z = \frac{\bar{X} - \mu}{\sigma/\sqrt{n}} = \frac{\bar{X} - 24.7}{3.9/\sqrt{36}}$$

is approximately $N(0, 1)$. Note that Z is a linear function of $\bar{X}$.

(c) You use formula (11.6) to find

$$z = \frac{23.79 - 24.7}{3.9/6} = -1.40$$

Thus (Fig. 11-5),

$$\begin{aligned} P(\bar{X} \le 23.79) &\approx P(Z \le -1.40) \\ &= P(Z > 1.40) \\ &= .5000 - P(0 \le Z \le 1.40) \\ &= .5000 - .4192 = .0808 \end{aligned}$$

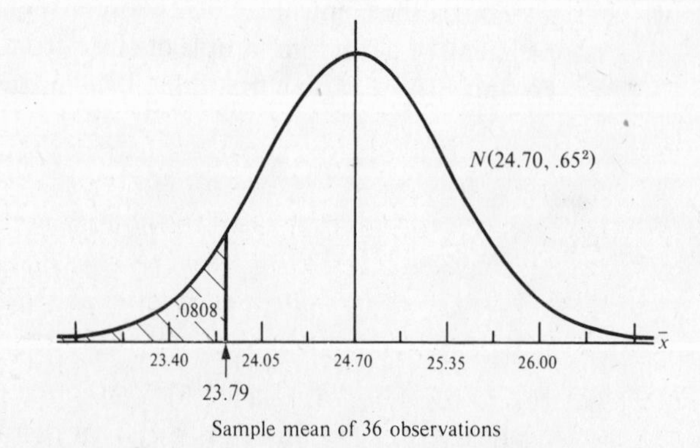

Figure 11-5

EXAMPLE 11-10 Let $\bar{X}$ be the sample mean of a random sample of size $n = 25$ from a distribution that has a mean of $\mu = 6.047$ and a standard deviation of $\sigma = .02$. **(a)** Give the approximate distribution of $\bar{X}$, and **(b)** the standard error of the mean. **(c)** Find $P(\bar{X} \le 6.036)$, approximately.

Solution

(a) By point 2, the approximate distribution of $\bar{X}$ is $N(6.047, (.02)^2/25)$.
(b) The standard error of the mean is $\sigma/\sqrt{n} = .02/5 = .004$.
(c) By formula (11.6),

$$z = \frac{6.036 - 6.047}{.02/5} = -2.75$$

Thus (Fig. 11-6),

$$P(\bar{X} \le 6.036) \approx P(Z \le -2.75)$$
$$= .5000 - P(0 \le Z \le 2.75)$$
$$= .5000 = .4970 = .0030$$

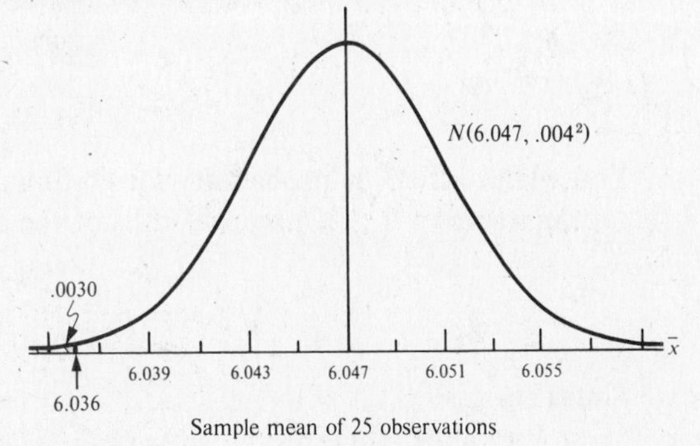

Figure 11-6

EXAMPLE 11-11 Let X equal the weight of a "20-g" mint. Suppose that the mean of X is $\mu = 21.53$ and the variance of X is $\sigma^2 = 1/3$. Let Y equal the total weight of $n = 48$ mints that are selected at random. **(a)** Give the approximate distribution of Y. **(b)** Give a function of Y that is approximately $N(0, 1)$. **(c)** Find $P(Y \le 1025)$, approximately.

Solution

(a) By point 2, the approximate distribution of Y is $N[(48)(21.53), (48)(1/3)] = N(1033.44, 16)$.
(b) By formula (11.5),

$$Z = \frac{Y - (48)(21.53)}{\sqrt{48(1/3)}} = \frac{Y - 1033.4}{4}$$

is approximately $N(0, 1)$.
(c) By formula (11.7),

$$z = \frac{1025 - (48)(21.53)}{\sqrt{48(1/3)}}$$
$$= \frac{1025 - 1033.44}{\sqrt{16}} = -2.11$$

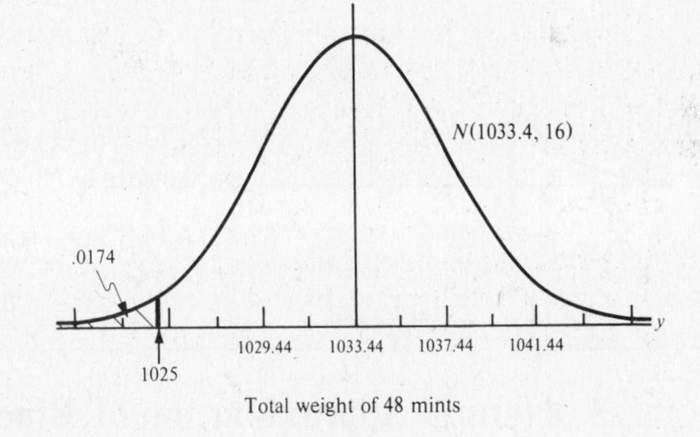

Figure 11-7

Thus (Fig. 11-7),

$$P(Y \le 1025) \approx P(Z \le -2.11)$$
$$= .5000 - .4826 = .0174$$

EXAMPLE 11-12 Let X equal the outcome when you roll a 4-sided die. Then $\mu = 2.5$ and $\sigma^2 = 1.25$ are the mean and variance of X. Suppose your employer, who is a gambler, agrees to donate money to your alma mater equal to the sum of the outcomes when you roll the 4-sided die 45 times. Let Y equal the sum of the 45 consecutive rolls of the die. **(a)** Give the approximate distribution of Y. **(b)** Find $P(100.5 \le Y \le 130.5)$, approximately. **(c)** Would you rather roll the die to determine your employer's contribution or obtain an outright contribution of \$121.50 for your alma mater?

Solution

(a) By point 2, the approximate distribution of Y is

$$N(n\mu, n\sigma^2) = N[(45)(2.5), (45)(1.25)] = N(112.5, 56.25)$$

(b) By formula (11.7),

$$z = \frac{100.5 - 112.5}{\sqrt{56.25}} = -1.60 \qquad z = \frac{130.5 - 112.5}{\sqrt{56.25}} = 2.40$$

Thus (Fig. 11-8),

$$\begin{aligned} P(100.5 \le Y \le 130.5) &\approx P(-1.60 \le Z \le 2.40) \\ &= P(-1.60 \le Z \le 0) + P(0 \le Z \le 2.40) \\ &= .4452 + .4918 = .9370 \end{aligned}$$

(c) You want to find the probability for getting more than \$121.50 when rolling the die—that is, for getting a sum of $Y > 121.5$ on 45 rolls of the die. By (11.7),

$$z = \frac{121.5 - 112.5}{\sqrt{56.25}} = 1.20$$

Thus (Fig. 11-9),

$$\begin{aligned} P(Y > 121.5) &\approx P(Z > 1.20) \\ &= .5000 - .3849 \\ &= .1151 \end{aligned}$$

Since you have only an 11.5% chance of receiving more than \$121.50, you would probably prefer to receive a check for \$121.50 instead of rolling the die for the contribution.

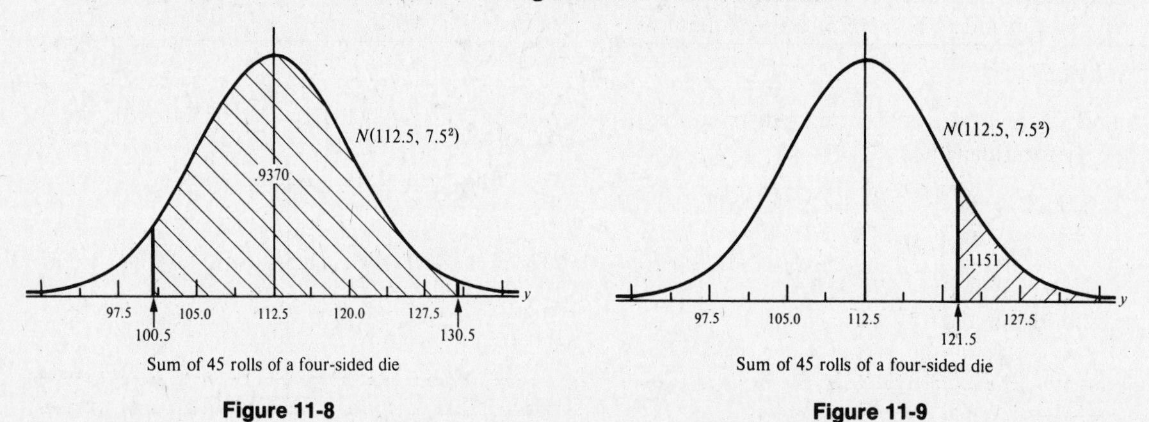

Figure 11-8 **Figure 11-9**

11-3. Normal Approximation of Binomial Probabilities

Let $X_1, X_2, \ldots, X_n$ be a random sample of n Bernoulli random variables. That is, X_i is $b(1, p)$ for $i = 1, 2, \ldots, n$. Recall that $\mu = 1p = p$ and $\sigma^2 = 1p(1-p) = p(1-p) = pq$. Let $Y = \sum_{i=1}^{n} X_i$. Applying formula (11.5) with $\mu = p$ and $\sigma^2 = p(1-p)$, the Central Limit Theorem says that

$$Z = \frac{Y - n\mu}{\sqrt{n}\,\sigma} = \frac{Y - np}{\sqrt{n}\sqrt{p(1-p)}}$$

has a distribution that is approximately $N(0, 1)$. Recall, however, that the exact distribution of Y, the sum of n Bernoulli trials, is $b(n, p)$ and that the mean of Y is $\mu = np$ and the variance of Y is $np(1-p)$. We thus have the following theorem.

THEOREM: If Y has a binomial distribution $b(n, p)$, then the distribution of

$$Z = \frac{Y - np}{\sqrt{np(1-p)}} = \frac{Y - np}{\sqrt{npq}} \qquad \textbf{(11.8)}$$

is approximately $N(0, 1)$ provided that $np \ge 5$ and $n(1-p) \ge 5$.

Several points must be made in connection with this theorem:

1. We often use X rather than Y for a binomial random variable. If X is a binomial random variable $b(n, p)$, X has a mean of $\mu = np$ and a variance of $\sigma^2 = np(1-p)$. Formula (11.8) states that if you subtract its mean np from a binomial random variable X and divide this difference by the standard deviation $\sqrt{np(1-p)}$, the resulting random variable is approximately $N(0, 1)$.
2. From formula (11.8) we can say that if the distribution of X is $b(n, p)$, then the approximate distribution of X is $N[np, np(1-p)]$ provided that $np \geq 5$ and $n(1-p) \geq 5$.
3. The reason for stating that $np \geq 5$ and $n(1-p) \geq 5$ is to ensure that n is sufficiently large so that the approximation is good. Most statisticians accept this rule-of-thumb for using the normal distribution to approximate binomial probabilities.
4. Because we are using the standard normal distribution, which is continuous, to approximate probabilities for a binomial distribution, which is discrete, we make what is called a *half-unit correction for continuity*. This is done by adjusting the probability region by .5 at each included end-point of the uncorrected region. So, for example, if X is $b(n, p)$, to find $P(X \leq k)$, approximately, where k is an integer, we first note that

$$P(X \leq k) = P(X \leq k + .5)$$

since a binomial random variable always equals an integer. Then find the value of

$$z = \frac{k + .5 - np}{\sqrt{np(1-p)}} \tag{11.9}$$

It follows that

$$P(X \leq k) = P(X \leq k + .5) \approx P(Z \leq z)$$

where z is given by (11.9) and the latter probability is found in Table 4 in the Appendix.

Similarly, to find $P(k \leq X)$, approximately, first make a half-unit *decrease* in the left-hand limit, $P(k \leq X) = P(k - .5 \leq X)$. Then find the value of

$$z = \frac{k - .5 - np}{\sqrt{np(1-p)}} \tag{11.10}$$

It follows that

$$P(k \leq X) = P(k - .5 \leq X) \approx P(z \leq Z)$$

where z is given by (11.10) and the latter probability is found in Table 4 in the Appendix.

EXAMPLE 11-13 Let the distribution of X be $b(16, \frac{1}{2})$. **(a)** Sketch the graph of the probability histogram for the distribution of X. Superimpose over this graph the pdf for the normal distribution $N[np, np(1-p)] = N[16(\frac{1}{2}), 16(\frac{1}{2})(\frac{1}{2})] = N(8, 4)$. **(b)** Find $P(9 \leq X \leq 12)$ using the Binomial Probabilities Table (Table 2 in the Appendix). Shade the rectangles in Fig. 11-10 that correspond to this probability. **(c)** The area under the pdf between 8.5 and 12.5 for the normal distribution $N(8, 4)$ gives an approximation of $P(9 \leq X \leq 12)$. Find this normal approximation.

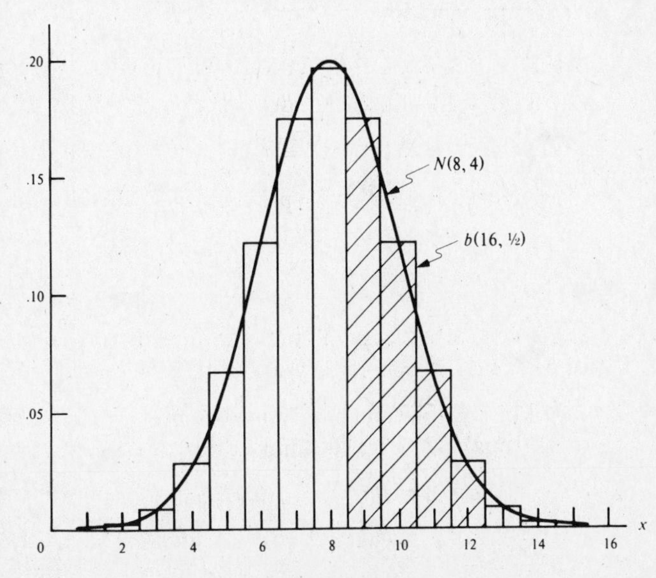

Figure 11-10

Solution

(a) To construct the probability histogram in Fig. 11-10, use the binomial probabilities from Table 2 in the Appendix with $n = 16$ and $p = .5$. The normal probability curve will intersect each rectangle close to its midpoint.

(b) From Table 2,

$$P(9 \le X \le 12) = P(X = 9, 10, 11, 12)$$
$$= .1746 + .1222 + .0667 + .0278$$
$$= .3913$$

(c) First make the half-unit correction for continuity. Both 9 and 12 are included in the desired probability region (note the less than *or* equal to signs), so the corrected region becomes 8.5 to 12.5. Now apply formulas (11.10) and (11.9).

$$z = \frac{8.5 - 16(\frac{1}{2})}{\sqrt{16(\frac{1}{2})(\frac{1}{2})}} \qquad z = \frac{12.5 - 16(\frac{1}{2})}{\sqrt{16(\frac{1}{2})(\frac{1}{2})}}$$
$$= \frac{8.5 - 8}{\sqrt{4}} = .25 \qquad = \frac{12.5 - 8}{\sqrt{4}} = 2.25$$

Thus,

$$P(9 \le X \le 12) = P(8.5 \le X \le 12.5)$$
$$\approx P(.25 \le Z \le 2.25)$$
$$= .4878 - .0987 = .3891$$

You see that the answers to **(b)** and **(c)** both equal 0.39 when rounded off to 2 decimal places. Thus, the normal approximation to binomial probabilities is quite good in this case.

EXAMPLE 11-14 For each question on a 48-question "multiple guess" test there are four possible answers, of which exactly 1 is correct. Let X equal the number of correct answers if you guess on each question. **(a)** How is X distributed? **(b)** Use the normal approximation to find the probability that you correctly guess *at most* 10 answers; that is, find $P(X \le 10)$, approximately. **(c)** Use the normal approximation to find $P(X > 16)$, approximately.

Solution

(a) If you treat each of the 48 questions as a Bernoulli trial with $p = \frac{1}{4}$ as the probability for success, then the distribution of X is $b(48, \frac{1}{4})$. That is, X has a binomial distribution with $n = 48$ and $p = \frac{1}{4}$.

(b) You are going to use the normal approximation, so first make the half-unit correction for continuity. Since 10 is an included end-point, the probability is extended to $X \le 10.5$. Now apply formula (11.9):

$$z = \frac{10.5 - 48(\frac{1}{4})}{\sqrt{48(\frac{1}{4})(\frac{3}{4})}}$$
$$= \frac{10.5 - 12}{\sqrt{9}}$$
$$= -.50$$

Finally, you obtain the approximate probability (Fig. 11-11):

$$P(X \le 10) \approx P(Z \le -.50)$$
$$= .5000 - .1915 = .3085$$

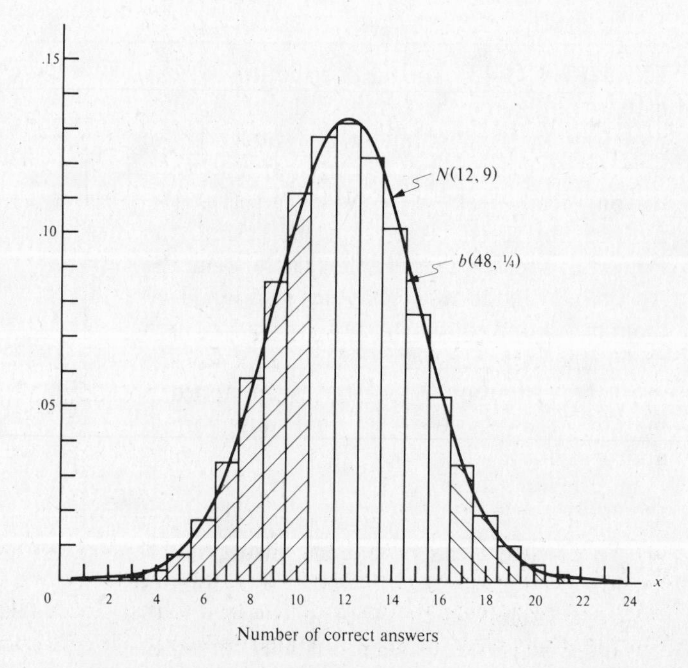

Figure 11-11

(c) $P(X > 16) = P(16 < X) = P(17 \le X)$

The half-unit correction for continuity gives us

$$P(17 \le X) = P(16.5 \le X)$$

You can see that the *integers* that satisfy the inequality on the left (17, 18, ..., 48) also satisfy the inequality on the right after the half-unit correction has been made. *Make sure that this always holds true.* Applying (11.10), you find

$$z = \frac{16.5 - 12}{\sqrt{9}} = 1.50$$

Now you find the approximate probability:

$$P(X > 16) = P(16.5 \le X) \approx P(1.50 \le Z) = .5000 - .4332 = .0668$$

EXAMPLE 11-15 Among gifted seventh-graders who scored very high on a mathematics exam, approximately 20% ($\frac{1}{5}$) were left-handed or ambidextrous. Let X equal the number of left-handed or ambidextrous students among a random sample of $n = 25$ gifted seventh-graders. **(a)** How is X distributed? **(b)** Use the normal approximation to find the probability that there are less than 10 left-handed or ambidextrous seventh-grade students in a random sample of 25; that is, find $P(X < 10)$, approximately.

Solution

(a) If you assume that these are 25 Bernoulli trials, then X is $b(25, \frac{1}{5})$.
(b) Note that $P(X < 10) = P(X \le 9)$. So you can apply formula (11.9) to find

$$z = \frac{9.5 - 25(\frac{1}{5})}{\sqrt{25(\frac{1}{5})(\frac{4}{5})}} = \frac{9.5 - 5}{\sqrt{4}} = 2.25$$

Therefore, the approximate probability is (Fig. 11-12)

$$P(X < 10) \approx P(Z \le 2.25)$$
$$= .5000 + .4878 = .9878$$

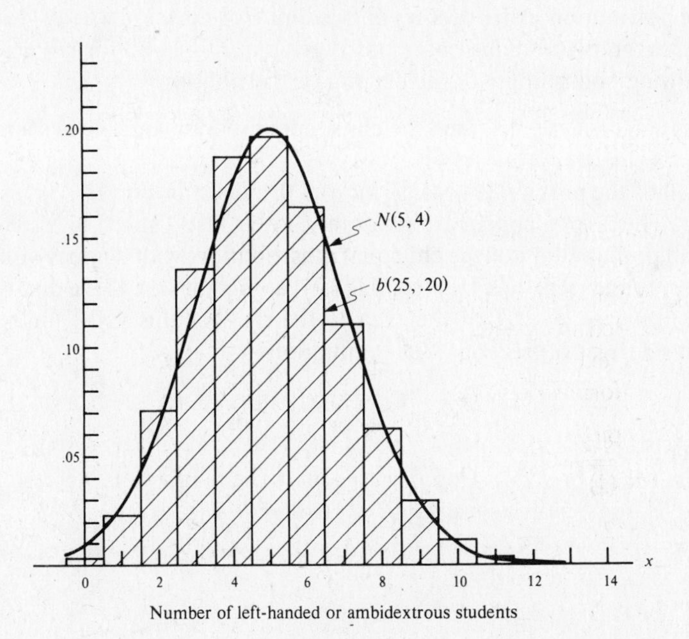

Figure 11-12

EXAMPLE 11-16 In the casino game of roulette, the probability for winning with a bet on red is $p = \frac{18}{38}$. After winning the State lottery, you feel lucky and decide to play roulette and gamble your winnings by placing 1000 consecutive bets on red. Let X equal the number of winning bets out of the 1000 that you place. **(a)** How is X distributed? **(b)** Use the normal approximation to find the probability that you will win more bets than you lose; that is, find $P(X > 500)$, approximately.

Solution

(a) If you treat each bet as a Bernoulli trial, X is $b(1000, \frac{18}{38})$.
(b) You note that $P(X > 500) = P(500 < X) = P(501 \le X)$. Using formula (11.10),

$$z = \frac{500.5 - 1000(\frac{18}{38})}{\sqrt{1000(\frac{18}{38})(\frac{20}{38})}} = \frac{26.816}{15.789} = 1.70$$

Thus
$$P(X > 500) = P(500.5 \le X) \approx P(1.70 \le Z) = .5000 - .4554 = .0446$$

The probability for winning more bets than you lose is very slim. Better save your money!

11-4. The Chi-Square Distribution

There are three distributions that play an important role in **statistical inference**, which is the process of drawing conclusions about an entire population based on sample observations. They are especially important when sampling from normal distributions. The *chi-square distribution* is discussed in this section. The Student's t and the F distributions will be discussed in the remaining sections of this chapter.

A. The chi-square distribution

A chi-square pdf is positive only when x is positive. It is a skewed curve and there is a different curve for each positive integer r. The integer r is called the number of degrees of freedom and we say that X has a chi-square distribution with r degrees of freedom. For notation we say that the distribution of X is $\chi^2(r)$. A typical graph of a chi-square pdf is given in Fig. 11-13 for $r = 6$ degrees of freedom.

The following points apply to the chi-square distribution:

1. If the distribution of X is $\chi^2(r)$, then the chi-square random variable has a mean of $\mu = r$ and a variance of $\sigma^2 = 2r$.
2. Notice that all of the probability occurs for $x > 0$, as seen in Fig. 11-13.
3. To find probabilities for a chi-square random variable, use Table 5 in the Appendix. This table gives right-tail probabilities for the chi-square distribution, with degrees of freedom listed down the left-hand side and right-tail area (probability) across the top. If the distribution of X is $\chi^2(r)$, then $\chi^2_\alpha(r)$ is a number such that the area to the right of this value for a chi-square random variable with r degrees of freedom is α. In notation,

$$P[X > \chi^2_\alpha(r)] = \alpha$$

See Figure 11-14.
4. Many authors use df or $d.f.$ instead of r to denote the number of degrees of freedom.
5. In place of $\chi^2_\alpha(r)$, some authors use χ^2_α, $\chi^2(\alpha, r)$, or $\chi^2(\alpha, df)$.

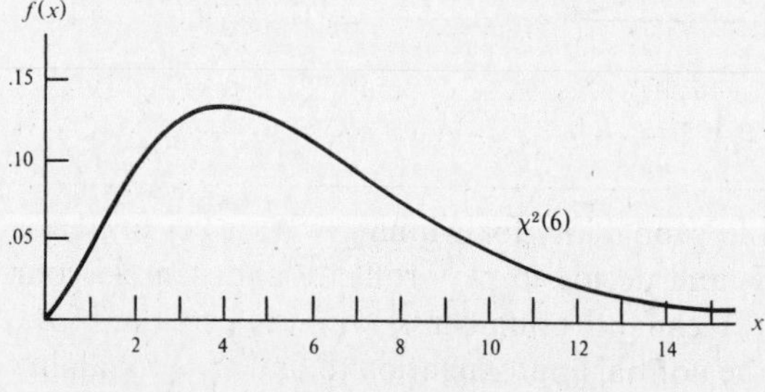

Figure 11-13

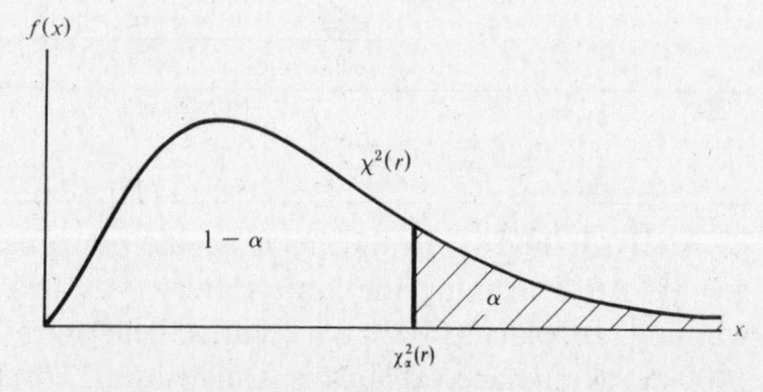

Figure 11-14

EXAMPLE 11-17 Let X be a random variable that has a chi-square distribution with 6 degrees of freedom, $\chi^2(6)$. Give **(a)** the mean of X, and **(b)** the variance of X. Find the values of **(c)** $\chi^2_{.05}(6)$ and **(d)** $\chi^2_{.90}(6)$.

Solution

(a) The mean of any chi-square random variable is equal to the number of degrees of freedom. Thus the mean of X is

$$\mu = r = 6$$

(b) The variance of X is equal to twice the number of degrees of freedom:

$$\sigma^2 = 2r = (2)(6) = 12$$

For **(c)** and **(d)** we reproduce a part of Table 5 from the Appendix.

r \ α	.95	.90	.10	.05
6	1.64	[2.20]	10.64	[12.59]
7	2.17	2.83	12.02	14.07
8	2.73	3.49	13.36	15.51
9	3.33	4.17	14.68	16.92

(c) Here you want to find a value such that the probability for obtaining an observation greater than this value for a chi-square random variable with $r = 6$ is .05. Look down the r column to the row $r = 6$, and then across to the column headed by .05 to read (see Fig. 11-15)

$$\chi^2_{.05}(6) = 12.59$$

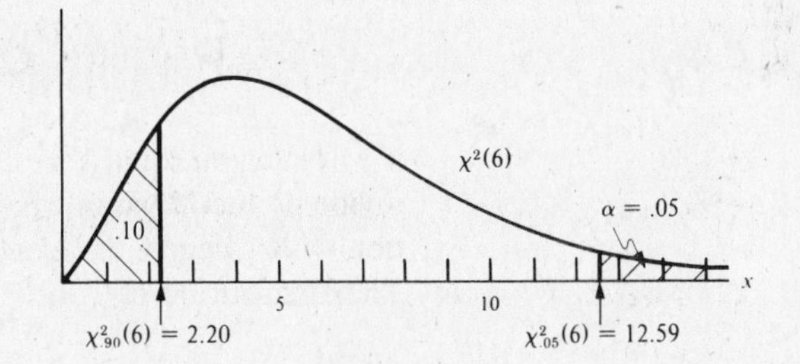

Figure 11-15

(d) In this case, you are looking for the value such that 90% of the probability is to its right. If you look down the r column to the row $r = 6$ and then across to the column headed by .90, you read (see Fig. 11-15)

$$\chi^2_{.90}(6) = 2.20$$

EXAMPLE 11-18 Let the distribution of X be $\chi^2(9)$. Find **(a)** $\chi^2_{.10}(9)$, **(b)** $\chi^2_{.90}(9)$, **(c)** $P(X \leq 14.68)$, **(d)** $P(X \leq 4.17)$, and **(e)** $P(4.17 \leq X \leq 14.68)$.

Solution

(a) Turning to Table 5 in the Appendix, read down the r column to $r = 9$, and across to the column headed by .10 to find (Fig. 11-16)

$$\chi^2_{.10}(9) = 14.68$$

(b) Read down to $r = 9$ in Table 5 and across to the column headed by .90 to find

$$\chi^2_{.90}(9) = 4.17$$

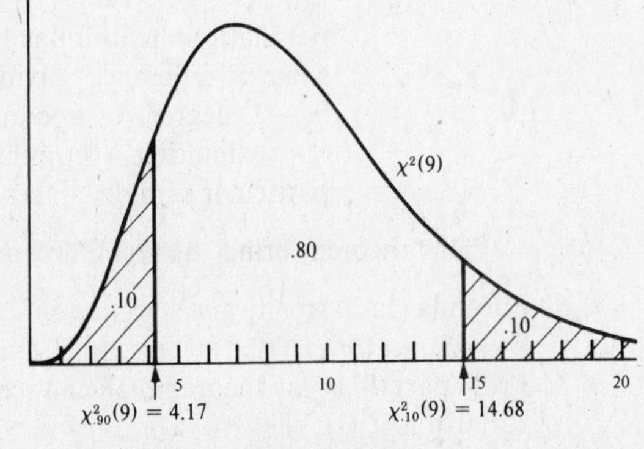

Figure 11-16

(c) Look across the row $r = 9$ in Table 5 until you find the value 14.68. This corresponds to an upper-tail probability of $\alpha = .10$. Therefore,

$$P(X \leq 14.68) = 1 - P(X > 14.68)$$
$$= 1 - .10 = .90$$

(d) Looking across the row $r = 9$ in Table 5, you find that the value 4.17 corresponds to an upper-tail area of $\alpha = .90$. Therefore,

$$P(X \leq 4.17) = 1 - P(X > 4.17) = 1 - .90 = .10$$

(e) Using the results you obtained in **(c)** and **(d)**, you find

$$P(4.17 \leq X \leq 14.68) = P(X \leq 14.68) - P(X < 4.17) = .90 - .10 = .80$$

The following theorem illustrates how the chi-square distribution is used in connection with random samples from normal distributions.

THEOREM: Let $X_1, X_2, \ldots, X_n$ be the items of a random sample of size n from a normal distribution $N(\mu, \sigma^2)$.

(a) For each i, the distribution of

$$Z_i = \frac{X_i - \mu}{\sigma} \tag{11.11}$$

is $N(0, 1)$. That is, each individual item can be transformed into a standard normal random variable by subtracting the mean and dividing the difference by the standard deviation.

(b) For each i, the distribution of

$$Z_i^2 = \frac{(X_i - \mu)^2}{\sigma^2} \tag{11.12}$$

is $\chi^2(1)$. If you square the difference between the ith item and the mean and then divide by the variance, the resulting random variable has a chi-square distribution with 1 degree of freedom.

(c) The distribution of

$$W = \frac{\sum_{i=1}^{n} (X_i - \mu)^2}{\sigma^2} = \frac{\sum (X - \mu)^2}{\sigma^2} \tag{11.13}$$

is $\chi^2(n)$. This part of the theorem says that the *sum* of the squared differences for all n items in the random sample divided by the variance has a chi-square distribution with n degrees of freedom.

(d) The distribution of

$$\frac{(n-1)S^2}{\sigma^2} = \frac{\sum_{i=1}^{n} (X_i - \bar{X})^2}{\sigma^2} = \frac{\sum (X - \bar{X})^2}{\sigma^2} \tag{11.14}$$

is $\chi^2(n - 1)$. Formula (11.14) is very similar to (11.13), except that here the population mean μ has been replaced by the sample mean $\bar{X}$. Now the sum of squared differences divided by the variance has a chi-square distribution with $(n - 1)$ degrees of freedom—a degree of freedom has been "lost," since we are approximating a population parameter (μ) with a value estimated from the particular sample chosen ($\bar{X}$).

This theorem brings up the following additional points about the chi-square distribution:

6. Formula (11.11) really gives no new information, since we already knew that the distribution of $Z = (X - \mu)/\sigma$ is $N(0, 1)$ when the distribution of X is $N(\mu, \sigma^2)$.

7. From part **(b)** of the theorem you can see that the distribution of the square of a $N(0, 1)$ random variable is $\chi^2(1)$.

8. Notice that the number of degrees of freedom in part **(c)** is equal to the number of terms in the sum—that is, the number of summands.

9. Notice again that when the distribution mean μ in part **(c)** is replaced by the sample mean $\bar{X}$ in part **(d)**, that 1 degree of freedom is lost.

EXAMPLE 11-19 Let $X_1, X_2, \ldots, X_7$ be a random sample of size 7 from a normal distribution $N(1027, 100)$. Let S^2 be the sample variance of this random sample. **(a)** Give the distribution of $6S^2/100$. Find **(b)** $P(6S^2/100 > 12.59)$ and **(c)** $P(6S^2/100 \leq 1.64)$.

Solution

(a) Since $n = 7$, $(n - 1) = 6$, and you can use formula (11.14) to find that the distribution of

$$\frac{(n-1)S^2}{\sigma^2} = \frac{6S^2}{100}$$

is $\chi^2(n - 1) = \chi^2(6)$.

(b) Using Table 5 in the Appendix, read across the row $r = 6$ until you find the number 12.59. This value is in the column headed by .05. Thus,

$$P\left(\frac{6S^2}{100} > 12.59\right) = .05$$

(c) In Table 5 read across the row $r = 6$ to find 1.64. This is in the column headed by .95. Thus,

$$P\left(\frac{6S^2}{100} > 1.64\right) = .95$$

It follows that

$$P\left(\frac{6S^2}{100} \leq 1.64\right) = 1 - P\left(\frac{6S^2}{100} > 1.64\right)$$
$$= 1 - .95 = .05$$

EXAMPLE 11-20 Let $X_1, X_2, \ldots, X_{10}$ be a random sample of size 10 from the normal distribution $N(456.2, 5.76)$. Let S^2 be the sample variance of this random sample. **(a)** Give the distribution of $9S^2/5.76$. **(b)** Find $P(9S^2/5.76 > 16.92)$.

Solution

(a) By formula (11.14), the distribution of $9S^2/5.76$ is $\chi^2(9)$.

(b) Using Table 5, look across the row $r = 9$ to find 16.92, which is in the column headed by .05. Thus

$$P\left(\frac{9S^2}{5.76} > 16.92\right) = .05$$

11-5. Student's *t* Distribution

A distribution that occurs very frequently in applications is the *t distribution.* A ***t*** **(or Student's *t*) random variable** can be defined as follows:

THEOREM: Let the distribution of U be $\chi^2(r)$ and the distribution of Z be $N(0, 1)$, with U and Z independent. Then the random variable T where

$$T = \frac{Z}{\sqrt{U/r}} \qquad \textbf{(11.15)}$$

has a t distribution with r degrees of freedom.

Notice the following points:

1. As the number of degrees of freedom increases, the graph of the pdf of T becomes closer to the pdf for the standard normal distribution. This is illustrated in Fig. 11-17.
2. If T has a t distribution with r degrees of freedom, then the mean of T is $\mu = 0$ when $r \geq 2$ (the distribution is symmetric about zero), and the variance of T is $\sigma^2 = r/(r - 2)$ when $r \geq 3$.

To find probabilities for the t (or Student's t) distribution, use Table 6 in the Appendix. The notation used is similar to that for the chi-square distribution. If T has a t distribution with r degrees of freedom, then $t_\alpha(r)$ is a number such that the probability that T obtains a value greater than this number is α. In symbols,

$$P[T > t_\alpha(r)] = \alpha$$

You can also say that the right-tail probability is equal to α. See Fig. 11-18.

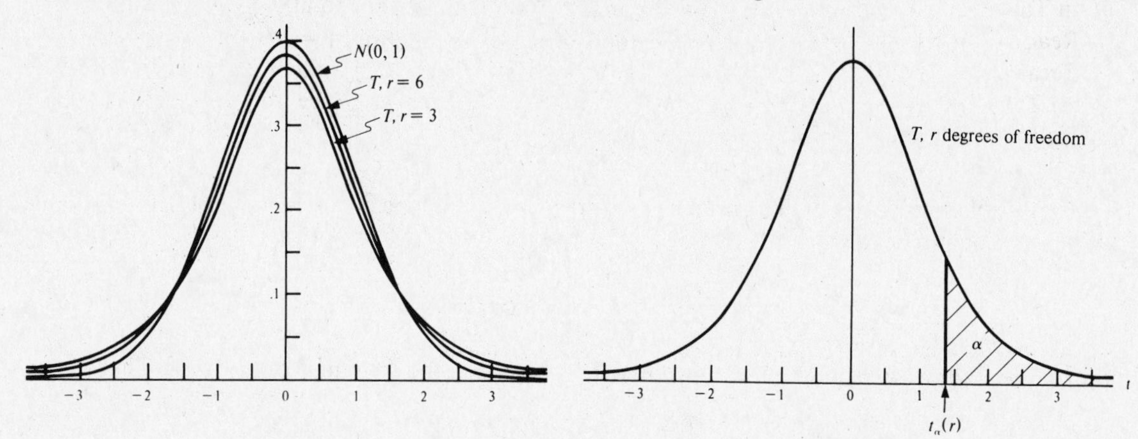

Figure 11-17

Figure 11-18

EXAMPLE 11-21 Let T have a t distribution with $r = 6$ degrees of freedom. Give **(a)** the mean and **(b)** the variance of T. Find the values of **(c)** $t_{.05}(6)$, **(d)** $t_{.01}(6)$, and **(e)** $-t_{.05}(6) = t_{.95}(6)$.

Solution

(a) Point 2 states that every t random variable for $r \geq 2$ has a mean of $\mu = 0$.
(b) Point 2 also gives the variance of a t random variable for $r \geq 3$ by

$$\sigma^2 = \frac{r}{r-2} = \frac{6}{4} = 1.5$$

For **(c)** and **(d)** we reproduce a part of Table 6 from the Appendix.

r \ α	.10	.05	.025	.01
6	1.44	1.94	2.45	3.14
7	1.42	1.89	2.36	3.00
8	1.40	1.86	2.31	2.90
9	1.38	1.83	2.26	2.82

(c) To use this table, you find the row $r = 6$ and then look across to the column headed by $\alpha = .05$ to read $t_{.05}(6) = 1.94$.
(d) Look across the row $r = 6$ to the column headed by $\alpha = .01$, to find that $t_{.01}(6) = 3.14$.
(e) Because of the symmetry of the pdf for the t distribution, the number that has a .05 probability to the left of it is just the negative of the number that has a .05 probability to the right of it (see Fig. 11-19). Thus,

$$-t_{.05}(6) = -1.94 = t_{.95}(6)$$

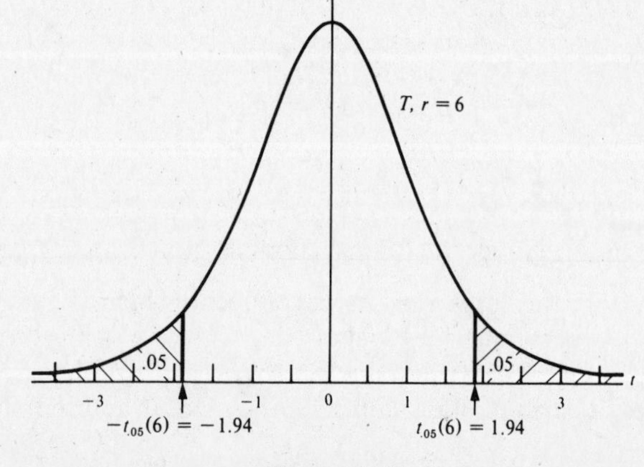

Figure 11-19

EXAMPLE 11-22 Let T have a t distribution with $r = 9$ degrees of freedom. Find **(a)** $t_{.025}(9)$, **(b)** $t_{.10}(9)$, **(c)** $-t_{.025}(9)$, and **(d)** $P(-2.26 \le T \le 2.26)$.

Solution

(a) In Table 6, read across the row $r = 9$ to the column headed by .025 to find $t_{.025}(9) = 2.26$.
(b) Read across the row $r = 9$ to the column headed by .10 to find $t_{.10}(9) = 1.38$.
(c) Because the t distribution is symmetric about zero, the answer here is the negative of the answer to **(a)**. Thus, $-t_{.025}(9) = -2.26 = t_{.975}(9)$. Notice that the probability to the *left* of -2.26 is .025 and to the *right* of 2.26 is .025 (see Fig. 11-20).

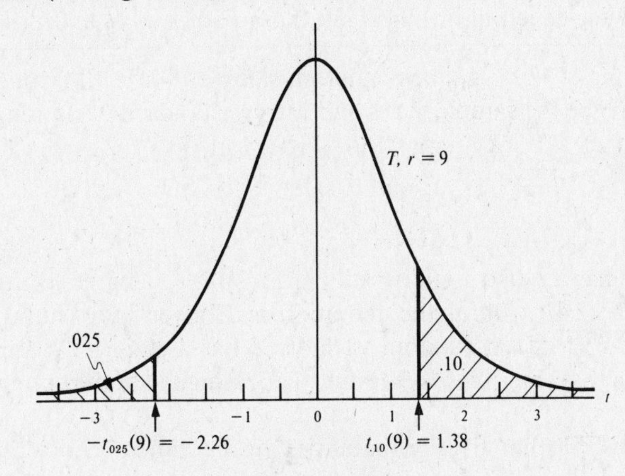

Figure 11-20

(d) In **(a)** you found that $P(T > 2.26) = .025$. So you know that $P(T \le 2.26) = 1 - .025 = .975$. And from **(c)** you know that $P(T \le -2.26) = .025$. Combining these results you find

$$P(-2.26 \le T \le 2.26) = P(T \le 2.26) - P(T \le -2.26)$$
$$= .975 - .025 = .95$$

You could also solve this problem by noticing that

$$P(-2.26 < T < 2.26) = 1 - P(T < -2.26) - P(T > 2.26) = 1 - .025 - .025 = .950$$

The next theorem illustrates how to use the t distribution in connection with random samples from normal distributions.

THEOREM: Let $X_1, X_2, \ldots, X_n$ be a random sample of size n from a normal distribution $N(\mu, \sigma^2)$.

(a) The distribution of

$$Z = \frac{\bar{X} - \mu}{\sigma/\sqrt{n}} \tag{11.16}$$

is $N(0, 1)$.

(b) The distribution of

$$U = \frac{(n-1)S^2}{\sigma^2} \tag{11.17}$$

is $\chi^2(n-1)$.

(c) From (11.15), (11.16), and (11.17), it follows that

$$T = \frac{Z}{\sqrt{\dfrac{U}{n-1}}} = \frac{\dfrac{\bar{X} - \mu}{\sigma/\sqrt{n}}}{\sqrt{\dfrac{(n-1)S^2}{\sigma^2(n-1)}}} = \frac{\bar{X} - \mu}{S/\sqrt{n}} \tag{11.18}$$

has a t distribution with $r = n - 1$ degrees of freedom.

EXAMPLE 11-23 Let $X_1, X_2, \ldots, X_7$ be a random sample of size 7 from a normal distribution $N(1027, 100)$. Let $\bar{X}$ and S be the sample mean and sample standard deviation of this random sample. **(a)** Give the distribution of $T = (\bar{X} - \mu)/(S/\sqrt{7})$. Find **(b)** $P(T > 3.14)$ and **(c)** $P(T < -1.94)$.

Solution

(a) Using formula (11.18), you see that T has a t distribution with $r = 7 - 1 = 6$ degrees of freedom.
(b) In Table 6, look across row $r = 6$ until you find the number 3.14. You see that this is in the column headed by $\alpha = .01$. Thus, $P(T > 3.14) = .01$.
(c) In Table 6, in the row $r = 6$, you see that 1.94 is in the column headed by $\alpha = .05$. Thus the probability to the right of 1.94 or to the left of -1.94 is .05 so, $P(T < -1.94) = .05$. (See Fig. 11-19.)

EXAMPLE 11-24 Let $X_1, X_2, \ldots, X_{10}$ be a random sample of size 10 from the normal distribution $N(456.2, 5.76)$. Let $\bar{X}$ and S be the sample mean and sample standard deviation of this random sample. Give the distribution **(a)** of $T = (\bar{X} - 456.2)/(S/\sqrt{10})$, and **(b)** of $Z = (\bar{X} - 456.2)/(2.4/\sqrt{10})$. Find **(c)** $P(T > 1.83)$ and **(d)** $P(Z > 1.83)$.

Solution

(a) By formula (11.18), T has a t distribution with $r = 10 - 1 = 9$ degrees of freedom.
(b) Since $Z = (X - 456.2)/(2.4/\sqrt{10})$ fits the definition of Z given in formula (11.16), with $\mu = 456.2$, $\sigma = \sqrt{5.76} = 2.4$, and $n = 10$, the random variable Z has a standard normal distribution $N(0, 1)$.
(c) Using Table 6, read across the row $r = 9$ to find 1.83, which is in the column headed by .05. Thus, $P(T > 1.83) = .05$.
(d) Here you must use the table that gives probabilities for the standard normal distribution (Table 4 in the Appendix):

$$P(Z > 1.83) = .5000 - P(0 \le Z \le 1.83) = .5000 - .4664 = .0336$$

Parts **(c)** and **(d)** illustrate that the t distribution has more probability in the tails than the normal distribution does, since the probability that T is greater than some value is *more* than the probability that Z is greater than the same value. (Look again at Fig. 11-17).

11-6. The *F* Distribution

An ***F* random variable** can be defined as follows:

THEOREM: Let the distribution of U be $\chi^2(r_1)$, the distribution of V be $\chi^2(r_2)$, with U and V independent. Then

$$F = \frac{U/r_1}{V/r_2} \tag{11.19}$$

has an F distribution with r_1 and r_2 degrees of freedom. That is, an F random variable is the ratio of two independent chi-square random variables, each divided by their respective degrees of freedom.

Notice the following points about an F distribution:

1. The order in which the degrees of freedom are given for an F random variable is important. The number of degrees of freedom for the numerator, r_1, is listed first and the number of degrees of freedom for the denominator, r_2, is listed second.
2. The mean of the F distribution is

$$\mu = \frac{r_2}{r_2 - 2}$$

and the variance is

$$\sigma^2 = \frac{2r_2^2(r_1 + r_2 - 2)}{r_1(r_2 - 2)^2(r_2 - 4)}$$

3. The F distribution is skewed, and all of the probability occurs for positive values of F. See Fig. 11-21.
4. As r_2 increases, the mean of F approaches 1.

To find probabilities associated with the F distribution, use Table 7 in the Appendix. If F has an F distribution with r_1 and r_2 degrees of freedom, let $F_\alpha(r_1, r_2)$ be a number such that the probability that F obtains a value greater than this number is α. In symbols,

$$P[F > F_\alpha(r_1, r_2)] = \alpha$$

That is, the probability in the right tail is α. See Fig. 11-22 for an illustration. (Other authors may use the notation F_α or $F(r_1, r_2, \alpha)$ rather than $F_\alpha(r_1, r_2)$ to denote the right-tail probability.)

Because you need both r_1 and r_2 to find probabilities for the F distribution, the tabulation for F probabilities is more involved than that for other types of probabilities you have studied. Table 7 in the Appendix is in three parts: Table 7a gives right-tail probabilities for $\alpha = .01$, Table 7b for $\alpha = .025$, and Table 7c for $\alpha = .05$. (These tables will be sufficient for most of your applications. Computer programs are available to find probabilities that are not available in tables.) You first must select the appropriate table according to the chosen α level; then find the desired value at the intersection of the r_2 row and the r_1 column of the table. (Notice that numerator degrees of freedom are given across the top and denominator degrees of freedom are listed down the side of the table.)

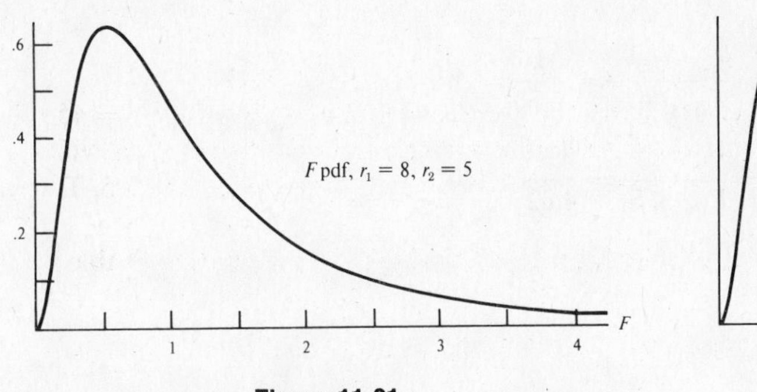

Figure 11-21

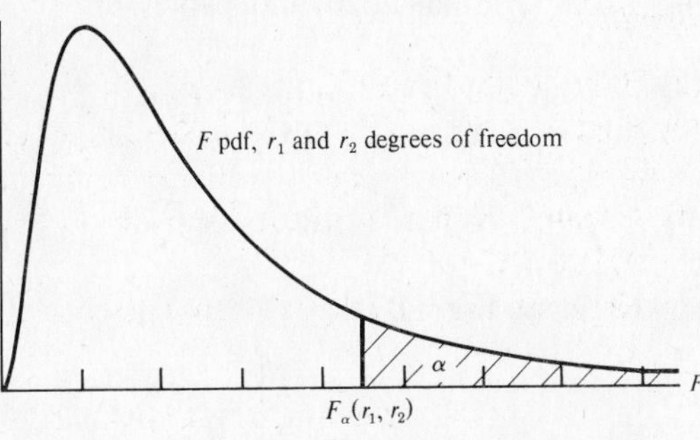

Figure 11-22

EXAMPLE 11-25 Let F have an F distribution with $r_1 = 5$ and $r_2 = 8$ degrees of freedom. Find **(a)** $F_{.01}(5, 8)$, **(b)** $F_{.025}(5, 8)$, and **(c)** $F_{.05}(5, 8)$.

Solution

(a) Because $\alpha = .01$, you will use Table 7a to find the desired value. In Table 7a read *across* to the column $r_1 = 5$ and *down* to the row $r_2 = 8$ to find $F_{.01}(5, 8) = 6.63$.

(b) Using Table 7b (since $\alpha = .025$), read across to the column $r_1 = 5$ and down to the row $r_2 = 8$ to find $F_{.025}(5, 8) = 4.82$.

(c) Using Table 7c (since $\alpha = .05$), read across to the column $r_1 = 5$ and down to row $r_2 = 8$ to find $F_{.05}(5, 8) = 3.69$. (See Fig. 11-23.)

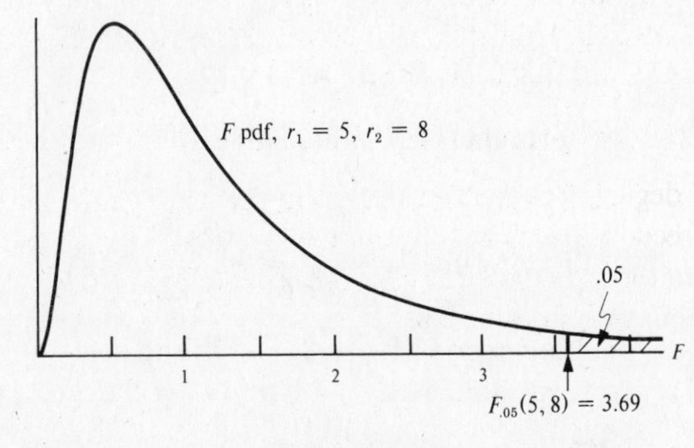

Figure 11-23

Since Table 7 is restricted to right-tail probabilities of $\alpha = .01$, .025, and .05, you will find the following formula very useful:

$$F_{1-\alpha}(r_1, r_2) = \frac{1}{F_\alpha(r_2, r_1)} \tag{11.20}$$

You can use this formula to find F values for *right-tail probabilities* of $1 - \alpha = .99$, .975, and .95, or for *left-tail probabilities* of $\alpha = .01, .025$, and .05. (You know that if the right-tail probability is $1 - \alpha$, the left-tail probability is α.) Notice that in formula (11.20) the degrees of freedom on the right side of the equation are reversed—(r_2, r_1) instead of (r_1, r_2). And remember that the subscripts in formula (11.20) are right-tail probabilities, even though one of those subscripts is $1 - \alpha$.

EXAMPLE 11-26 Let F have an F distribution with $r_1 = 5$ and $r_2 = 8$ degrees of freedom. Find **(a)** $F_{.99}(5, 8)$, **(b)** $F_{.975}(5, 8)$, and **(c)** $F_{.95}(5, 8)$.

Solution Here you are being asked to find F values, given the right-tail probability $1 - \alpha$.

(a) By using formula (11.20), and then Table 7a:

$$F_{.99}(5, 8) = \frac{1}{F_{.01}(8, 5)} = \frac{1}{10.29} = .097$$

(b) From formula (11.20) and Table 7b:

$$F_{.975}(5, 8) = \frac{1}{F_{.025}(8, 5)} = \frac{1}{6.76} = .148$$

(c)

$$F_{.95}(5, 8) = \frac{1}{F_{.05}(8, 5)} = \frac{1}{4.82} = .207$$

(Also see Fig. 11-24.)

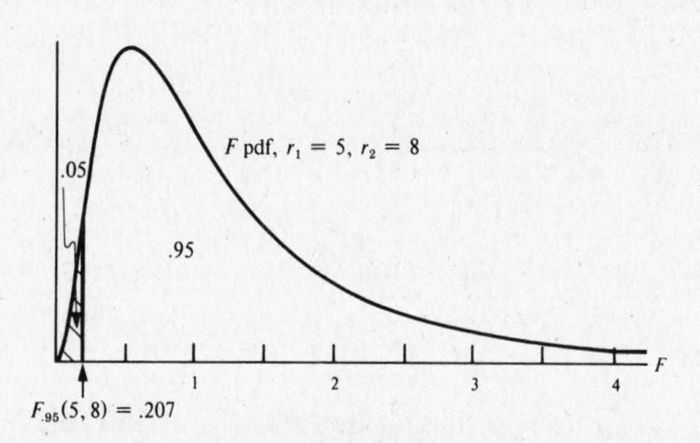

Figure 11-24

EXAMPLE 11-27 Let F have an F distribution with $r_1 = 6$ and $r_2 = 9$ degrees of freedom. Find **(a)** $F_{.025}(6, 9)$, **(b)** $F_{.975}(6, 9)$, **(c)** $P(.181 \leq F \leq 4.32)$, and **(d)** constants a and b so that $P(a \leq F \leq b) = .90$.

Solution

(a) Since $\alpha = .025$, use Table 7b:

$$F_{.025}(6, 9) = 4.32$$

(b) Here $1 - \alpha = .975$, so you use formula (11.20) and Table 7b to find the F value that goes with this right-tail probability.

$$F_{.975}(6, 9) = \frac{1}{F_{.025}(9, 6)} = \frac{1}{5.52} = .181$$

(c) You found in **(b)** that the probability to the left of .181 is .025 and in **(a)** that the probability to the right of 4.32 is .025 for an F random variable with $r_1 = 6$ and $r_2 = 9$ degrees of freedom. Thus

$$P(.181 \leq F \leq 4.32) = 1 - .025 - .025 = .95$$

(d) You can select a and b so that the left-tail and the right-tail probabilities are each equal to .05. In Table 7c you find

$$a = F_{.95}(6, 9) = \frac{1}{F_{.05}(9, 6)} = \frac{1}{4.10} = .244$$

and $$b = F_{.05}(6, 9) = 3.37$$

Thus, $$P(.244 \leq F \leq 3.37) = 1 - .05 - .05 = .90$$

The next theorem illustrates how the F distribution is used in connection with random samples from two normal distributions.

THEOREM: Let S_x^2 be the sample variance of a random sample of size n from a normal distribution $N(\mu_x, \sigma^2)$. Let S_y^2 be the sample variance of a random sample of size m from an independent normal distribution $N(\mu_y, \sigma^2)$.

(a) The distribution of

$$\frac{(n-1)S_x^2}{\sigma^2} \tag{11.21}$$

is $\chi^2(n-1)$.

(b) The distribution of

$$\frac{(m-1)S_y^2}{\sigma^2} \tag{11.22}$$

is $\chi^2(m-1)$.

(c) Thus,

$$F = \frac{\dfrac{(n-1)S_x^2}{\sigma^2(n-1)}}{\dfrac{(m-1)S_y^2}{\sigma^2(m-1)}} = \frac{S_x^2}{S_y^2} \tag{11.23}$$

has an F distribution with $r_1 = n - 1$ and $r_2 = m - 1$ degree of freedom, since it is the ratio of two chi-square random variables, each divided by their respective degrees of freedom. [See formula (11.19)]. Notice that formula (11.23) states that the ratio of two sample variances from independent normal populations with equal population variances has an F distribution.

EXAMPLE 11-28 Let S_x^2 be the sample variance of a random sample of size $n = 7$ from a normal distribution $N(21.48, .31)$. Let S_y^2 be the sample variance of a random sample of size $m = 10$ from an independent normal distribution $N(22.17, .31)$. **(a)** Give the distribution of S_x^2/S_y^2. **(b)** Find $P(S_x^2/S_y^2 > 3.37)$.

Solution

(a) S_x^2 and S_y^2 are sample variances of random samples from normal populations and each has a population variance of .31. Thus by formula (11.23), $F = S_x^2/S_y^2$ has an F distribution with $r_1 = 7 - 1 = 6$ and $r_2 = 10 - 1 = 9$ degrees of freedom.

(b) Look in Tables 7a, b, and c under $r_1 = 6$ and $r_2 = 9$ degrees of freedom. You'll find 3.37 in Table 7c, where $\alpha = .05$. Thus, $P(S_x^2/S_y^2 > 3.37) = .05$.

SUMMARY

1. A *random sample* $X_1, X_2, \ldots, X_n$ is a set of n independent and identically distributed random variables.
2. The probability distribution of the mean of a random sample, $\bar{X}$, is called the *sampling distribution of the mean.*
3. The *mean* of the distribution of $\bar{X}$ is $\mu_{\bar{x}} = \mu$.
4. The *variance* of the distribution of $\bar{X}$ is $\sigma_{\bar{x}}^2 = \sigma^2/n$.
5. The standard deviation of the distribution of $\bar{X}$, also known as the *standard error of the mean,* is $\sigma_{\bar{x}} = \sigma/\sqrt{n}$.
6. When $\bar{X}$ is the mean of a random sample from a normal distribution $N(\mu, \sigma^2)$, the distribution of $Z = (\bar{X} - \mu)/(\sigma/\sqrt{n})$ is $N(0, 1)$.

7. For a random sample from any distribution with mean μ and finite variance σ^2, the approximate distribution of $Z = (\bar{X} - \mu)/(\sigma/\sqrt{n})$ is $N(0, 1)$ when n is sufficiently large.
8. When n is sufficiently large, the approximate distribution of $Z = (Y - n\mu)/(\sigma\sqrt{n})$ is $N(0, 1)$, where Y is the sum of the items of the random sample.
9. If Y has a binomial distribution $b(n, p)$, then the distribution of $Z = (Y - np)/\sqrt{npq}$ is approximately $N(0, 1)$, provided that $np \geq 5$ and $n(1 - p) = nq \geq 5$.
10. If S^2 is the sample variance of a random sample of size n from a normal distribution $N(\mu, \sigma^2)$, then the distribution of $(n - 1)S^2/\sigma^2$ is $\chi^2(n - 1)$.
11. If $\bar{X}$ and S are the sample mean and sample standard deviation of a random sample of size n from a normal distribution $N(\mu, \sigma^2)$, then $T = (\bar{X} - \mu)/(S/\sqrt{n})$ has a t distribution with $r = n - 1$ degrees of freedom.
12. If S_x^2 and S_y^2 are the sample variances of random samples of sizes n and m from the independent normal distributions $N(\mu_x, \sigma^2)$ and $N(\mu_y, \sigma^2)$, respectively, then $F = S_x^2/S_y^2$ has an F distribution with $r_1 = n - 1$ and $r_2 = m - 1$ degrees of freedom.

RAISE YOUR GRADES

Can you . . . ?

☑ tell the difference between a population distribution and a sampling distribution of a statistic
☑ define a random sample
☑ find the mean, variance, and standard deviation of $\bar{X}$
☑ define the standard error of the mean
☑ give the distribution of $\bar{X}$ when sampling from normal populations
☑ state the Central Limit Theorem in terms of $\bar{X}$, the mean of the X's
☑ state the Central Limit Theorem in terms of Y, the sum of the X's
☑ use the normal distribution to approximate binomial probabilities
☑ use Table 5 for finding chi-square probabilities
☑ give the distribution of $(n - 1)S^2/\sigma^2$ when sampling from normal populations
☑ use Table 6 for finding t probabilities
☑ give the distribution of $(\bar{X} - \mu)/(S/\sqrt{n})$ when sampling from normal populations
☑ use Table 7 for finding F probabilities
☑ give the distribution of S_x^2/S_y^2 when taking samples of sizes n and m from independent normal populations that have equal variances

RAPID REVIEW

1. A set of n independent and identically distributed random variables is called a ____________ ____________.
2. Let $\bar{X}$ be the sample mean of a random sample of size 9 from a distribution that has a mean of $\mu = 75$ and a variance of $\sigma^2 = 36$. Then, for the distribution of $\bar{X}$, the mean is **(a)** ____________, the variance is **(b)** ____________, and the standard deviation is **(c)** ____________.
3. If $\bar{X}$ is the mean of a random sample of size $n = 25$ from the normal distribution $N(63, 100)$, the distribution of $\bar{X}$ is ____________.
4. If $\bar{X}$ is the mean of a random sample of size $n > 30$ from a distribution with mean μ and variance σ^2, then the distribution of ____________ is approximately $N(0, 1)$.
5. If the distribution of X is $b(100, 1/2)$, then the distribution of ____________ is approximately $N(0, 1)$.
6. If the distribution of X is $\chi^2(9)$, then $\mu =$ **(a)** ____________ and $\sigma^2 =$ **(b)** ____________.
7. If the distribution of X is $\chi^2(9)$, $\chi^2_{.05}(9) =$ **(a)** ____________ and $\chi^2_{.95}(9) =$ **(b)** ____________.
8. If S^2 is the sample variance of a random sample of size 21 from the normal distribution $N(63, 100)$, then the distribution of $20S^2/100 = S^2/5$ is **(a)** ____________ and $P(S^2/5 > 28.41) =$ **(b)** ____________.

9. For the t distribution, $t_{.05}(20) =$ **(a)** ______________ and $-t_{.025}(20) =$ **(b)** ______________.

10. If $\bar{X}$ and S are the mean and standard deviation of a random sample of size 21 from the normal distribution $N(63, 100)$, then $T = (\bar{X} - 63)/(S/\sqrt{21})$ has a t distribution with **(a)** ______________ degrees of freedom and $P(T > 1.33) =$ **(b)** ______________.

11. For the F distribution, $F_{.025}(3, 15) =$ **(a)** ______________ and $F_{.975}(13, 4) =$ **(b)** ______________.

12. If S_x^2 and S_y^2 are random samples of sizes $n = 6$ and $m = 11$ from independent normal distributions that have the same variance, then $P(S_x^2/S_y^2 > 5.64) =$ ______________.

Answers **(1)** random sample **(2) (a)** 75 **(b)** 4 **(c)** 2 **(3)** $N(63, 4)$ **(4)** $(\bar{X} - \mu)/(\sigma/\sqrt{n})$ **(5)** $(X - 50)/5$ **(6) (a)** 9 **(b)** 18 **(7) (a)** 16.92 **(b)** 3.33 **(8) (a)** $\chi^2(20)$ **(b)** .10 **(9) (a)** 1.72 **(b)** −2.09 **(10) (a)** 20 **(b)** .10 **(11) (a)** 4.15 **(b)** .25 **(12)** .01

SOLVED PROBLEMS

Random Samples: The Mean, Variance, and Standard Deviation of $\bar{X}$

PROBLEM 11-1 Let $X_1, X_2, \ldots, X_{16}$ be a random sample of size $n = 16$ from the normal distribution $N(110, 100)$. Let $\bar{X}$ be the sample mean of this random sample. For the sampling distribution of $\bar{X}$, give **(a)** the mean, **(b)** the variance, and **(c)** the standard error of the mean.

Solution

(a) By formula (11.1),

$$\mu_{\bar{x}} = \mu = 110$$

(b) By formula (11.2),

$$\sigma_{\bar{x}}^2 = \frac{\sigma^2}{n} = \frac{100}{16} = 6.25$$

(c) By formula (11.3),

$$\sigma_{\bar{x}} = \frac{\sigma}{\sqrt{n}} = \frac{10}{\sqrt{16}} = 2.5$$

PROBLEM 11-2 Let X equal the outcome when a fair 6-sided die is rolled. Then $\mu = 7/2 = 3.5$ and $\sigma^2 = 35/12$. Let $\bar{X}$ be the mean of a random sample of 35 rolls of this die. Give the values of **(a)** $\mu_{\bar{x}}$, **(b)** $\sigma_{\bar{x}}^2$, and **(c)** $\sigma_{\bar{x}}$.

Solution

(a) By formula (11.1), $\mu_{\bar{x}} = 3.5$.
(b) By formula (11.2), $\sigma_{\bar{x}}^2 = (35/12)/35 = 1/12$.
(c) By formula (11.3), $\sigma_{\bar{x}} = \sqrt{1/12} = .289$.

PROBLEM 11-3 Let $\bar{X}$ be the sample mean of a random sample of size $n = 18$ from the distribution with pdf $f(x) = 1 - x/2$ for $0 \le x \le 2$. For this distribution $\mu = 2/3$ and $\sigma^2 = 2/9$. Give the values of **(a)** $\mu_{\bar{x}}$, **(b)** $\sigma_{\bar{x}}^2$, and **(c)** $\sigma_{\bar{x}}$.

Solution

(a) $\mu_{\bar{x}} = 2/3$
(b) $\sigma_{\bar{x}}^2 = (2/9)/18 = 1/81$
(c) $\sigma_{\bar{x}} = 1/9$

PROBLEM 11-4 Let $\bar{X}$ be the sample mean when a 4-sided die is rolled $n = 45$ times. The mean for each roll is $\mu = 2.5$; the variance is $\sigma^2 = 5/4$. (See Example 11-2.) Give the values of **(a)** $\mu_{\bar{x}}$ and **(b)** the standard error of the mean.

Solution

(a) $\mu_{\bar{x}} = 2.5$
(b) The standard error of the mean is the standard deviation of the distribution of $\bar{X}$. Thus by formula (11.3), $\sigma_{\bar{x}} = \sqrt{(5/4)/45} = 1/6$.

PROBLEM 11-5 Let $\bar{X}$ be the sample mean for a random sample of muscle-strength measurements of 64 freshman males. Assume that for a population of freshman males, $\mu = 48.72$ and $\sigma = 11.44$ are the mean and standard deviation of their muscle strengths. Give the values of **(a)** $\mu_{\bar{x}}$ and **(b)** the standard error of the mean.

Solution

(a) $\mu_{\bar{x}} = 48.72$
(b) $\sigma_{\bar{x}} = 11.44/8 = 1.43$

The Distribution of the Sample Mean

PROBLEM 11-6 Let $\bar{X}$ be the sample mean of a random sample of size $n = 16$ from the normal distribution $N(50.025, 17.64)$. Give **(a)** the distribution of $\bar{X}$, and **(b)** the standard error of the mean. **(c)** Find $P(\bar{X} \leq 52.797)$.

Solution

(a) The mean and variance are given by formulas (11.1) and (11.2) and you are told that $\bar{X}$ has a normal distribution. So, the distribution of $\bar{X}$ is $N(50.025, 17.64/16)$.
(b) By formula (11.3), the standard error of the mean is

$$\sigma_{\bar{x}} = \frac{\sigma}{\sqrt{n}} = \sqrt{\frac{17.64}{16}} = 1.05$$

(c) You first find the standard z value by subtracting the mean and dividing by the standard deviation (see Example 11-6).

$$z = \frac{52.797 - 50.025}{1.05} = 2.64$$

Then you can find the desired probability (see Fig. 11-25):

$$P(\bar{X} \leq 52.797) = P(Z \leq 2.64)$$
$$= .5000 + .4959 = .9959$$

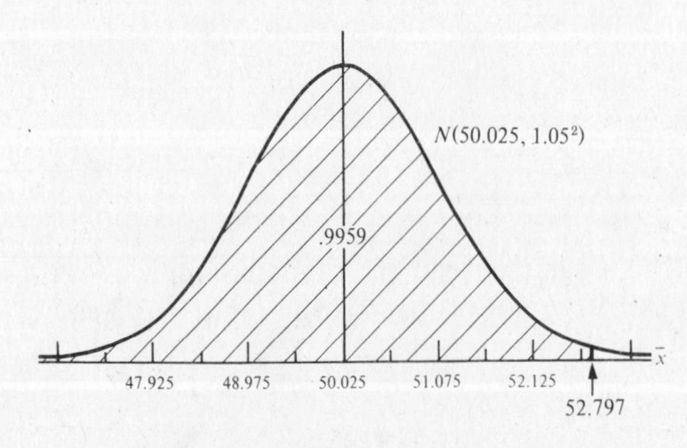

Figure 11-25

PROBLEM 11-7 Using the information given in Problem 11-1, find $P(107.3 \leq \bar{X} \leq 112.9)$.

Solution From Problem 11-1, $\bar{X}$ is $N(110, 6.25)$ and $\sigma/\sqrt{n} = 2.5$, so,

$$z = \frac{107.3 - 110}{2.5} \qquad z = \frac{112.9 - 100}{2.5}$$

$$= -1.08 \qquad\qquad = 1.16$$

The probability you want to find is

$$P(107.3 \leq \bar{X} \leq 112.9) = P(-1.08 \leq Z \leq 1.16)$$
$$= .3599 + .3770 = .7369$$

(See Fig. 11-26.)

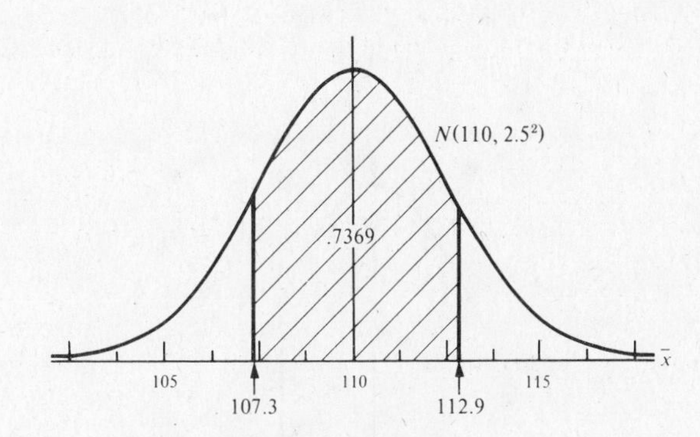

Figure 11-26

PROBLEM 11-8 Assume that the length of life of a light bulb has a normal distribution $N(1500, 38.1^2)$. Let $\bar{X}$ be the sample mean of the length of life of 9 bulbs that are selected at random. (**a**) Give the distribution of $\bar{X}$. (**b**) Give the standard error of the mean. (**c**) Find $P(\bar{X} \leq 1482.22)$.

Solution

(**a**) The distribution of $\bar{X}$ is $N(1500, 38.1^2/9)$.
(**b**) The standard error of the mean is $\sigma_{\bar{x}} = 38.1/\sqrt{9} = 12.7$.
(**c**)

$$z = \frac{1482.22 - 1500}{12.7}$$
$$= -1.40$$

Thus,

$$P(\bar{X} \leq 1482.22) = P(Z \leq -1.40) = P(Z \geq 1.40)$$
$$= .5000 - .4192 = .0808$$

PROBLEM 11-9 Yet Y equal the total length of life for the 9 light bulbs described in Problem 11-8. Give (**a**) the distribution of Y, and (**b**) the standard deviation of the distribution of Y. (**c**) Find $P(Y > 13{,}694.31)$.

Solution

(**a**) $N(9 \cdot 1500, 9 \cdot 38.1^2) = N(13{,}500, 13{,}064.49)$
(**b**) $\sigma_y = \sqrt{13{,}064.49} = 114.3$
(**c**) Therefore,

$$z = \frac{13{,}694.31 - 13{,}500}{114.3}$$
$$= 1.70$$

$$P(Y > 13{,}694.31) = P(Z > 1.70)$$
$$= .5000 - .4554 = .0446$$

(See Fig. 11-27.)

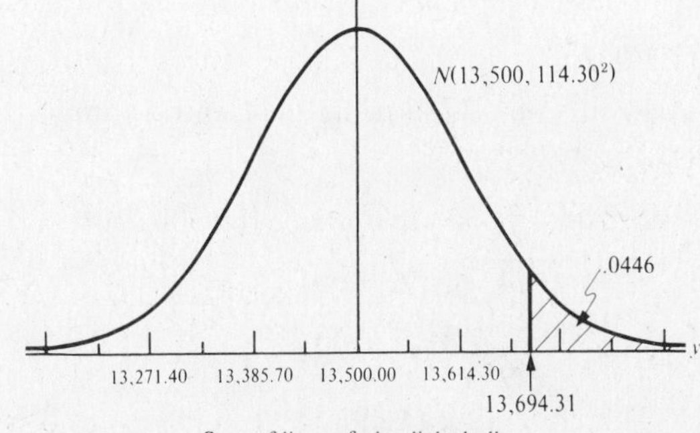

Figure 11-27

PROBLEM 11-10 Assume that the weight of a box of a certain type of soap powder has a mean of $\mu = 6.047$ g and a standard deviation of $\sigma = .02$. Let $\bar{X}$ be the sample mean of a random sample of 16 packages of soap. Find $P(6.037 \leq \bar{X} \leq 6.052)$, approximately.

Solution You are not told whether the population from which the sample is taken is normally distributed. However, by the Central Limit Theorem, $\bar{X}$ has an approximate normal distribution, so you can compute z and find the desired probability, approximately. By formula (11.6), $z = \dfrac{c - \mu}{\sigma/\sqrt{n}}$,

$$z = \frac{6.037 - 6.047}{.02/4} = -2.00 \qquad z = \frac{6.052 - 6.047}{.02/4} = 1.00$$

Therefore (see Fig. 11-28),

$$P(6.037 \leq \bar{X} \leq 6.052) \approx P(-2.00 \leq Z \leq 1.00) = .4772 + .3413 = .8185$$

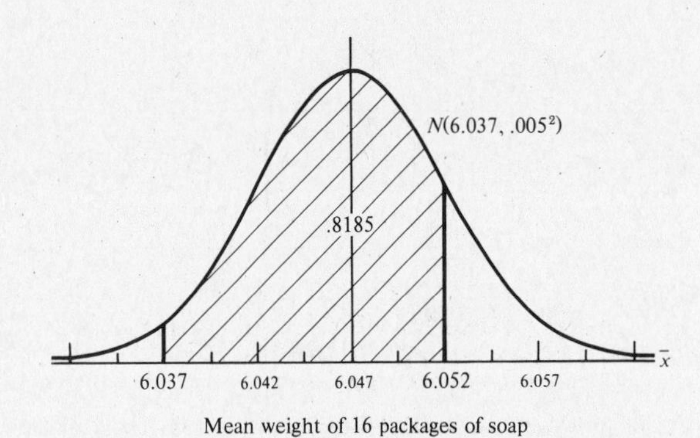

Figure 11-28

PROBLEM 11-11 Let X equal the maximal oxygen intake in milliliters of oxygen per minute per kilogram of weight of a human on a treadmill. Assume that for a particular population the mean of X is $\mu = 54.030$ and the standard deviation of X is $\sigma = 5.8$. Let $\bar{X}$ be the sample mean of a random sample of size $n = 47$. Find $P(52.761 \leq \bar{X} \leq 54.453)$, approximately.

Solution Applying the Central Limit Theorem, and using formula (11.6), you find

$$z = \frac{52.761 - 54.030}{5.8/\sqrt{47}} = -1.50 \qquad z = \frac{54.453 - 54.030}{5.8/\sqrt{47}} = .50$$

Therefore,

$$P(52.761 \leq \bar{X} \leq 54.453) \approx P(-1.50 \leq Z \leq .50) = .4332 + .1915 = .6247$$

PROBLEM 11-12 Let X equal the weight of a "3-pound" bag of apples when weighed on a particular scale. Assume that the mean of X is $\mu = 3.197$ and the standard deviation of X is $\sigma = .169$. Let Y equal the sum of the weights of $n = 32$ bags of apples selected at random. Give (**a**) the approximate distribution of Y, and (**b**) the standard deviation of the distribution of Y. (**c**) Find $P(Y \geq 100)$, approximately.

Solution

(**a**) From point 2 following the Central Limit Theorem:

$$N(32 \cdot 3.197, 32 \cdot .169^2) = N(102.304, .914)$$

(**b**) $\sigma_y = \sqrt{.914} = .956$

(**c**) $z = \dfrac{100 - 102.304}{.956} = -2.41$

Therefore (Fig. 11-29),

$$P(Y \geq 100) \approx P(Z \geq -2.41) = P(Z \leq 2.41)$$

$$= .5000 + .4920 = .9920$$

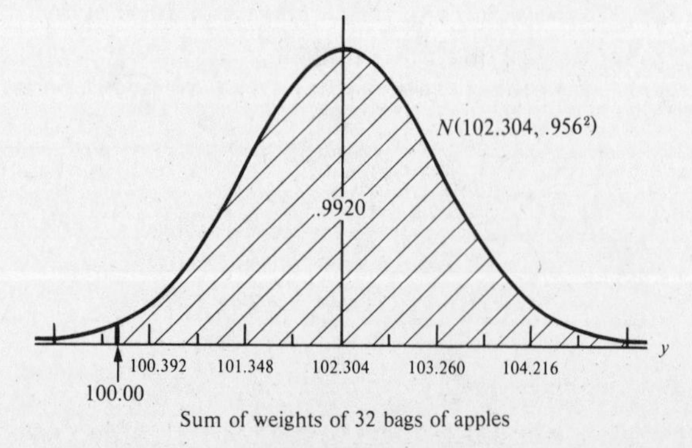

Figure 11-29

Normal Approximation of Binomial Probabilities

PROBLEM 11-13 If X is $b(100, .1)$, use the normal distribution to find $P(9 \le X \le 14)$, approximately.

Solution You first note that $P(9 \le X \le 14) = P(8.5 \le X \le 14.5)$. That is, with the half-unit correction for continuity, the region for which you want to find the probability is $8.5 \le X \le 14.5$ (see Fig. 11-30). By formulas (11.10) and (11.9),

$$z = \frac{k - .5 - np}{\sqrt{np(1-p)}} = \frac{8.5 - 100(.1)}{\sqrt{100(.1)(.9)}} = \frac{8.5 - 10}{\sqrt{9}} = -.50 \qquad z = \frac{k + .5 - np}{\sqrt{np(1-p)}} = \frac{14.5 - 100(.1)}{\sqrt{100(.1)(.9)}} = \frac{14.5 - 10}{\sqrt{9}} = 1.50$$

Therefore,

$$P(9 \le X \le 14) = P(8.5 \le X \le 14.5)$$
$$\approx P(-.50 \le Z \le 1.50) = .1915 + .4332 = .6247$$

PROBLEM 11-14 Let X equal the number of correct guesses on a 100-question true–false test so that the probability for guessing correctly is $p = 1/2$. Thus X is $b(100, 1/2)$. Use the normal approximation to find $P(X \le 45)$, approximately.

Solution Use formula (11.9), making the half-unit correction for continuity:

$$z = \frac{45.5 - 100(.5)}{\sqrt{100(.5)(.5)}} = -.90$$

Therefore (Fig. 11-31),

$$P(X \le 45) = P(X \le 45.5)$$
$$\approx P(Z \le -.90) = .5000 - .3159 = .1841$$

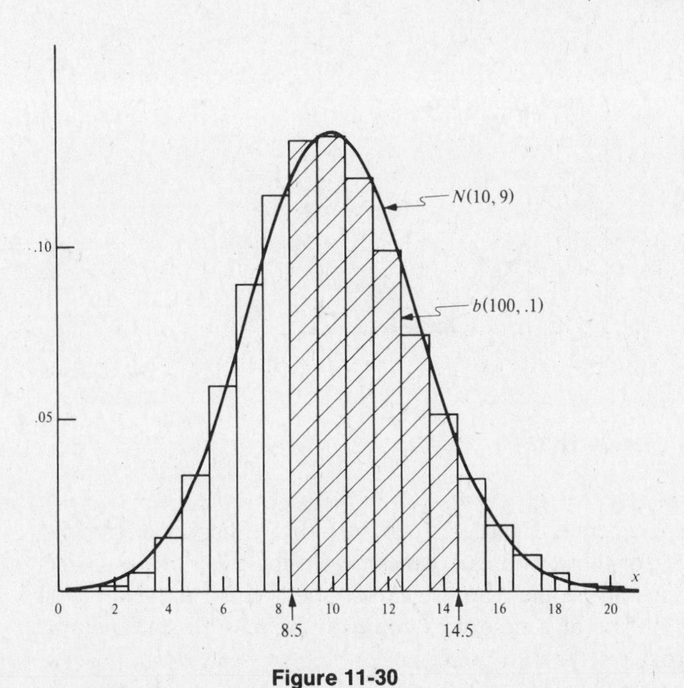

Figure 11-30

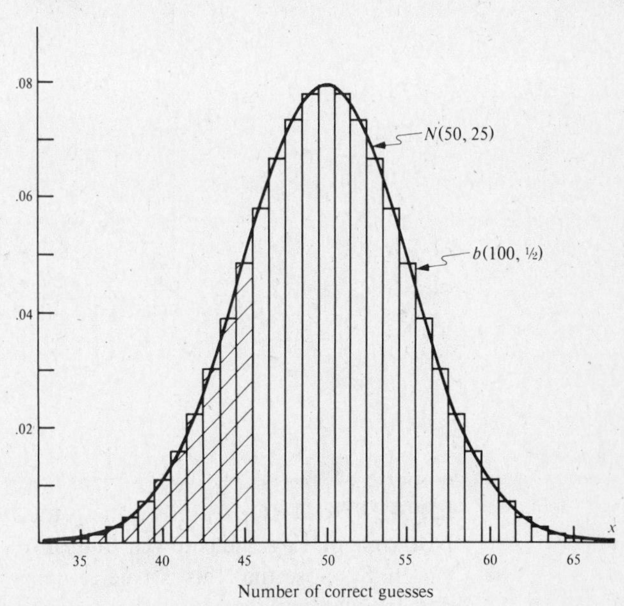

Figure 11-31

PROBLEM 11-15 Suppose that 20% of the American public believes that we could survive a nuclear war. Let Y equal the number of people in a random sample of size 25 who answer "yes" to the question, "Do you believe that we could survive a nuclear war?" Use the Central Limit Theorem to determine the probability that there are anywhere from 6 to 9 people in a sample of 25 who answer "yes"; that is, find $P(6 \leq Y \leq 9)$, approximately.

Solution We see that the distribution of Y is $b(25, .20)$, since $n = 25$ and the probability for a "yes" is .20. Use formulas (11.10) and (11.9) (using y's in place of the x's), and incorporate the half-unit corrections for continuity:

$$z = \frac{5.5 - 5}{\sqrt{4}} = .25 \qquad z = \frac{9.5 - 5}{\sqrt{4}} = 2.25$$

Therefore (Fig. 11-32),

$$\begin{aligned} P(6 \leq Y \leq 9) &= P(5.5 \leq Y \leq 9.5) \\ &\approx P(.25 \leq Z \leq 2.25) \\ &= .4878 - .0987 = .3891 \end{aligned}$$

The exact probability, using $b(25, .20)$, is

$$\begin{aligned} P(Y = 6, 7, 8, 9) &= .1633 + .1108 + .0623 + .0294 \\ &= .3658 \end{aligned}$$

So the normal approximation is close to this exact probability, although you see that the normal distribution does over-approximate the exact probability in this case. Notice in Fig. 11-32 that the area under the normal curve between 5.5 and 9.5 is larger than the area of the 4 rectangles in the binomial probability histogram.

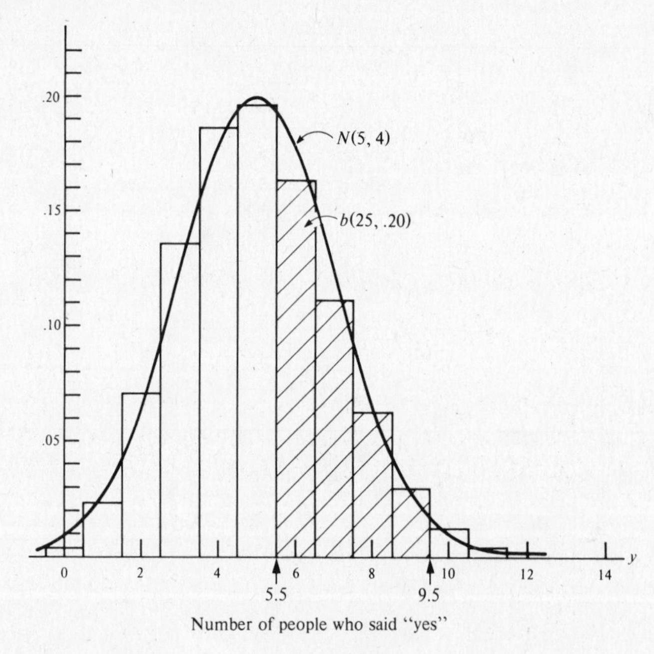

Figure 11-32

PROBLEM 11-16 A study done at the Department of Health Care of the Elderly, Sherwood Hospital, Nottingham, England, showed that only 25% of those who used canes were using canes of the correct length. Suppose that this is true in general for all people who use canes. Let X equal the number of persons who have canes of the correct length out of a random sample of size $n = 108$. (**a**) Give the distribution of X. (**b**) Find $P(X < 32)$, approximately.

Solution

(**a**) The distribution of X is $b(108, .25)$ since $n = 108$ and the probability for "success" (for having a cane of the correct length) is .25.

(b) $P(X < 32) = P(X \leq 31)$
By formula (11.9):

$$z = \frac{31.5 - (108)25}{\sqrt{108(.25)(.75)}} = \frac{31.5 - 27}{4.5} = 1.00$$

Therefore,

$$\begin{aligned} P(X \leq 31) &= P(X \leq 31.5) \\ &\approx P(Z \leq 1) \\ &= .5000 + .3413 = .8413 \end{aligned}$$

PROBLEM 11-17 Let p equal the proportion of London residents who prefer that police on general duty carry only whistles and truncheons, not firearms, except when hunting for violent criminals. Assume that $p = .83$. Let X equal the number of such London residents among a random sample of size $n = 55$. **(a)** Give the distribution of X. **(b)** Find $P(X \leq 50)$, approximately.

Solution

(a) The distribution of X is $b(55, .83)$.

(b) $z = \dfrac{50.5 - 55(.83)}{\sqrt{55(.83)(.17)}} = \dfrac{50.5 - 45.65}{\sqrt{7.7605}} = 1.74$

Therefore,

$$\begin{aligned} P(X \leq 50) &= P(X \leq 50.5) \\ &\approx P(Z \leq 1.74) \\ &= .5000 + .4591 = .9591 \end{aligned}$$

The Chi-Square Distribution

PROBLEM 11-18 Let the distribution of X be $\chi^2(17)$; that is, X has a chi-square distribution with 17 degrees of freedom. Give **(a)** the mean and **(b)** the variance of X. Find the values of **(c)** $\chi^2_{.05}(17)$, **(d)** $\chi^2_{.95}(17)$, and **(e)** $P(7.56 \leq X \leq 30.19)$.

Solution

(a) The mean is equal to the number of degrees of freedom, so $\mu = 17$.
(b) The variance is equal to twice the number of degrees of freedom, so $\sigma^2 = (2)(17) = 34$.
(c) Using Table 5 in the Appendix, find the row $r = 17$ and read across to the column headed by .05 (Fig. 11-33):

$$\chi^2_{.05}(17) = 27.59$$

(d) Using Table 5, go across row $r = 17$ to the column headed by .95 (Fig. 11-33):

$$\chi^2_{.95}(17) = 8.67$$

(e) Go across row $r = 17$ to find the numbers 7.56 and 30.19. Since 7.56 is in the column headed by .975, the probability to the right of 7.56 is .975. The number 30.19 is in the column headed by .025, so the probability to the right of 30.19 is .025. Thus,

$$\begin{aligned} P(7.56 \leq X \leq 30.19) &= .975 - .025 \\ &= .950 \end{aligned}$$

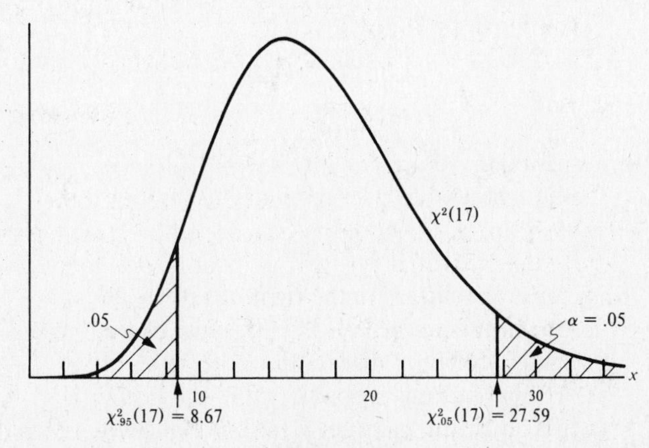

Figure 11-33

PROBLEM 11-19 Let $X_1, X_2, \ldots, X_{18}$ be a random sample of size $n = 18$ from a normal distribution $N(112, 121)$. Let S^2 be the sample variance of this random sample. **(a)** Give the distribution of $17S^2/121$. Find **(b)** $P(17S^2/121 > 30.19)$ and **(c)** $P(17S^2/121 \leq 7.56)$.

Solution

(a) By formula (11.14), the distribution of $\dfrac{(n-1)S^2}{\sigma^2} = \dfrac{17S^2}{121}$ is $\chi^2(18-1) = \chi^2(17)$.

(b) From Table 5 in the Appendix and Fig. 11-34:

$$P\left(\frac{17S^2}{121} > 30.19\right) = .025$$

(c) $P\left(\dfrac{17S^2}{121} > 7.56\right) = .975$

so that
$$P\left(\frac{17S^2}{121} \leq 7.56\right) = 1 - .975 = .025$$

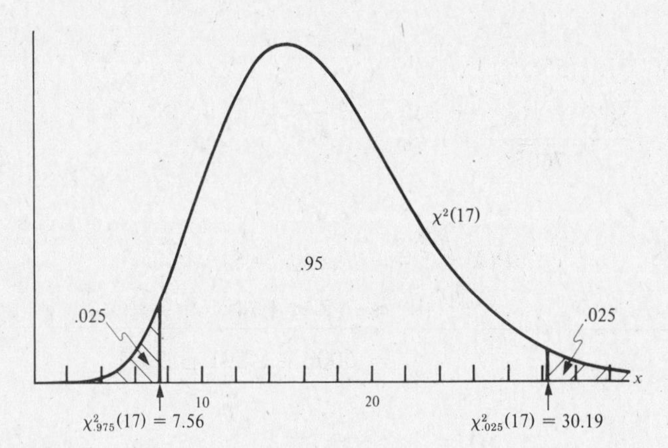

Figure 11-34

Student's *t* Distribution

PROBLEM 11-20 Let T have a t distribution with $r = 17$ degrees of freedom. Give **(a)** the mean and **(b)** the variance of T. Find the values of **(c)** $t_{.025}(17)$ and **(d)** $t_{.95}(17) = -t_{.05}(17)$.

Solution

(a) From point 2, for every t random variable (with $r \geq 2$), $\mu = 0$, since the t distribution is symmetric about zero.

(b) Also from point 2, the variance of a t distribution (for $r \geq 3$) is given by

$$\sigma^2 = \frac{r}{r-2} = \frac{17}{17-2} = \frac{17}{15} = 1.133$$

(c) In Table 6, look across row $r = 17$ to the column headed by .025 to find (Fig. 11-35)

$$t_{.025}(17) = 2.11$$

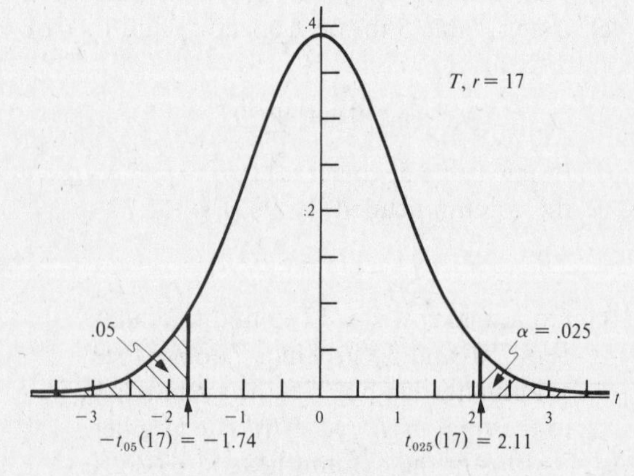

Figure 11-35

(d) If .95 probability is to the right of t, then .05 probability is to the left. Also, because of the symmetry of the t distribution, the value of t with a left-tail probability of .05 is the negative of the value of t with a right-tail probability of .05. Therefore, in Table 6 look across the row $r = 17$ to the column headed by .05 to read 1.74. You now know that

$$t_{.95}(17) = -t_{.05}(17) = -1.74$$

PROBLEM 11-21 Let $X_1, X_2, \ldots, X_{18}$ be a random sample of size $n = 18$ from a normal distribution $N(112, 121)$. Let $\bar{X}$ and S be the sample mean and sample standard deviation of this random sample, respectively. **(a)** Give the distribution of T. Find **(b)** $P(T > 2.11)$ and **(c)** $P(T \leq -1.74)$.

Solution

(a) Using formula (11.18), $T = (\bar{X} - \mu)/(S/\sqrt{18})$, you see that T has a t distribution with $r = 18 - 1 = 17$ degrees of freedom.

(b) From Table 6 and Fig. 11-35:

$$P(T > 2.11) = .025$$

(c) From Table 6 we find that $P(T > 1.74) = .05$. By the symmetry of the t distribution it follows that $P(T \leq -1.74) = .05$.

The *F* Distribution

PROBLEM 11-22 Let F have an F distribution with $r_1 = 7$ and $r_2 = 11$ degrees of freedom. Find **(a)** $F_{.01}(7, 11)$, **(b)** $F_{.025}(7, 11)$, and **(c)** $F_{.05}(7, 11)$.

Solution

(a) Since $\alpha = .01$, the appropriate F table is Table 7a. Read down column $r_1 = 7$ to row $r_2 = 11$ to find $F_{.01}(7, 11) = 4.89$.

(b) Here $\alpha = .025$, so use Table 7b. Read down column $r_1 = 7$ to row $r_2 = 11$ to find (Fig. 11-36) $F_{.025}(7, 11) = 3.76$.

(c) Here Table 7c is the appropriate F table. Read down column $r_1 = 7$ to row $r_2 = 11$ to find $F_{.05}(7, 11) = 3.01$.

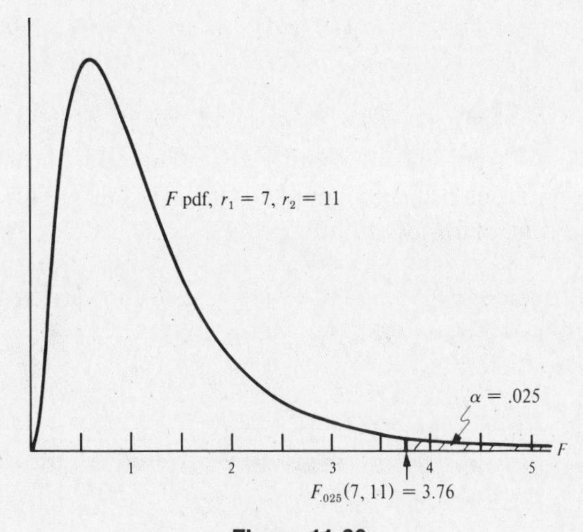

Figure 11-36

PROBLEM 11-23 Let F have an F distribution with $r_1 = 7$ and $r_2 = 4$ degrees of freedom. Find **(a)** $F_{.99}(7, 4)$, **(b)** $F_{.975}(7, 4)$, and **(c)** $F_{.95}(7, 4)$.

Solution

(a) Using Table 7a and formula (11.20),

$$F_{1-\alpha} = \frac{1}{F_\alpha(r_2, r_1)}$$

$$F_{.99}(7, 4) = \frac{1}{F_{.01}(4, 7)} = \frac{1}{7.85} = .127$$

(b) Using Table 7b and (11.20),

$$F_{.975}(7,4) = \frac{1}{F_{.025}(4,7)} = \frac{1}{5.52} = .181$$

(c) Using Table 7c and (11.20),

$$F_{.95}(7,4) = \frac{1}{F_{.05}(4,7)} = \frac{1}{4.12} = .243$$

(See Fig. 11-37.)

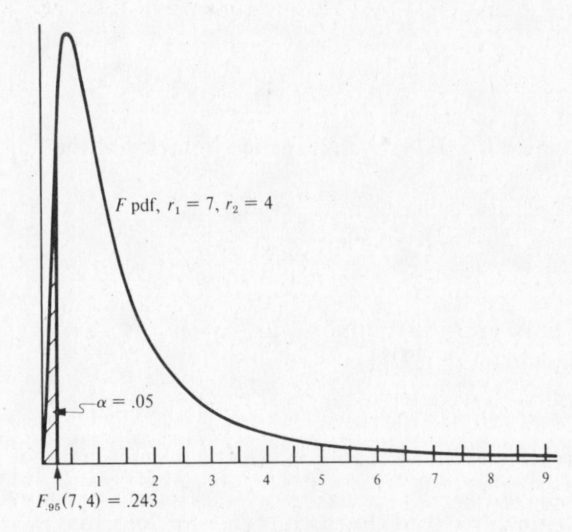

Figure 11-37

PROBLEM 11-24 Let S_x^2 be the sample variance of a random sample of size $n = 8$ from a normal distribution $N(115, 121)$. Let S_y^2 be the sample variance of a random sample of size $m = 5$ from an independent normal distribution $N(110, 121)$. **(a)** Give the distribution of S_x^2/S_y^2. **(b)** Find $P(S_x^2/S_y^2 > 9.07)$.

Solution

(a) You learned in the final theorem of this chapter that the ratio of two sample variances from independent normal distributions with equal population variances has an F distribution. That is, by formula (11.23), $F = S_x^2/S_y^2$ has an F distribution with $r_1 = 8 - 1 = 7$ and $r_2 = 5 - 1 = 4$ degrees of freedom.

(b) Look in Tables 7a, b, and c under $r_1 = 7$ and $r_2 = 4$ degrees of freedom to find the F value 9.07. You find it in Table 7b, where $\alpha = .025$, so $P(S_x^2/S_y^2 > 9.07) = .025$.

Supplementary Problems

PROBLEM 11-25 Let X equal the age of a customer at a particular record shop. Assume that $\mu = 27.5$ and $\sigma = 8.5$. Let $\bar{X}$ be the sample mean age in a random sample of 100 customers. Give the values of **(a)** $\mu_{\bar{x}}$, **(b)** $\sigma_{\bar{x}}^2$, and **(c)** $\sigma_{\bar{x}}$. **(d)** How is $\bar{X}$ distributed, approximately?

Answer (a) 27.5 (b) $8.5^2/100$ (c) $8.5/10 = .85$ (d) $N(27.5, 8.5^2/100)$

PROBLEM 11-26 Let X equal the gas mileage (in miles per gallon) for a particular car model. Suppose that the distribution of X is $N(56.3, 4)$. Let $\bar{X}$ be the sample mean of a random sample of 9 observations of X. **(a)** Give the distribution of $\bar{X}$. **(b)** Give the value of the standard error of the mean.

Answer (a) $N(56.3, 4/9)$ (b) $\sigma_{\bar{x}} = 2/3$

PROBLEM 11-27 In Problem 11-25, find $P(\bar{X} \le 25.8)$, approximately.

Answer .0228

PROBLEM 11-28 Let X equal the size of a deposit made at a particular bank. Assume that X has a mean of $\mu = 500$, a variance of $\sigma^2 = 640{,}000$, and a standard deviation of $\sigma = 800$. Let Y equal the sum of 25 deposits that are selected at random. **(a)** Give the approximate distribution of Y. **(b)** Find $P(Y > 19{,}080)$, approximately.

Answer **(a)** $N(12{,}500, 16{,}000{,}000)$ **(b)** .05

PROBLEM 11-29 Suppose that an egg laid by a moorhen will hatch 80% of the time. In a group of 49 eggs selected at random, let X equal the number of eggs that hatch. **(a)** How is X distributed? **(b)** Find $P(39 \le X \le 45)$, approximately.

Answer **(a)** $b(49, .80)$ **(b)** .5865

PROBLEM 11-30 Give the values of **(a)** $\chi^2_{.01}(23)$, **(b)** $\chi^2_{.975}(23)$, **(c)** $\chi^2_{.90}(23)$, and **(d)** $\chi^2_{.10}(23)$.

Answer **(a)** 41.64 **(b)** 11.69 **(c)** 14.85 **(d)** 32.01

PROBLEM 11-31 Give the values of **(a)** $t_{.01}(23)$, **(b)** $t_{.90}(23)$, and **(c)** $-t_{.05}(23)$.

Answer **(a)** 2.50 **(b)** -1.32 **(c)** -1.71

PROBLEM 11-32 Give the values of **(a)** $F_{.01}(12, 15)$, **(b)** $F_{.99}(12, 15)$, **(c)** $F_{.025}(15, 12)$, and **(d)** $F_{.975}(15, 12)$.

Answer **(a)** 3.67 **(b)** .249 **(c)** 3.18 **(d)** .338

EXAM 2

1. In 1982 developing countries imported 100 million tons of cereals: 60 million tons in wheat, 30 million in coarse grains, and 10 million in rice. Suppose that it is possible to select 1 ton of a particular grain at random. **(a)** Describe the sample space. **(b)** Let $A = \{\text{Rice}\}$, and find $P(A)$.

2. A fair, 20-sided die is rolled once, so that the sample space is $S = \{1, 2, 3, \ldots, 20\}$. Let event $A = \{5, 10, 15, 20\}$, event $B = \{6, 12, 18\}$, and event $C = \{4, 8, 12, 16, 20\}$. **(a)** Which pair of these events are mutually exclusive? **(b)** Give the values of $P(A)$, $P(C)$, and $P(A \cap C)$. **(c)** Use your answers to **(b)** to find $P(A \cup C)$.

Problems 3–15 are based on the following data:

Some coins that are approximately the same size are the French 1 franc, the German 1 deutsche mark, the Italian 200 lire, the Canadian quarter, and the United States quarter. After a three-week trip, a traveler ended up with some of each of these coins in a coin purse.

3. Suppose that there is exactly 1 of each of the 5 coins in the coin purse. Give the number of ordered samples of size 3 that can be taken from this set of 5 coins when sampling **(a)** with replacement, and **(b)** without replacement. **(c)** Give the number of permutations of these 5 coins.

4. Suppose that in the coin purse there are 4 coins from each of the 5 countries. How many sets of 3 coins can be selected out of the 20 coins?

5. Suppose that there are 6 coins from each of the 5 countries. Select one coin at random. Let E be the event that it belongs to a European country, and let D be the event that it is a deutsche mark. Find $P(D \mid E)$.

6. Suppose that there are 6 coins from each of the 5 countries. If 2 coins are selected at random, one at a time without replacement, give the probability that they are **(a)** both French francs, **(b)** both from the same country.

7. Suppose that in the coin purse there are 2 coins from each of the 5 countries. You select a coin at random, note whether or not it is a European coin, and then return it to the purse. You repeat this selection with replacement 3 times. Give the probability that **(a)** all 3 coins are European, **(b)** exactly 2 of the 3 coins are European.

8. There are 20 coins in the coin purse: 6 are from France, 5 from Germany, 4 from Italy, 3 from Canada, and 2 from the United States. Select one coin at random, and let the random variable X equal 1, 2, 3, 4, or 5 if the coin is from France, Germany, Italy, Canada, or the United States, respectively. **(a)** Define the probability function of X. **(b)** Draw a probability histogram.

9. **(a)** Define and **(b)** graph the distribution function for the distribution in Problem 8.

10. Find **(a)** the mean, **(b)** the variance, and **(c)** the standard deviation for the distribution in Problem 8.

11. Suppose that the coin purse contains 1 coin from each of the 5 countries. Select 1 coin at random, and call the trial a success if it is a European coin. Let the random variable X equal 1 for success and 0 for failure. **(a)** Define the probability function of X. **(b)** Find the mean and **(c)** the variance of X.

12. The coin purse contains 1 coin from each of the 5 countries. Select a random sample of size 9, one at a time with replacement. Let X equal the number of European coins in this sample. **(a)** How is X distributed? **(b)** Find $P(X = 4)$.

13. The coin purse contains 1 coin from each of the 5 countries. Select a random sample of size 8, one at a time with replacement. Let X equal the number of Canadian coins in the sample. Give **(a)** the distribution, **(b)** the mean, and **(c)** the variance of X.

14. The coin purse contains 7 German and 9 French coins. Let X equal the number of German coins among 4 coins that are selected at random without replacement. **(a)** Define the probability function of X. **(b)** Give the mean of X. **(c)** Find $P(X = 3)$ using formula (6.7) to evaluate the combinatorials.

15. The coin purse contains 5 Italian, 3 French, and 2 German coins. A random sample of size 9 is selected, one at a time with replacement. Let X_1, X_2, and X_3 equal the numbers of Italian, French, and German coins, respectively, in the sample. Find the probability that in a sample of 9 coins there are 4 Italian, 3 French, and 2 German coins; that is, find $P(X_1 = 4, X_2 = 3, X_3 = 2)$.

16. Let X have a Poisson distribution with $\lambda = 4.2$. Give **(a)** the mean and **(b)** the variance of X. **(c)** Find $P(X = 5)$.

17. Let the pdf of the continuous random variable X be defined by $f(x) = 1/2 = .5, 0 \le x \le 2$. **(a)** Find $P(0 \le X \le .8)$. **(b)** Find $P(0 \le X \le x)$ for $0 \le x \le 2$. **(c)** Define the distribution function of X.

18. Let Z have a standard normal distribution $N(0, 1)$. Find **(a)** $P(Z < -2.17)$, **(b)** $P(0 \le Z \le .84)$, and **(c)** $P(-2.17 \le Z \le .84)$.

19. Give the values of **(a)** $z_{.025}$ and **(b)** $-z_{.05}$.

20. Let X equal the diameter (in millimeters) of the head of a fetus between the 16th and 25th weeks of pregnancy. Assume that the distribution of X is $N(51.4, 77.44)$. Find $P(44.8 \le X \le 62.4)$.

21. Let $\bar{X}$ be the mean of a random sample of size $n = 16$ from the normal distribution $N(51.4, 77.44)$. Give the values of **(a)** $\mu_{\bar{x}}$, the mean of the distribution of $\bar{X}$, **(b)** $\sigma_{\bar{x}}^2$, the variance of the distribution of $\bar{X}$, and **(c)** the standard error of the mean. **(d)** How is $\bar{X}$ distributed?

22. Let X equal the time (in minutes) it takes to serve a customer at a bank. Assume that the mean and variance of X are $\mu = 3.9$ and $\sigma^2 = 16$. Let $\bar{X}$ be the sample mean of a random sample of $n = 25$ service times. Find $P(\bar{X} \le 2.5)$, approximately.

23. A cross between a bar-eyed female fruit fly and a normal-eyed male fruit fly is expected to yield a 3:1 ratio of normal to bar; i.e., the probability that an offspring has normal eyes is $p = .75$. Let X equal the number of offspring with normal eyes in a random sample of size $n = 48$. **(a)** How is X distributed? **(b)** Find $P(X \le 40)$, approximately.

24. Give the values of **(a)** $\chi^2_{.05}(23)$ and **(b)** $\chi^2_{.90}(23)$.

25. Give the values of **(a)** $t_{.05}(13)$ and **(b)** $t_{.90}(13)$.

26. Give the values of **(a)** $F_{.01}(7, 4)$ and **(b)** $F_{.99}(7, 4)$.

Answers to Exam 2

1. (Section 6-1)

(a) The sample space is the set of all the possible outcomes when 1 ton of grain is selected at random: $S = \{\text{Wheat, Coarse grain, Rice}\}$.

(b) Assuming that there are 100 million equally likely events, of which 10 million belong to A, $P(A) = 10/100 = .1$.

2. (Section 6-2)

(a) Because $A \cap B = \varnothing$, A and B are mutually exclusive.

(b) With a fair die each outcome is equally likely; thus,

$$P(A) = \tfrac{4}{20} \quad P(C) = \tfrac{5}{20} \quad P(A \cap C) = \tfrac{1}{20}$$

(c) Using formula (6.2), you find

$$\begin{aligned} P(A \cup C) &= P(A) + P(C) - P(A \cap C) \\ &= \tfrac{4}{20} + \tfrac{5}{20} - \tfrac{1}{20} = \tfrac{8}{20} \end{aligned}$$

3. (Section 6-3)

(a) When sampling with replacement, you have 5 choices each time a coin is selected; thus $5 \cdot 5 \cdot 5 = 5^3 = 125$.
(b) When selected items aren't replaced, the number of possible outcomes is $5 \cdot 4 \cdot 3 = 60$.
(c) The number of permutations of 5 objects is $5! = 5 \cdot 4 \cdot 3 \cdot 2 \cdot 1 = 120$.

4. (Section 6-3) Using formula (6.7): $\binom{n}{r} = \dfrac{n!}{r!(n-r)!}$

$$\binom{20}{3} = \frac{20!}{3!17!} = \frac{20 \cdot 19 \cdot 18 \cdot 17!}{3 \cdot 2 \cdot 1 \cdot 17!} = 1140$$

5. (Section 6-4) Using formula (6.8):

$$P(D \mid E) = \frac{P(D \cap E)}{P(E)} = \frac{P(D)}{P(E)} = \frac{6/30}{18/30} = \frac{1}{3}$$

6. (Section 6-4)

(a) $A \cap B$ is the event that both coins are francs. By formula (6.9):

$$P(A \cap B) = P(A)P(B \mid A) = \frac{6}{30} \cdot \frac{5}{29} = \frac{1}{29}$$

(b) The probability that both coins are from France is $1/29$. The probability that both come from any one of the other countries is $1/29$ as well. Thus $\frac{1}{29} + \frac{1}{29} + \frac{1}{29} + \frac{1}{29} + \frac{1}{29} = \frac{5}{29}$.

7. (Section 6-4)

(a) Assuming that the trials are independent, and letting E be the event that a European coin is selected, $P(EEE) = \frac{6}{10} \cdot \frac{6}{10} \cdot \frac{6}{10} = \frac{216}{1000}$.
(b) You could select 2 European coins on the first two trials, on the first and third trials, or on the last two trials. Thus,

$$P(EEE') + P(EE'E) + P(E'EE) = \frac{6}{10} \cdot \frac{6}{10} \cdot \frac{4}{10} + \frac{6}{10} \cdot \frac{4}{10} \cdot \frac{6}{10} + \frac{4}{10} \cdot \frac{6}{10} \cdot \frac{6}{10}$$

$$= 3\left(\frac{6}{10}\right)^2\left(\frac{4}{10}\right) = 3\left(\frac{144}{1000}\right) = \frac{432}{1000}$$

where E' stands for the event that a coin is not European.

8. (Section 7-2)

(a) You can define the probability function either using the formula $f(x) = \dfrac{7-x}{20}$, $x = 1, 2, 3, 4, 5$, or by the following table:

x	$f(x)$
1	6/20
2	5/20
3	4/20
4	3/20
5	2/20
	20/20

(b) See Fig. E2-1.

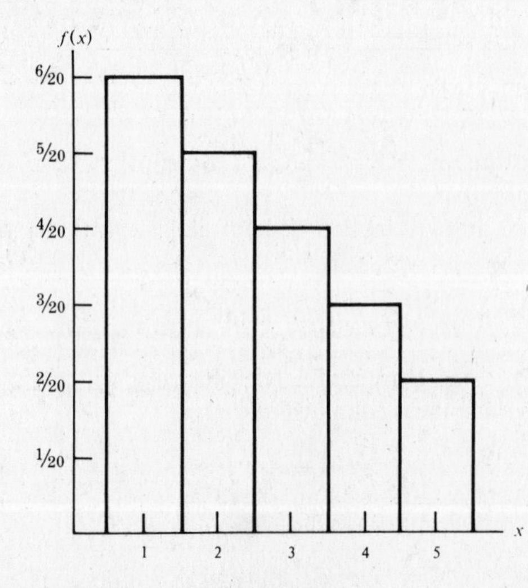

Figure E2-1

9. (Section 7-3)

(a) The distribution function is

$$F(x) = \begin{cases} 0, & x < 1 \\ 6/20, & 1 \le x < 2 \\ 11/20, & 2 \le x < 3 \\ 15/20, & 3 \le x < 4 \\ 18/20, & 4 \le x < 5 \\ 20/20, & 5 \le x \end{cases}$$

(b) See Fig. E2-2.

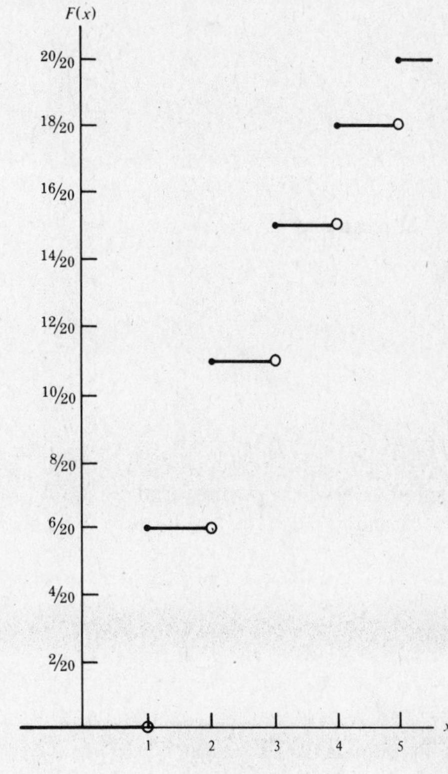

Figure E2-2

10. (Section 7-4)

(a) By formula (7.1), $\mu = \sum xf(x) = \frac{50}{20} = 2.5$
(b) By formula (7.3), $\sigma^2 = \sum x^2 f(x) - \mu^2 = 8 - (2.5)^2 = 1.75$
(c) By formula (7.4), $\sigma = \sqrt{\sigma^2} = \sqrt{1.75} = 1.32$

11. (Sections 7-2 and 7-4)

(a) Three of the coins are from Europe, so $P(X = 1) = 3/5$. The probability function of X is

$$f(0) = P(X = 0) = \tfrac{2}{5}$$

$$f(1) = P(X = 1) = \tfrac{3}{5}$$

(b) The mean of X is $\mu = \sum xf(x) = 0(\frac{2}{5}) + 1(\frac{3}{5}) = \frac{3}{5}$
(c) The variance of X is $\sigma^2 = \sum x^2 f(x) - \mu^2$

$$= 0^2(\tfrac{2}{5}) + 1^2(\tfrac{3}{5}) - (\tfrac{3}{5})^2$$

$$= \tfrac{3}{5} - \tfrac{9}{25} = \tfrac{15}{25} - \tfrac{9}{25} = \tfrac{6}{25}$$

12. (Section 8-2)

(a) You're observing $n = 9$ independent Bernoulli (success–failure) trials with $p = .6$, the probability for success on each trial. Thus X has a binomial distribution $b(9, .6)$.
(b) From the binomial probability table (Table 2 in the Appendix), $P(X = 4) = .1672$.

13. (Section 8-3)

(a) Since there are $n = 8$ trials and the probability for success $p = 1/5$ on each trial, X has a binomial distribution $b(8, \frac{1}{5})$.
(b) The mean of X is $\mu = np = 8(\frac{1}{5}) = 1.6$.
(c) The variance of X is $\sigma^2 = np(1 - p) = 8(\frac{1}{5})(\frac{4}{5}) = \frac{32}{25} = 1.28$.

14. (Section 9-1)

(**a**) X has a hypergeometric distribution, since you're selecting a random sample of $r = 4$ objects from a collection of $n = n_1 + n_2$ objects, where $n_1 = 7$ and $n_2 = 9$. Therefore, by formula (9.1), the probability function of X is

$$f(x) = \frac{\binom{n_1}{x}\binom{n_2}{r-x}}{\binom{n}{r}} = \frac{\binom{7}{x}\binom{9}{4-x}}{\binom{16}{4}}, \quad x = 0, 1, 2, 3, 4$$

(**b**) The mean of X is $\mu = \frac{rn_1}{n} = \frac{4 \cdot 7}{16} = \frac{28}{16} = 1.75$.

(**c**)

$$P(X = 3) = f(3)$$

$$= \frac{\binom{7}{3}\binom{9}{1}}{\binom{16}{4}} = \frac{35 \cdot 9}{1820} = .173$$

15. (Section 9-3) The random variables X_1, X_2, and X_3 have a multinomial distribution with $n = 9$, $p_1 = .5$, $p_2 = .3$, and $p_3 = .2$. Thus, by formula (9.6),

$$P(X_1 = 4, X_2 = 3, X_3 = 2) = \frac{9!}{4!3!2!}(.5)^4(.3)^3(.2)^2$$

$$= 1260(.0625)(.027)(.04) = .085$$

16. (Section 9-2)

(**a**) By formula (9.6), the mean is $\mu = \lambda = 4.2$.
(**b**) By formula (9.6), the variance is $\sigma^2 = \lambda = 4.2$
(**c**) From the Poisson probability table (Table 3 in the Appendix), $P(X = 5) = .1633$.

17. (Section 10-1)

(**a**) This probability is given by the area of a rectangle with a base of length = .8 and a height = .5. Thus, $P(0 \leq X \leq .8) = (.8)(.5) = .4$.
(**b**) This probability is given by the area of a rectangle with a base of length x and a height of .5. Thus, $P(0 \leq X \leq x) = x(.5) = x/2 \quad (0 \leq x \leq 2)$.
(**c**) The answer to (**b**) defines the distribution function for x between 0 and 2. There is no probability outside this interval. Thus the distribution function is

$$F(x) = \begin{cases} 0, & x < 0 \\ x/2, & 0 \leq x < 2 \\ 1, & 2 \leq x \end{cases}$$

18. (Section 10-2)

(**a**) In the normal probability table (Table 4 in the Appendix), locate row $z = 2.1$ and then look across to column .07. You find $P(Z < -2.17) = P(Z > 2.17) = .5000 - P(0 \leq Z \leq 2.17) = .5000 - .4850 = .0150$.
(**b**) In Table 4 of the Appendix, look in row $z = .8$ and column .04 to find $P(0 \leq Z \leq .84) = .2995$.
(**c**) Using the answers to (**a**) and (**b**), you find

$$P(-2.17 \leq Z \leq .84) = P(-2.17 \leq Z < 0) + P(0 \leq Z \leq .84)$$

$$= .4850 + .2995 = .7845$$

19. (Section 10-2)

(**a**) Here, .025 is α, the probability to the right of the z value. To use Table 4 to find z, you must know $(.5000 - \alpha)$, the probability between 0 and Z. Since $(.5000 - .0250) = .4750$, you can look up this probability in Table 4: $z_{.025} = 1.96$.
(**b**) Looking up the value $(.5000 - \alpha) = (.5000 - .0500) = .4500$ in Table 4, you find that this probability is the average of the probabilities for $z = 1.64$ and $z = 1.65$. Averaging these z values, you find $z_{.05} = 1.645$. Thus $-z_{.05} = -1.645$.

20. (Section 10-3) $z = \dfrac{44.8 - 51.4}{8.8}$ $\quad z = \dfrac{62.4 - 51.4}{8.8}$

$= -.75 \qquad = 1.25$

Thus,
$$P(44.8 \leq X \leq 62.4) = P(-.75 \leq Z \leq 1.25)$$
$$= P(-.75 \leq Z \leq 0) + P(0 < Z \leq 1.25)$$
$$= .2734 + .3944 = .6678$$

21. (Sections 11-1 and 11-2)

(a) By formula (11.1), the mean of $\bar{X}$ equals the distribution mean. Thus, $\mu_{\bar{x}} = \mu = 51.4$.

(b) By formula (11.2), the variance of $\bar{X}$ equals the distribution variance divided by n. Thus, $\sigma_{\bar{x}}^2 = \dfrac{\sigma^2}{n} = \dfrac{77.44}{16} = 4.84$.

(c) The standard error of the mean is the standard deviation of the distribution of $\bar{X}$. Thus, by formula (11.3),

$$\sigma_{\bar{x}} = \sqrt{\frac{\sigma^2}{n}} = \sqrt{\frac{77.44}{16}} = \frac{8.8}{4} = 2.2$$

(d) The distribution of $\bar{X}$ is normal with mean and variance as found in **(a)** and **(b)**. Thus the distribution of $\bar{X}$ is $N(51.4, 4.84)$.

22. (Section 11-2) Standardize to find the z value using formula (11.6). Following the Central Limit Theorem:

$$z = \frac{c - \mu}{\sigma/\sqrt{n}} = \frac{2.5 - 3.9}{4/\sqrt{25}} = -1.75$$

Therefore, $P(\bar{X} \leq 2.5) \approx P(Z \leq -1.75)$

$$= P(Z \geq 1.75) = P(Z \geq 0) - P(0 \leq Z < 1.75) = .5000 - .4599 = .0401$$

23. (Section 11-3)

(a) Assuming independent trials, X has a binomial distribution $b(48, .75)$.

(b) Make the half-unit correction for continuity, so that $P(X \leq 40.5)$, and use formula (11.9):

$$z = \frac{k + .5 - np}{np(1 - p)} = \frac{40.5 - 48(.75)}{\sqrt{48(.75)(.25)}} = \frac{40.5 - 36}{\sqrt{9}} = 1.50$$

Therefore,
$$P(X \leq 40) \approx P(Z \leq 1.50)$$
$$= P(Z < 0) + P(0 \leq Z \leq 1.50) = .5000 + .4332 = .9332$$

24. (Section 11-4)

(a) Use the chi-square table (Table 5 in the Appendix) to find these values for a chi-square distribution. Looking in row $r = 23$ and column .05, you find $\chi^2_{.05}(23) = 35.17$.

(b) Looking in row $r = 23$ and column .90 of Table 5, you find $\chi^2_{.90}(23) = 14.85$.

25. (Section 11-5)

(a) Use the t table (Table 6 in the Appendix) to find critical values for the t distribution. Looking in row $r = 13$ and column .05, you find $t_{.05}(13) = 1.77$.

(b) Looking in row $r = 13$ and column .10 of Table 6, you find $t_{.10}(13) = 1.35$. Thus, because of the symmetry of the t distribution, $t_{.90}(13) = -t_{.10}(13) = -1.35$.

26. (Section 11-6)

(a) In the F table listing α values of .01 (Table 7a in the Appendix), locate column $r_1 = 7$ and row $r_2 = 4$. You'll find $F_{.01}(7, 4) = 14.98$.

(b) By formula (11.12), $F_{.99}(7, 4) = 1/F_{.01}(4, 7)$. Therefore, looking in column $r_1 = 4$ and row $r_2 = 7$ in Table 7a, you find

$$F_{.99}(7, 4) = \frac{1}{F_{.01}(4, 7)} = \frac{1}{7.85} = .127$$

APPENDIX

TABLE 1: **Binomial Coefficients**
TABLE 2: **Binomial Probabilities**
TABLE 3: **Poisson Probabilities**
TABLE 4: **Probabilities for the Standard Normal Distribution**
TABLE 5: **Chi-Square Critical Values**
TABLE 6: **Student's *t* Critical Values**
TABLE 7: ***F* Distribution Critical Values**
TABLE 8: **Critical Values for the Correlation Coefficient**
TABLE 9: **Random Numbers on the Interval (0, 1)**

TABLE 1: Binomial Coefficients*

n	$\binom{n}{0}$	$\binom{n}{1}$	$\binom{n}{2}$	$\binom{n}{3}$	$\binom{n}{4}$	$\binom{n}{5}$	$\binom{n}{6}$	$\binom{n}{7}$	$\binom{n}{8}$	$\binom{n}{9}$	$\binom{n}{10}$	$\binom{n}{11}$	$\binom{n}{12}$	$\binom{n}{13}$
0	1													
1	1	1												
2	1	2	1											
3	1	3	3	1										
4	1	4	6	4	1									
5	1	5	10	10	5	1								
6	1	6	15	20	15	6	1							
7	1	7	21	35	35	21	7	1						
8	1	8	28	56	70	56	28	8	1					
9	1	9	36	84	126	126	84	36	9	1				
10	1	10	45	120	210	252	210	120	45	10	1			
11	1	11	55	165	330	462	462	330	165	55	11	1		
12	1	12	66	220	495	792	924	792	495	220	66	12	1	
13	1	13	78	286	715	1,287	1,716	1,716	1,287	715	286	78	13	1
14	1	14	91	364	1,001	2,002	3,003	3,432	3,003	2,002	1,001	364	91	14
15	1	15	105	455	1,365	3,003	5,005	6,435	6,435	5,005	3,003	1,365	455	105
16	1	16	120	560	1,820	4,368	8,008	11,440	12,870	11,440	8,008	4,368	1,820	560
17	1	17	136	680	2,380	6,188	12,376	19,448	24,310	24,310	19,448	12,376	6,188	2,380
18	1	18	153	816	3,060	8,568	18,564	31,824	43,758	48,620	43,758	31,824	18,564	8,568
19	1	19	171	969	3,876	11,628	27,132	50,388	75,582	92,378	92,378	75,582	50,388	27,132
20	1	20	190	1,140	4,845	15,504	38,760	77,520	125,970	167,960	184,756	167,960	125,970	77,520
21	1	21	210	1,330	5,985	20,349	54,264	116,280	203,490	293,930	352,716	352,716	293,930	203,490
22	1	22	231	1,540	7,315	26,334	74,613	170,544	319,770	497,420	646,646	705,432	646,646	497,420
23	1	23	253	1,771	8,855	33,649	100,947	245,157	490,314	817,190	1,144,066	1,352,078	1,352,078	1,144,066
24	1	24	276	2,024	10,626	42,504	134,596	346,104	735,471	1,307,504	1,961,256	2,496,144	2,704,156	2,496,144
25	1	25	300	2,300	12,650	53,130	177,100	480,700	1,081,575	2,042,975	3,268,760	4,457,400	5,200,300	5,200,300
26	1	26	325	2,600	14,950	65,780	230,230	657,800	1,562,275	3,124,550	5,311,735	7,726,160	9,657,700	10,400,600

* For $r > 13$ you may use the identity $\binom{n}{r} = \binom{n}{n-r}$.

TABLE 2: Binomial Probabilities

$$P(X = x) = f(x) = \frac{n!}{x!(n-x)!} p^x(1-p)^{n-x}, \quad x = 0, 1, \ldots, n$$

						p						
n	x	.1	.2	.25	.3	.4	.5	.6	.7	.75	.8	.9
1	0	.9000	.8000	.7500	.7000	.6000	.5000	.4000	.3000	.2500	.2000	.1000
	1	.1000	.2000	.2500	.3000	.4000	.5000	.6000	.7000	.7500	.8000	.9000
2	0	.8100	.6400	.5625	.4900	.3600	.2500	.1600	.0900	.0625	.0400	.0100
	1	.1800	.3200	.3750	.4200	.4800	.5000	.4800	.4200	.3750	.3200	.1800
	2	.0100	.0400	.0625	.0900	.1600	.2500	.3600	.4900	.5625	.6400	.8100
3	0	.7290	.5120	.4219	.3430	.2160	.1250	.0640	.0270	.0156	.0080	.0010
	1	.2430	.3840	.4219	.4410	.4320	.3750	.2880	.1890	.1406	.0960	.0270
	2	.0270	.0960	.1406	.1890	.2880	.3750	.4320	.4410	.4219	.3840	.2430
	3	.0010	.0080	.0156	.0270	.0640	.1250	.2160	.3430	.4219	.5120	.7290
4	0	.6561	.4096	.3164	.2401	.1296	.0625	.0256	.0081	.0039	.0016	.0001
	1	.2916	.4096	.4219	.4116	.3456	.2500	.1536	.0756	.0469	.0256	.0036
	2	.0486	.1536	.2109	.2646	.3456	.3750	.3456	.2646	.2109	.1536	.0486
	3	.0036	.0256	.0469	.0756	.1536	.2500	.3456	.4116	.4219	.4096	.2916
	4	.0001	.0016	.0039	.0081	.0256	.0625	.1296	.2401	.3164	.4096	.6561
5	0	.5905	.3277	.2373	.1681	.0778	.0313	.0102	.0024	.0010	.0003	.0000
	1	.3281	.4096	.3955	.3602	.2592	.1563	.0768	.0283	.0146	.0064	.0005
	2	.0729	.2048	.2637	.3087	.3456	.3125	.2304	.1323	.0879	.0512	.0081
	3	.0081	.0512	.0879	.1323	.2304	.3125	.3456	.3087	.2637	.2048	.0729
	4	.0005	.0064	.0146	.0283	.0768	.1563	.2592	.3602	.3955	.4096	.3281
	5	.0000	.0003	.0010	.0024	.0102	.0313	.0778	.1681	.2373	.3277	.5905
6	0	.5314	.2621	.1780	.1176	.0467	.0156	.0041	.0007	.0002	.0001	.0000
	1	.3543	.3932	.3560	.3025	.1866	.0938	.0369	.0102	.0044	.0015	.0001
	2	.0984	.2458	.2966	.3241	.3110	.2344	.1382	.0595	.0330	.0154	.0012
	3	.0146	.0819	.1318	.1852	.2765	.3125	.2765	.1852	.1318	.0819	.0146
	4	.0012	.0154	.0330	.0595	.1382	.2344	.3110	.3241	.2966	.2458	.0984
	5	.0001	.0015	.0044	.0102	.0369	.0938	.1866	.3025	.3560	.3932	.3543
	6	.0000	.0001	.0002	.0007	.0041	.0156	.0467	.1176	.1780	.2621	.5314
7	0	.4783	.2097	.1335	.0824	.0280	.0078	.0016	.0002	.0001	.0000	.0000
	1	.3720	.3670	.3115	.2471	.1306	.0547	.0172	.0036	.0013	.0004	.0000
	2	.1240	.2753	.3115	.3177	.2613	.1641	.0774	.0250	.0115	.0043	.0002
	3	.0230	.1147	.1730	.2269	.2903	.2734	.1935	.0972	.0577	.0287	.0026
	4	.0026	.0287	.0577	.0972	.1935	.2734	.2903	.2269	.1703	.1147	.0230
	5	.0002	.0043	.0115	.0250	.0774	.1641	.2613	.3177	.3115	.2753	.1240
	6	.0000	.0004	.0013	.0036	.0172	.0547	.1306	.2471	.3115	.3670	.3720
	7	.0000	.0000	.0001	.0002	.0016	.0078	.0280	.0824	.1335	.2097	.4783
8	0	.4305	.1678	.1001	.0576	.0168	.0039	.0007	.0001	.0000	.0000	.0000
	1	.3826	.3355	.2670	.1977	.0896	.0312	.0079	.0012	.0004	.0001	.0000
	2	.1488	.2936	.3115	.2965	.2090	.1094	.0413	.0100	.0038	.0011	.0000
	3	.0331	.1468	.2076	.2541	.2787	.2188	.1239	.0467	.0231	.0092	.0004
	4	.0046	.0459	.0865	.1361	.2322	.2734	.2322	.1361	.0865	.0459	.0046
	5	.0004	.0092	.0231	.0467	.1239	.2188	.2787	.2541	.2076	.1468	.0331
	6	.0000	.0011	.0038	.0100	.0413	.1094	.2090	.2965	.3115	.2936	.1488
	7	.0000	.0001	.0004	.0012	.0079	.0312	.0896	.1977	.2670	.3355	.3826
	8	.0000	.0000	.0000	.0001	.0007	.0039	.0168	.0576	.1001	.1678	.4305
9	0	.3874	.1342	.0751	.0404	.0101	.0020	.0003	.0000	.0000	.0000	.0000
	1	.3874	.3020	.2253	.1556	.0605	.0176	.0035	.0004	.0001	.0000	.0000
	2	.1722	.3020	.3003	.2668	.1612	.0703	.0212	.0039	.0012	.0003	.0000
	3	.0446	.1762	.2336	.2668	.2508	.1641	.0743	.0210	.0087	.0028	.0001
	4	.0074	.0661	.1168	.1715	.2508	.2461	.1672	.0735	.0389	.0165	.0008
	5	.0008	.0165	.0389	.0735	.1672	.2461	.2508	.1715	.1168	.0661	.0074
	6	.0001	.0028	.0087	.0210	.0743	.1641	.2508	.2668	.2336	.1762	.0446
	7	.0000	.0003	.0012	.0039	.0212	.0703	.1612	.2668	.3003	.3020	.1722
	8	.0000	.0000	.0001	.0004	.0035	.0176	.0605	.1556	.2253	.3020	.3874
	9	.0000	.0000	.0000	.0000	.0003	.0020	.0101	.0404	.0751	.1342	.3874

TABLE 2 (*continued*)

		p										
n	*x*	.1	.2	.25	.3	.4	.5	.6	.7	.75	.8	.9
10	0	.3487	.1074	.0563	.0282	.0060	.0010	.0001	.0000	.0000	.0000	.0000
	1	.3874	.2684	.1877	.1211	.0403	.0098	.0016	.0001	.0000	.0000	.0000
	2	.1937	.3020	.2816	.2335	.1209	.0439	.0106	.0014	.0004	.0001	.0000
	3	.0574	.2013	.2503	.2668	.2150	.1172	.0425	.0090	.0031	.0008	.0000
	4	.0112	.0881	.1460	.2001	.2508	.2051	.1115	.0368	.0162	.0055	.0001
	5	.0015	.0264	.0584	.1029	.2007	.2461	.2007	.1029	.0584	.0264	.0015
	6	.0001	.0055	.0162	.0368	.1115	.2051	.2508	.2001	.1460	.0881	.0112
	7	.0000	.0008	.0031	.0090	.0425	.1172	.2150	.2668	.2503	.2013	.0574
	8	.0000	.0001	.0004	.0014	.0106	.0439	.1209	.2335	.2816	.3020	.1937
	9	.0000	.0000	.0000	.0001	.0016	.0098	.0403	.1211	.1877	.2684	.3874
	10	.0000	.0000	.0000	.0000	.0001	.0010	.0060	.0282	.0563	.1074	.3487
11	0	.3138	.0859	.0422	.0198	.0036	.0005	.0000	.0000	.0000	.0000	.0000
	1	.3835	.2362	.1549	.0932	.0266	.0054	.0007	.0000	.0000	.0000	.0000
	2	.2131	.2953	.2581	.1998	.0887	.0269	.0052	.0005	.0001	.0000	.0000
	3	.0710	.2215	.2581	.2568	.1774	.0806	.0234	.0037	.0011	.0002	.0000
	4	.0158	.1107	.1721	.2201	.2365	.1611	.0701	.0173	.0064	.0017	.0000
	5	.0025	.0388	.0803	.1321	.2207	.2256	.1471	.0566	.0268	.0097	.0003
	6	.0003	.0097	.0268	.0566	.1471	.2256	.2207	.1321	.0803	.0388	.0025
	7	.0000	.0017	.0064	.0173	.0701	.1611	.2365	.2201	.1721	.1107	.0158
	8	.0000	.0002	.0011	.0037	.0234	.0806	.1774	.2568	.2581	.2215	.0710
	9	.0000	.0000	.0001	.0005	.0052	.0269	.0887	.1998	.2581	.2953	.2131
	10	.0000	.0000	.0000	.0000	.0007	.0054	.0266	.0932	.1549	.2362	.3835
	11	.0000	.0000	.0000	.0000	.0000	.0005	.0036	.0198	.0422	.0859	.3138
12	0	.2824	.0687	.0317	.0138	.0022	.0002	.0000	.0000	.0000	.0000	.0000
	1	.3766	.2062	.1267	.0712	.0174	.0029	.0003	.0000	.0000	.0000	.0000
	2	.2301	.2835	.2323	.1678	.0639	.0161	.0025	.0002	.0000	.0000	.0000
	3	.0852	.2362	.2581	.2397	.1419	.0537	.0125	.0015	.0004	.0001	.0000
	4	.0213	.1329	.1936	.2311	.2128	.1208	.0420	.0078	.0024	.0005	.0000
	5	.0038	.0532	.1032	.1585	.2270	.1934	.1009	.0291	.0115	.0033	.0000
	6	.0005	.0155	.0401	.0792	.1766	.2256	.1766	.0792	.0401	.0155	.0005
	7	.0000	.0033	.0115	.0291	.1009	.1934	.2270	.1585	.1032	.0532	.0038
	8	.0000	.0005	.0024	.0078	.0420	.1208	.2128	.2311	.1936	.1329	.0213
	9	.0000	.0001	.0004	.0015	.0125	.0537	.1419	.2397	.2581	.2362	.0852
	10	.0000	.0000	.0000	.0002	.0025	.0161	.0639	.1678	.2323	.2835	.2301
	11	.0000	.0000	.0000	.0000	.0003	.0029	.0174	.0712	.1267	.2062	.3766
	12	.0000	.0000	.0000	.0000	.0000	.0002	.0022	.0138	.0317	.0687	.2824
13	0	.2542	.0550	.0238	.0097	.0013	.0001	.0000	.0000	.0000	.0000	.0000
	1	.3672	.1787	.1029	.0540	.0113	.0016	.0001	.0000	.0000	.0000	.0000
	2	.2448	.2680	.2059	.1388	.0453	.0095	.0012	.0001	.0000	.0000	.0000
	3	.0997	.2457	.2517	.2181	.1107	.0349	.0065	.0006	.0001	.0000	.0000
	4	.0277	.1535	.2097	.2337	.1845	.0873	.0243	.0034	.0009	.0001	.0000
	5	.0055	.0691	.1258	.1803	.2214	.1571	.0656	.0142	.0047	.0011	.0000
	6	.0008	.0230	.0559	.1030	.1968	.2095	.1312	.0442	.0186	.0058	.0001
	7	.0001	.0058	.0186	.0442	.1312	.2095	.1968	.1030	.0559	.0230	.0008
	8	.0000	.0011	.0047	.0142	.0656	.1571	.2214	.1803	.1258	.0691	.0055
	9	.0000	.0001	.0009	.0034	.0243	.0873	.1845	.2337	.2097	.1535	.0277
	10	.0000	.0000	.0001	.0006	.0065	.0349	.1107	.2181	.2517	.2457	.0997
	11	.0000	.0000	.0000	.0001	.0012	.0095	.0453	.1388	.2059	.2680	.2448
	12	.0000	.0000	.0000	.0000	.0001	.0016	.0113	.0540	.1029	.1787	.3672
	13	.0000	.0000	.0000	.0000	.0000	.0001	.0013	.0097	.0238	.0550	.2542
14	0	.2288	.0440	.0178	.0068	.0008	.0001	.0000	.0000	.0000	.0000	.0000
	1	.3559	.1539	.0832	.0407	.0073	.0009	.0001	.0000	.0000	.0000	.0000
	2	.2570	.2501	.1802	.1134	.0317	.0056	.0005	.0000	.0000	.0000	.0000
	3	.1142	.2501	.2402	.1943	.0845	.0222	.0033	.0002	.0000	.0000	.0000
	4	.0349	.1720	.2202	.2290	.1549	.0611	.0136	.0014	.0003	.0000	.0000
	5	.0078	.0860	.1468	.1963	.2066	.1222	.0408	.0066	.0018	.0003	.0000
	6	.0013	.0322	.0734	.1262	.2066	.1833	.0918	.0232	.0082	.0020	.0000
	7	.0002	.0092	.0280	.0618	.1574	.2095	.1574	.0618	.0280	.0092	.0002

TABLE 2 (*continued*)

		p										
n	*x*	.1	.2	.25	.3	.4	.5	.6	.7	.75	.8	.9
	8	.0000	.0020	.0082	.0232	.0918	.1833	.2066	.1262	.0734	.0322	.0013
	9	.0000	.0003	.0018	.0066	.0408	.1222	.2066	.1963	.1468	.0860	.0078
	10	.0000	.0000	.0003	.0014	.0136	.0611	.1549	.2290	.2202	.1720	.0349
	11	.0000	.0000	.0000	.0002	.0033	.0222	.0845	.1943	.2402	.2501	.1142
	12	.0000	.0000	.0000	.0000	.0005	.0056	.0317	.1134	.1802	.2501	.2570
	13	.0000	.0000	.0000	.0000	.0001	.0009	.0073	.0407	.0832	.1539	.3559
	14	.0000	.0000	.0000	.0000	.0000	.0001	.0008	.0068	.0178	.0440	.2288
15	0	.2059	.0352	.0134	.0047	.0005	.0000	.0000	.0000	.0000	.0000	.0000
	1	.3432	.1319	.0668	.0305	.0047	.0005	.0000	.0000	.0000	.0000	.0000
	2	.2669	.2309	.1559	.0916	.0219	.0032	.0003	.0000	.0000	.0000	.0000
	3	.1285	.2501	.2252	.1700	.0634	.0139	.0016	.0001	.0000	.0000	.0000
	4	.0428	.1876	.2252	.2186	.1268	.0417	.0074	.0006	.0001	.0000	.0000
	5	.0105	.1032	.1651	.2061	.1859	.0916	.0245	.0030	.0007	.0001	.0000
	6	.0019	.0430	.0917	.1472	.2066	.1527	.0612	.0116	.0034	.0007	.0000
	7	.0003	.0138	.0393	.0811	.1771	.1964	.1181	.0348	.0131	.0035	.0000
	8	.0000	.0035	.0131	.0348	.1181	.1964	.1771	.0811	.0393	.0138	.0003
	9	.0000	.0007	.0034	.0116	.0612	.1527	.2066	.1472	.0917	.0430	.0019
	10	.0000	.0001	.0007	.0030	.0245	.0916	.1859	.2061	.1651	.1032	.0105
	11	.0000	.0000	.0001	.0006	.0074	.0417	.1268	.2186	.2252	.1876	.0428
	12	.0000	.0000	.0000	.0001	.0016	.0139	.0634	.1700	.2252	.2501	.1285
	13	.0000	.0000	.0000	.0000	.0003	.0032	.0219	.0916	.1559	.2309	.2669
	14	.0000	.0000	.0000	.0000	.0000	.0005	.0047	.0305	.0668	.1319	.3432
	15	.0000	.0000	.0000	.0000	.0000	.0000	.0005	.0047	.0134	.0352	.2059
16	0	.1853	.0281	.0100	.0033	.0003	.0000	.0000	.0000	.0000	.0000	.0000
	1	.3294	.1126	.0535	.0228	.0030	.0002	.0000	.0000	.0000	.0000	.0000
	2	.2745	.2111	.1336	.0732	.0150	.0018	.0001	.0000	.0000	.0000	.0000
	3	.1423	.2463	.2079	.1465	.0468	.0085	.0008	.0000	.0000	.0000	.0000
	4	.0514	.2001	.2252	.2040	.1014	.0278	.0040	.0002	.0000	.0000	.0000
	5	.0137	.1201	.1802	.2099	.1623	.0667	.0142	.0013	.0002	.0000	.0000
	6	.0028	.0550	.1101	.1649	.1983	.1222	.0392	.0056	.0014	.0002	.0000
	7	.0004	.0197	.0524	.1010	.1889	.1746	.0840	.0185	.0058	.0012	.0000
	8	.0001	.0055	.0197	.0487	.1417	.1964	.1417	.0487	.0197	.0055	.0001
	9	.0000	.0012	.0058	.0185	.0840	.1746	.1889	.1010	.0524	.0197	.0004
	10	.0000	.0002	.0014	.0056	.0392	.1222	.1983	.1649	.1101	.0550	.0028
	11	.0000	.0000	.0002	.0013	.0142	.0667	.1623	.2099	.1802	.1201	.0137
	12	.0000	.0000	.0000	.0002	.0040	.0278	.1014	.2040	.2252	.2001	.0514
	13	.0000	.0000	.0000	.0000	.0008	.0085	.0468	.1465	.2079	.2463	.1423
	14	.0000	.0000	.0000	.0000	.0001	.0018	.0150	.0732	.1336	.2111	.2745
	15	.0000	.0000	.0000	.0000	.0000	.0002	.0030	.0228	.0535	.1126	.3294
	16	.0000	.0000	.0000	.0000	.0000	.0000	.0003	.0033	.0100	.0281	.1853
20	0	.1216	.0115	.0032	.0008	.0000	.0000	.0000	.0000	.0000	.0000	.0000
	1	.2702	.0576	.0211	.0068	.0005	.0000	.0000	.0000	.0000	.0000	.0000
	2	.2852	.1369	.0669	.0278	.0031	.0002	.0000	.0000	.0000	.0000	.0000
	3	.1901	.2054	.1339	.0716	.0123	.0011	.0000	.0000	.0000	.0000	.0000
	4	.0898	.2182	.1897	.1304	.0350	.0046	.0003	.0000	.0000	.0000	.0000
	5	.0319	.1746	.2023	.1789	.0746	.0148	.0013	.0000	.0000	.0000	.0000
	6	.0089	.1091	.1686	.1916	.1244	.0370	.0049	.0002	.0000	.0000	.0000
	7	.0020	.0545	.1124	.1643	.1659	.0739	.0146	.0010	.0002	.0000	.0000
	8	.0004	.0222	.0609	.1144	.1797	.1201	.0355	.0039	.0008	.0001	.0000
	9	.0001	.0074	.0271	.0654	.1597	.1602	.0710	.0120	.0030	.0005	.0000
	10	.0000	.0020	.0099	.0308	.1171	.1762	.1171	.0308	.0099	.0020	.0000
	11	.0000	.0005	.0030	.0120	.0710	.1602	.1597	.0654	.0271	.0074	.0001
	12	.0000	.0001	.0008	.0039	.0355	.1201	.1797	.1144	.0609	.0222	.0004
	13	.0000	.0000	.0002	.0010	.0146	.0739	.1659	.1643	.1124	.0545	.0020
	14	.0000	.0000	.0000	.0002	.0049	.0370	.1244	.1916	.1686	.1091	.0089
	15	.0000	.0000	.0000	.0000	.0013	.0148	.0746	.1789	.2023	.1746	.0319
	16	.0000	.0000	.0000	.0000	.0003	.0046	.0350	.1304	.1897	.2182	.0898
	17	.0000	.0000	.0000	.0000	.0000	.0011	.0123	.0716	.1339	.2054	.1901

TABLE 2 (*continued*)

		p										
n	*x*	.1	.2	.25	.3	.4	.5	.6	.7	.75	.8	.9
	18	.0000	.0000	.0000	.0000	.0000	.0002	.0031	.0278	.0669	.1369	.2852
	19	.0000	.0000	.0000	.0000	.0000	.0000	.0005	.0068	.0211	.0576	.2702
	20	.0000	.0000	.0000	.0000	.0000	.0000	.0000	.0008	.0032	.0115	.1216
25	0	.0718	.0038	.0008	.0001	.0000	.0000	.0000	.0000	.0000	.0000	.0000
	1	.1994	.0236	.0063	.0014	.0000	.0000	.0000	.0000	.0000	.0000	.0000
	2	.2659	.0708	.0251	.0074	.0004	.0000	.0000	.0000	.0000	.0000	.0000
	3	.2265	.1358	.0641	.0243	.0019	.0001	.0000	.0000	.0000	.0000	.0000
	4	.1384	.1867	.1175	.0572	.0071	.0004	.0000	.0000	.0000	.0000	.0000
	5	.0646	.1960	.1645	.1030	.0199	.0016	.0000	.0000	.0000	.0000	.0000
	6	.0239	.1633	.1828	.1472	.0442	.0053	.0002	.0000	.0000	.0000	.0000
	7	.0072	.1108	.1654	.1712	.0800	.0143	.0009	.0000	.0000	.0000	.0000
	8	.0018	.0623	.1241	.1651	.1200	.0322	.0031	.0001	.0000	.0000	.0000
	9	.0004	.0294	.0781	.1336	.1511	.0609	.0088	.0004	.0000	.0000	.0000
	10	.0001	.0118	.0417	.0916	.1612	.0974	.0212	.0013	.0002	.0000	.0000
	11	.0000	.0040	.0189	.0536	.1465	.1328	.0434	.0042	.0007	.0001	.0000
	12	.0000	.0012	.0074	.0268	.1140	.1550	.0760	.0115	.0025	.0003	.0000
	13	.0000	.0003	.0025	.0115	.0760	.1550	.1140	.0268	.0074	.0012	.0000
	14	.0000	.0001	.0007	.0042	.0434	.1328	.1465	.0536	.0189	.0040	.0000
	15	.0000	.0000	.0002	.0013	.0212	.0974	.1612	.0916	.0417	.0118	.0001
	16	.0000	.0000	.0000	.0004	.0088	.0609	.1511	.1336	.0781	.0294	.0004
	17	.0000	.0000	.0000	.0001	.0031	.0322	.1200	.1651	.1241	.0623	.0018
	18	.0000	.0000	.0000	.0000	.0009	.0143	.0800	.1712	.1654	.1108	.0072
	19	.0000	.0000	.0000	.0000	.0002	.0053	.0442	.1472	.1828	.1633	.0239
	20	.0000	.0000	.0000	.0000	.0000	.0016	.0199	.1030	.1645	.1960	.0646
	21	.0000	.0000	.0000	.0000	.0000	.0004	.0071	.0572	.1175	.1867	.1384
	22	.0000	.0000	.0000	.0000	.0000	.0001	.0019	.0243	.0641	.1358	.2265
	23	.0000	.0000	.0000	.0000	.0000	.0000	.0004	.0074	.0251	.0708	.2659
	24	.0000	.0000	.0000	.0000	.0000	.0000	.0000	.0014	.0063	.0236	.1994
	25	.0000	.0000	.0000	.0000	.0000	.0000	.0000	.0001	.0008	.0038	.0718

TABLE 3: Poisson Probabilities

$P(X = x) = f(x) = \frac{\lambda^x e^{-\lambda}}{x!}$, $x = 0, 1, 2, \ldots$ where λ is the mean of a Poisson random variable

	λ									
x	.1	.2	.3	.4	.5	.6	.7	.8	.9	1.0
0	.9048	.8187	.7408	.6703	.6065	.5488	.4966	.4493	.4066	.3679
1	.0905	.1637	.2222	.2681	.3033	.3293	.3476	.3595	.3659	.3679
2	.0045	.0164	.0333	.0536	.0758	.0988	.1217	.1438	.1647	.1839
3	.0002	.0011	.0033	.0072	.0126	.0198	.0284	.0383	.0494	.0613
4	.0000	.0001	.0003	.0007	.0016	.0030	.0050	.0077	.0111	.0153
5	.0000	.0000	.0000	.0001	.0002	.0004	.0007	.0012	.0020	.0031
6	.0000	.0000	.0000	.0000	.0000	.0000	.0001	.0002	.0003	.0005
7	.0000	.0000	.0000	.0000	.0000	.0000	.0000	.0000	.0000	.0001

	λ									
x	1.1	1.2	1.3	1.4	1.5	1.6	1.7	1.8	1.9	2.0
0	.3329	.3012	.2725	.2466	.2231	.2019	.1827	.1653	.1496	.1353
1	.3662	.3614	.3543	.3452	.3347	.3230	.3106	.2975	.2842	.2707
2	.2014	.2169	.2303	.2417	.2510	.2584	.2640	.2678	.2700	.2707
3	.0738	.0867	.0998	.1128	.1255	.1378	.1496	.1607	.1710	.1804
4	.0203	.0260	.0324	.0395	.0471	.0551	.0636	.0723	.0812	.0902
5	.0045	.0062	.0084	.0111	.0141	.0176	.0216	.0260	.0309	.0361
6	.0008	.0012	.0018	.0026	.0035	.0047	.0061	.0078	.0098	.0120
7	.0001	.0002	.0003	.0005	.0008	.0011	.0015	.0020	.0027	.0034
8	.0000	.0000	.0001	.0001	.0001	.0002	.0003	.0005	.0006	.0009
9	.0000	.0000	.0000	.0000	.0000	.0000	.0001	.0001	.0001	.0002

	λ									
x	2.1	2.2	2.3	2.4	2.5	2.6	2.7	2.8	2.9	3.0
0	.1225	.1108	.1003	.0907	.0821	.0743	.0672	.0608	.0550	.0498
1	.2572	.2438	.2306	.2177	.2052	.1931	.1815	.1703	.1596	.1494
2	.2700	.2681	.2652	.2613	.2565	.2510	.2450	.2384	.2314	.2240
3	.1890	.1966	.2033	.2090	.2138	.2176	.2205	.2225	.2237	.2240
4	.0992	.1082	.1169	.1254	.1336	.1414	.1488	.1557	.1622	.1680
5	.0417	.0476	.0538	.0602	.0668	.0735	.0804	.0872	.0940	.1008
6	.0146	.0174	.0206	.0241	.0278	.0319	.0362	.0407	.0455	.0504
7	.0044	.0055	.0068	.0083	.0099	.0118	.0139	.0163	.0188	.0216
8	.0011	.0015	.0019	.0025	.0031	.0038	.0047	.0057	.0068	.0081
9	.0003	.0004	.0005	.0007	.0009	.0011	.0014	.0018	.0022	.0027
10	.0001	.0001	.0001	.0002	.0002	.0003	.0004	.0005	.0006	.0008
11	.0000	.0000	.0000	.0000	.0000	.0001	.0001	.0001	.0002	.0002
12	.0000	.0000	.0000	.0000	.0000	.0000	.0000	.0000	.0000	.0001

	λ									
x	3.1	3.2	3.3	3.4	3.5	3.6	3.7	3.8	3.9	4.0
0	.0450	.0408	.0369	.0334	.0302	.0273	.0247	.0224	.0202	.0183
1	.1397	.1304	.1217	.1135	.1057	.0984	.0915	.0850	.0789	.0733
2	.2165	.2087	.2008	.1929	.1850	.1771	.1692	.1615	.1539	.1465
3	.2237	.2226	.2209	.2186	.2158	.2125	.2087	.2046	.2001	.1954
4	.1734	.1781	.1823	.1858	.1888	.1912	.1931	.1944	.1951	.1954
5	.1075	.1140	.1203	.1264	.1322	.1377	.1429	.1477	.1522	.1563
6	.0555	.0608	.0662	.0716	.0771	.0826	.0881	.0936	.0989	.1042
7	.0246	.0278	.0312	.0348	.0385	.0425	.0466	.0508	.0551	.0595
8	.0095	.0111	.0129	.0148	.0169	.0191	.0215	.0241	.0269	.0298
9	.0033	.0040	.0047	.0056	.0066	.0076	.0089	.0102	.0116	.0132
10	.0010	.0013	.0016	.0019	.0023	.0028	.0033	.0039	.0045	.0053
11	.0003	.0004	.0005	.0006	.0007	.0009	.0011	.0013	.0016	.0019

TABLE 3 (*continued*)

	λ									
x	3.1	3.2	3.3	3.4	3.5	3.6	3.7	3.8	3.9	4.0
12	.0001	.0001	.0001	.0002	.0002	.0003	.0003	.0004	.0005	.0006
13	.0000	.0000	.0000	.0000	.0001	.0001	.0001	.0001	.0002	.0002
14	.0000	.0000	.0000	.0000	.0000	.0000	.0000	.0000	.0000	.0001

	λ									
x	4.1	4.2	4.3	4.4	4.5	4.6	4.7	4.8	4.9	5.0
0	.0166	.0150	.0136	.0123	.0111	.0101	.0091	.0082	.0074	.0067
1	.0679	.0630	.0583	.0540	.0500	.0462	.0427	.0395	.0365	.0337
2	.1393	.1323	.1254	.1188	.1125	.1063	.1005	.0948	.0894	.0842
3	.1904	.1852	.1798	.1743	.1687	.1631	.1574	.1517	.1460	.1404
4	.1951	.1944	.1933	.1917	.1898	.1875	.1849	.1820	.1789	.1755
5	.1600	.1633	.1662	.1687	.1708	.1725	.1738	.1747	.1753	.1755
6	.1093	.1143	.1191	.1237	.1281	.1323	.1362	.1398	.1432	.1462
7	.0640	.0686	.0732	.0778	.0824	.0869	.9014	.0959	.1002	.1044
8	.0328	.0360	.0393	.0428	.0463	.0500	.0537	.0575	.0614	.0653
9	.0150	.0168	.0188	.0209	.0232	.0255	.0280	.0307	.0334	.0363
10	.0061	.0071	.0081	.0092	.0104	.0118	.0132	.0147	.0164	.0181
11	.0023	.0027	.0032	.0037	.0043	.0049	.0056	.0064	.0073	.0082
12	.0008	.0009	.0011	.0014	.0016	.0019	.0022	.0026	.0030	.0034
13	.0002	.0003	.0004	.0005	.0006	.0007	.0008	.0009	.0011	.0013
14	.0001	.0001	.0001	.0001	.0002	.0002	.0003	.0003	.0004	.0005
15	.0000	.0000	.0000	.0000	.0001	.0001	.0001	.0001	.0001	.0002

	λ									
x	5.1	5.2	5.3	5.4	5.5	5.6	5.7	5.8	5.9	6.0
0	.0061	.0055	.0050	.0045	.0041	.0037	.0033	.0030	.0027	.0025
1	.0311	.0287	.0265	.0244	.0225	.0207	.0191	.0176	.0162	.0149
2	.0793	.0746	.0701	.0659	.0618	.0580	.0544	.0509	.0477	.0446
3	.1348	.1293	.1239	.1185	.1133	.1082	.1033	.0985	.0938	.0892
4	.1719	.1681	.1641	.1600	.1558	.1515	.1472	.1428	.1383	.1339
5	.1753	.1748	.1740	.1728	.1714	.1697	.1678	.1656	.1632	.1606
6	.1490	.1515	.1537	.1555	.1571	.1584	.1594	.1601	.1605	.1606
7	.1086	.1125	.1163	.1200	.1234	.1267	.1298	.1326	.1353	.1377
8	.0692	.0731	.0771	.0810	.0849	.0887	.0925	.0962	.0998	.1033
9	.0392	.0423	.0454	.0486	.0519	.0552	.0586	.0620	.0654	.0688
10	.0200	.0220	.0241	.0262	.0285	.0309	.0334	.0359	.0386	.0413
11	.0093	.0104	.0116	.0129	.0143	.0157	.0173	.0190	.0207	.0225
12	.0039	.0045	.0051	.0058	.0065	.0073	.0082	.0092	.0102	.0113
13	.0015	.0018	.0021	.0024	.0028	.0032	.0036	.0041	.0046	.0052
14	.0006	.0007	.0008	.0009	.0011	.0013	.0015	.0017	.0019	.0022
15	.0002	.0002	.0003	.0003	.0004	.0005	.0006	.0007	.0008	.0009
16	.0001	.0001	.0001	.0001	.0001	.0002	.0002	.0002	.0003	.0003
17	.0000	.0000	.0000	.0000	.0000	.0000	.0001	.0001	.0001	.0001

	λ									
x	6.1	6.2	6.3	6.4	6.5	6.6	6.7	6.8	6.9	7.0
0	.0022	.0020	.0018	.0017	.0015	.0014	.0012	.0011	.0010	.0009
1	.0137	.0126	.0116	.0106	.0098	.0090	.0082	.0076	.0070	.0064
2	.0417	.0390	.0364	.0340	.0318	.0296	.0276	.0258	.0240	.0223
3	.0848	.0806	.0765	.0726	.0688	.0652	.0617	.0584	.0552	.0521
4	.1294	.1249	.1205	.1162	.1118	.1076	.1034	.0992	.0952	.0912
5	.1579	.1549	.1519	.1487	.1454	.1420	.1385	.1349	.1314	.1277
6	.1605	.1601	.1595	.1586	.1575	.1562	.1546	.1529	.1511	.1490
7	.1399	.1418	.1435	.1450	.1462	.1472	.1480	.1486	.1489	.1490
8	.1066	.1099	.1130	.1160	.1188	.1215	.1240	.1263	.1284	.1304
9	.0723	.0757	.0791	.0825	.0858	.0891	.0923	.0954	.0985	.1014

TABLE 3 (*continued*)

	λ									
x	6.1	6.2	6.3	6.4	6.5	6.6	6.7	6.8	6.9	7.0
10	.0441	.0469	.0498	.0528	.0558	.0588	.0618	.0649	.0679	.0710
11	.0245	.0265	.0285	.0307	.0330	.0353	.0377	.0401	.0426	.0452
12	.0124	.0137	.0150	.0164	.0179	.0194	.0210	.0227	.0245	.0264
13	.0058	.0065	.0073	.0081	.0089	.0098	.0108	.0119	.0130	.0142
14	.0025	.0029	.0033	.0037	.0041	.0046	.0052	.0058	.0064	.0071
15	.0010	.0012	.0014	.0016	.0018	.0020	.0023	.0026	.0029	.0033
16	.0004	.0005	.0005	.0006	.0007	.0008	.0010	.0011	.0013	.0014
17	.0001	.0002	.0002	.0002	.0003	.0003	.0004	.0004	.0005	.0006
18	.0000	.0001	.0001	.0001	.0001	.0001	.0001	.0002	.0002	.0002
19	.0000	.0000	.0000	.0000	.0000	.0000	.0000	.0001	.0001	.0001

	λ									
x	7.1	7.2	7.3	7.4	7.5	7.6	7.7	7.8	7.9	8.0
0	.0008	.0007	.0007	.0006	.0006	.0005	.0005	.0004	.0004	.0003
1	.0059	.0054	.0049	.0045	.0041	.0038	.0035	.0032	.0029	.0027
2	.0208	.0194	.0180	.0167	.0156	.0145	.0134	.0125	.0116	.0107
3	.0492	.0464	.0438	.0413	.0389	.0366	.0345	.0324	.0305	.0286
4	.0874	.0836	.0799	.0764	.0729	.0696	.0663	.0632	.0602	.0573
5	.1241	.1204	.1167	.1130	.1094	.1057	.1021	.0986	.0951	.0916
6	.1468	.1445	.1420	.1394	.1367	.1339	.1311	.1282	.1252	.1221
7	.1489	.1486	.1481	.1474	.1465	.1454	.1442	.1428	.1413	.1396
8	.1321	.1337	.1351	.1363	.1373	.1382	.1388	.1392	.1395	.1396
9	.1042	.1070	.1096	.1121	.1144	.1167	.1187	.1207	.1224	.1241
10	.0740	.0770	.0800	.0829	.0858	.0887	.0914	.0941	.0967	.0993
11	.0478	.0504	.0531	.0558	.0585	.0613	.0640	.0667	.0695	.0722
12	.0283	.0303	.0323	.0344	.0366	.0388	.0411	.0434	.0457	.0481
13	.0154	.0168	.0181	.0196	.0211	.0227	.0243	.0260	.0278	.0296
14	.0078	.0086	.0095	.0104	.0113	.0123	.0134	.0145	.0157	.0169
15	.0037	.0041	.0046	.0051	.0057	.0062	.0069	.0075	.0083	.0090
16	.0016	.0019	.0021	.0024	.0026	.0030	.0033	.0037	.0041	.0045
17	.0007	.0008	.0009	.0010	.0012	.0013	.0015	.0017	.0019	.0021
18	.0003	.0003	.0004	.0004	.0005	.0006	.0006	.0007	.0008	.0009
19	.0001	.0001	.0001	.0002	.0002	.0002	.0003	.0003	.0003	.0004
20	.0000	.0000	.0001	.0001	.0001	.0001	.0001	.0001	.0001	.0002
21	.0000	.0000	.0000	.0000	.0000	.0000	.0000	.0000	.0001	.0001

	λ									
x	8.1	8.2	8.3	8.4	8.5	8.6	8.7	8.8	8.9	9.0
0	.0003	.0003	.0002	.0002	.0002	.0002	.0002	.0002	.0001	.0001
1	.0025	.0023	.0021	.0019	.0017	.0016	.0014	.0013	.0012	.0011
2	.0100	.0092	.0086	.0079	.0074	.0068	.0063	.0058	.0054	.0050
3	.0269	.0252	.0237	.0222	.0208	.0195	.0183	.0171	.0160	.0150
4	.0544	.0517	.0491	.0466	.0443	.0420	.0398	.0377	.0357	.0337
5	.0882	.0849	.0816	.0784	.0752	.0722	.0692	.0663	.0635	.0607
6	.1191	.1160	.1128	.1097	.1066	.1034	.1003	.0972	.0941	.0911
7	.1378	.1358	.1338	.1317	.1294	.1271	.1247	.1222	.1197	.1171
8	.1395	.1392	.1388	.1382	.1375	.1366	.1356	.1344	.1332	.1318
9	.1256	.1269	.1280	.1290	.1299	.1306	.1311	.1315	.1317	.1318
10	.1017	.1040	.1063	.1084	.1104	.1123	.1140	.1157	.1172	.1186
11	.0749	.0776	.0802	.0828	.0853	.0878	.0902	.0925	.0948	.0970
12	.0505	.0530	.0555	.0579	.0604	.0629	.0654	.0679	.0703	.0728
13	.0315	.0334	.0354	.0374	.0395	.0416	.0438	.0459	.0481	.0504
14	.0182	.0196	.0210	.0225	.0240	.0256	.0272	.0289	.0306	.0324
15	.0098	.0107	.0116	.0126	.0136	.0147	.0158	.0169	.0182	.0194
16	.0050	.0055	.0060	.0066	.0072	.0079	.0086	.0093	.0101	.0109
17	.0024	.0026	.0029	.0033	.0036	.0040	.0044	.0048	.0053	.0058

TABLE 3 (*continued*)

	λ									
x	8.1	8.2	8.3	8.4	8.5	8.6	8.7	8.8	8.9	9.0
18	.0011	.0012	.0014	.0015	.0017	.0019	.0021	.0024	.0026	.0029
19	.0005	.0005	.0006	.0007	.0008	.0009	.0010	.0011	.0012	.0014
20	.0002	.0002	.0002	.0003	.0003	.0004	.0004	.0005	.0005	.0006
21	.0001	.0001	.0001	.0001	.0001	.0002	.0002	.0002	.0002	.0003
22	.0000	.0000	.0000	.0000	.0001	.0001	.0001	.0001	.0001	.0001

	λ									
x	9.1	9.2	9.3	9.4	9.5	9.6	9.7	9.8	9.9	10
0	.0001	.0001	.0001	.0001	.0001	.0001	.0001	.0001	.0001	.0000
1	.0010	.0009	.0009	.0008	.0007	.0007	.0006	.0005	.0005	.0005
2	.0046	.0043	.0040	.0037	.0034	.0031	.0029	.0027	.0025	.0023
3	.0140	.0131	.0123	.0115	.0107	.0100	.0093	.0087	.0081	.0076
4	.0319	.0302	.0285	.0269	.0254	.0240	.0226	.0213	.0201	.0189
5	.0581	.0555	.0530	.0506	.0483	.0460	.0439	.0418	.0398	.0378
6	.0881	.0851	.0822	.0793	.0764	.0736	.0709	.0682	.0656	.0631
7	.1145	.1118	.1091	.1064	.1037	.1010	.0982	.0955	.0928	.0901
8	.1302	.1286	.1269	.1251	.1232	.1212	.1191	.1170	.1148	.1126
9	.1317	.1315	.1311	.1306	.1300	.1293	.1284	.1274	.1263	.1251
10	.1198	.1210	.1219	.1228	.1235	.1241	.1245	.1249	.1250	.1251
11	.0991	.1012	.1031	.1049	.1067	.1083	.1098	.1112	.1125	.1137
12	.0752	.0776	.0799	.0822	.0844	.0866	.0888	.0908	.0928	.0948
13	.0526	.0549	.0572	.0594	.0617	.0640	.0662	.0685	.0707	.0729
14	.0342	.0361	.0380	.0399	.0419	.0439	.0459	.0479	.0500	.0521
15	.0208	.0221	.0235	.0250	.0265	.0281	.0297	.0313	.0330	.0347
16	.0118	.0127	.0137	.0147	.0157	.0168	.0180	.0192	.0204	.0217
17	.0063	.0069	.0075	.0081	.0088	.0095	.0103	.0111	.0119	.0128
18	.0032	.0035	.0039	.0042	.0046	.0051	.0055	.0060	.0065	.0071
19	.0015	.0017	.0019	.0021	.0023	.0026	.0028	.0031	.0034	.0037
20	.0007	.0008	.0009	.0010	.0011	.0012	.0014	.0015	.0017	.0019
21	.0003	.0003	.0004	.0004	.0005	.0006	.0006	.0007	.0008	.0009
22	.0001	.0001	.0002	.0002	.0002	.0002	.0003	.0003	.0004	.0004
23	.0000	.0001	.0001	.0001	.0001	.0001	.0001	.0001	.0002	.0002
24	.0000	.0000	.0000	.0000	.0000	.0000	.0000	.0001	.0001	.0001

	λ									
x	11	12	13	14	15	16	17	18	19	20
0	.0000	.0000	.0000	.0000	.0000	.0000	.0000	.0000	.0000	.0000
1	.0002	.0001	.0000	.0000	.0000	.0000	.0000	.0000	.0000	.0000
2	.0010	.0004	.0002	.0001	.0000	.0000	.0000	.0000	.0000	.0000
3	.0037	.0018	.0008	.0004	.0002	.0001	.0000	.0000	.0000	.0000
4	.0102	.0053	.0027	.0013	.0006	.0003	.0001	.0001	.0000	.0000
5	.0224	.0127	.0070	.0037	.0019	.0010	.0005	.0002	.0001	.0001
6	.0411	.0255	.0152	.0087	.0048	.0026	.0014	.0007	.0004	.0002
7	.0646	.0437	.0281	.0174	.0104	.0060	.0034	.0018	.0010	.0005
8	.0888	.0655	.0457	.0304	.0194	.0120	.0072	.0042	.0024	.0013
9	.1085	.0874	.0661	.0473	.0324	.0213	.0135	.0083	.0050	.0029
10	.1194	.1048	.0859	.0663	.0486	.0341	.0230	.0150	.0095	.0058
11	.1194	.1144	.1015	.0844	.0663	.0496	.0355	.0245	.0164	.0106
12	.1094	.1144	.1099	.0984	.0829	.0661	.0504	.0368	.0259	.0176
13	.0926	.1056	.1099	.1060	.0956	.0814	.0658	.0509	.0378	.0271
14	.0728	.0905	.1021	.1060	.1024	.0930	.0800	.0655	.0514	.0387
15	.0534	.0724	.0885	.0989	.1024	.0992	.0906	.0786	.0650	.0516
16	.0367	.0543	.0719	.0866	.0960	.0992	.0963	.0884	.0772	.0646
17	.0237	.0383	.0550	.0713	.0847	.0934	.0963	.0936	.0863	.0760
18	.0145	.0256	.0397	.0554	.0706	.0830	.0909	.0936	.0911	.0844
19	.0084	.0161	.0272	.0409	.0557	.0699	.0814	.0887	.0911	.0888

TABLE 3 (*continued*)

	λ									
x	11	12	13	14	15	16	17	18	19	20
20	.0046	.0097	.0177	.0286	.0418	.0559	.0692	.0798	.0866	.0888
21	.0024	.0055	.0109	.0191	.0299	.0426	.0560	.0684	.0783	.0846
22	.0012	.0030	.0065	.0121	.0204	.0310	.0433	.0560	.0676	.0769
23	.0006	.0016	.0037	.0074	.0133	.0216	.0320	.0438	.0559	.0669
24	.0003	.0008	.0020	.0043	.0083	.0144	.0226	.0328	.0442	.0557
25	.0001	.0004	.0010	.0024	.0050	.0092	.0154	.0237	.0336	.0446
26	.0000	.0002	.0005	.0013	.0029	.0057	.0101	.0164	.0246	.0343
27	.0000	.0001	.0002	.0007	.0016	.0034	.0063	.0109	.0173	.0254
28	.0000	.0000	.0001	.0003	.0009	.0019	.0038	.0070	.0117	.0181
29	.0000	.0000	.0001	.0002	.0004	.0011	.0023	.0044	.0077	.0125
30	.0000	.0000	.0000	.0001	.0002	.0006	.0013	.0026	.0049	.0083
31	.0000	.0000	.0000	.0000	.0001	.0003	.0007	.0015	.0030	.0054
32	.0000	.0000	.0000	.0000	.0001	.0001	.0004	.0009	.0018	.0034
33	.0000	.0000	.0000	.0000	.0000	.0001	.0002	.0005	.0010	.0020
34	.0000	.0000	.0000	.0000	.0000	.0000	.0001	.0002	.0006	.0012
35	.0000	.0000	.0000	.0000	.0000	.0000	.0000	.0001	.0003	.0007
36	.0000	.0000	.0000	.0000	.0000	.0000	.0000	.0001	.0002	.0004
37	.0000	.0000	.0000	.0000	.0000	.0000	.0000	.0000	.0001	.0002
38	.0000	.0000	.0000	.0000	.0000	.0000	.0000	.0000	.0000	.0001
39	.0000	.0000	.0000	.0000	.0000	.0000	.0000	.0000	.0000	.0001

TABLE 4: Probabilities for the Standard Normal Distribution

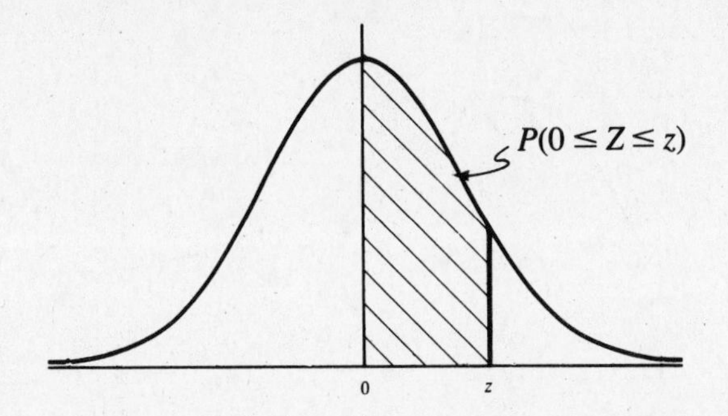

z	Second decimal place in z									
	.00	.01	.02	.03	.04	.05	.06	.07	.08	.09
.0	.0000	.0040	.0080	.0120	.0160	.0199	.0239	.0279	.0319	.0359
.1	.0398	.0438	.0478	.0517	.0557	.0596	.0636	.0675	.0714	.0753
.2	.0793	.0832	.0871	.0910	.0948	.0987	.1026	.1064	.1103	.1141
.3	.1179	.1217	.1255	.1293	.1331	.1368	.1406	.1443	.1480	.1517
.4	.1554	.1591	.1628	.1664	.1700	.1736	.1772	.1808	.1844	.1879
.5	.1915	.1950	.1985	.2019	.2054	.2088	.2123	.2157	.2190	.2224
.6	.2257	.2291	.2324	.2357	.2389	.2422	.2454	.2486	.2517	.2549
.7	.2580	.2611	.2642	.2673	.2704	.2734	.2764	.2794	.2823	.2852
.8	.2881	.2910	.2939	.2967	.2995	.3023	.3051	.3078	.3106	.3133
.9	.3159	.3186	.3212	.3238	.3264	.3289	.3315	.3340	.3365	.3389
1.0	.3413	.3438	.3461	.3485	.3508	.3531	.3554	.3577	.3599	.3621
1.1	.3643	.3665	.3686	.3708	.3729	.3749	.3770	.3790	.3810	.3830
1.2	.3849	.3869	.3888	.3907	.3925	.3944	.3962	.3980	.3997	.4015
1.3	.4032	.4049	.4066	.4082	.4099	.4115	.4131	.4147	.4162	.4177
1.4	.4192	.4207	.4222	.4236	.4251	.4265	.4279	.4292	.4306	.4319
1.5	.4332	.4345	.4357	.4370	.4382	.4394	.4406	.4418	.4429	.4441
1.6	.4452	.4463	.4474	.4484	.4495	.4505	.4515	.4525	.4535	.4545
1.7	.4554	.4564	.4573	.4582	.4591	.4599	.4608	.4616	.4625	.4633
1.8	.4641	.4649	.4656	.4664	.4671	.4678	.4686	.4693	.4699	.4706
1.9	.4713	.4719	.4726	.4732	.4738	.4744	.4750	.4756	.4761	.4767
2.0	.4772	.4778	.4783	.4788	.4793	.4798	.4803	.4808	.4812	.4817
2.1	.4821	.4826	.4830	.4834	.4838	.4842	.4846	.4850	.4854	.4857
2.2	.4861	.4864	.4868	.4871	.4875	.4878	.4881	.4884	.4887	.4890
2.3	.4893	.4896	.4898	.4901	.4904	.4906	.4909	.4911	.4913	.4916
2.4	.4918	.4920	.4922	.4925	.4927	.4929	.4931	.4932	.4934	.4936
2.5	.4938	.4940	.4941	.4943	.4945	.4946	.4948	.4949	.4951	.4952
2.6	.4953	.4955	.4956	.4957	.4959	.4960	.4961	.4962	.4963	.4964
2.7	.4965	.4966	.4967	.4968	.4969	.4970	.4971	.4972	.4973	.4974
2.8	.4974	.4975	.4976	.4977	.4977	.4978	.4979	.4979	.4980	.4981
2.9	.4981	.4982	.4982	.4983	.4984	.4984	.4985	.4985	.4986	.4986
3.0	.4987	.4987	.4987	.4988	.4988	.4989	.4989	.4989	.4990	.4990
3.1	.4990	.4991	.4991	.4991	.4992	.4992	.4992	.4992	.4993	.4993
3.2	.4993	.4993	.4994	.4994	.4994	.4994	.4994	.4995	.4995	.4995
3.3	.4995	.4995	.4995	.4996	.4996	.4996	.4996	.4996	.4996	.4997
3.4	.4997	.4997	.4497	.4997	.4997	.4997	.4997	.4997	.4997	.4998

TABLE 5: Chi-Square Critical Values

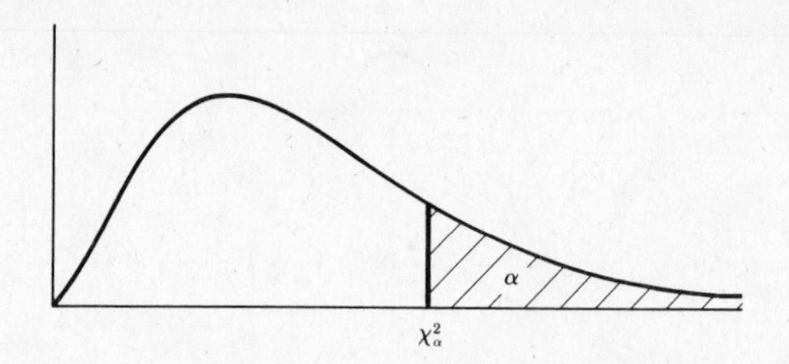

	α									
r	.995	.99	.975	.95	.90	.10	.05	.025	.01	.005
1	.00	.00	.00	.00	.02	2.71	3.84	5.02	6.63	7.88
2	.01	.02	.05	.10	.21	4.61	5.99	7.38	9.21	10.60
3	.07	.11	.22	.35	.58	6.25	7.81	9.35	11.34	12.84
4	.21	.30	.48	.71	1.06	7.78	9.49	11.14	13.28	14.86
5	.41	.55	.83	1.15	1.61	9.24	11.07	12.83	15.09	16.75
6	.68	.87	1.24	1.64	2.20	10.64	12.59	14.45	16.81	18.55
7	.99	1.24	1.69	2.17	2.83	12.02	14.07	16.01	18.48	20.28
8	1.34	1.65	2.18	2.73	3.49	13.36	15.51	17.54	20.09	21.96
9	1.73	2.09	2.70	3.33	4.17	14.68	16.92	19.02	21.67	23.59
10	2.16	2.56	3.25	3.94	4.87	15.99	18.31	20.48	23.21	25.19
11	2.60	3.05	3.82	4.57	5.58	17.28	19.68	21.92	24.72	26.76
12	3.07	3.57	4.40	5.23	6.30	18.55	21.03	23.34	26.22	28.30
13	3.57	4.11	5.01	5.89	7.04	19.81	22.36	24.74	27.69	29.82
14	4.07	4.66	5.63	6.57	7.79	21.06	23.68	26.12	29.14	31.32
15	4.60	5.23	6.26	7.26	8.55	22.31	25.00	27.49	30.58	32.80
16	5.14	5.81	6.91	7.96	9.31	23.54	26.30	28.85	32.00	34.27
17	5.70	6.41	7.56	8.67	10.09	24.77	27.59	30.19	33.41	35.72
18	6.26	7.01	8.23	9.39	10.86	25.99	28.87	31.53	34.81	37.16
19	6.84	7.63	8.91	10.12	11.65	27.20	30.14	32.85	36.19	38.58
20	7.43	8.26	9.59	10.85	12.44	28.41	31.41	34.17	37.57	40.00
21	8.03	8.90	10.28	11.59	13.24	29.62	32.67	35.48	38.93	41.40
22	8.64	9.54	10.98	12.34	14.04	30.81	33.92	36.78	40.29	42.80
23	9.26	10.20	11.69	13.09	14.85	32.01	35.17	38.08	41.64	44.18
24	9.89	10.86	12.40	13.85	15.66	33.20	36.42	39.36	42.98	45.56
25	10.52	11.52	13.12	14.61	16.47	34.38	37.65	40.65	44.31	46.93
26	11.16	12.20	13.84	15.38	17.29	35.56	38.89	41.92	45.64	48.29
27	11.81	12.88	14.57	16.15	18.11	36.74	40.11	43.19	46.96	49.65
28	12.46	13.56	15.31	16.93	18.94	37.92	41.34	44.46	48.28	50.99
29	13.12	14.26	16.05	17.71	19.77	39.09	42.56	45.72	49.59	52.34
30	13.79	14.95	16.79	18.49	20.60	40.26	43.77	46.98	50.89	53.67

TABLE 6: Student's *t* Critical Values*

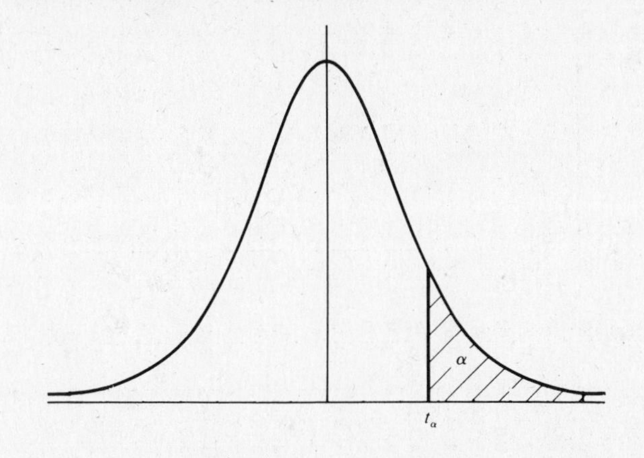

	α					
r	.25	.10	.05	.025	.01	.005
1	1.000	3.08	6.31	12.7	31.8	63.7
2	.816	1.89	2.92	4.30	6.97	9.92
3	.765	1.64	2.35	3.18	4.54	5.84
4	.741	1.53	2.13	2.78	3.75	4.60
5	.727	1.48	2.02	2.57	3.37	4.03
6	.718	1.44	1.94	2.45	3.14	3.71
7	.711	1.42	1.89	2.36	3.00	3.50
8	.706	1.40	1.86	2.31	2.90	3.36
9	.703	1.38	1.83	2.26	2.82	3.25
10	.700	1.37	1.81	2.23	2.76	3.17
11	.697	1.36	1.80	2.20	2.72	3.11
12	.695	1.36	1.78	2.18	2.68	3.05
13	.694	1.35	1.77	2.16	2.65	3.01
14	.692	1.35	1.76	2.14	2.62	2.98
15	.691	1.34	1.75	2.13	2.60	2.95
16	.690	1.34	1.75	2.12	2.58	2.92
17	.689	1.33	1.74	2.11	2.57	2.90
18	.688	1.33	1.73	2.10	2.55	2.88
19	.688	1.33	1.73	2.09	2.54	2.86
20	.687	1.33	1.72	2.09	2.53	2.85
21	.686	1.32	1.72	2.08	2.52	2.83
22	.686	1.32	1.72	2.07	2.51	2.82
23	.685	1.32	1.71	2.07	2.50	2.81
24	.685	1.32	1.71	2.06	2.49	2.80
25	.684	1.32	1.71	2.06	2.49	2.79
26	.684	1.32	1.71	2.06	2.48	2.78
27	.684	1.31	1.70	2.05	2.47	2.77
28	.683	1.31	1.70	2.05	2.47	2.76
29	.683	1.31	1.70	2.05	2.46	2.76
z_α	.674	1.28	1.645	1.96	2.33	2.58

* For df $\geq$ 30, the critical value t_α is approximated by z_α, given in the bottom row of table.

TABLE 7: *F* Distribution Critical Values*
Part A: Values of $F_{.01}$

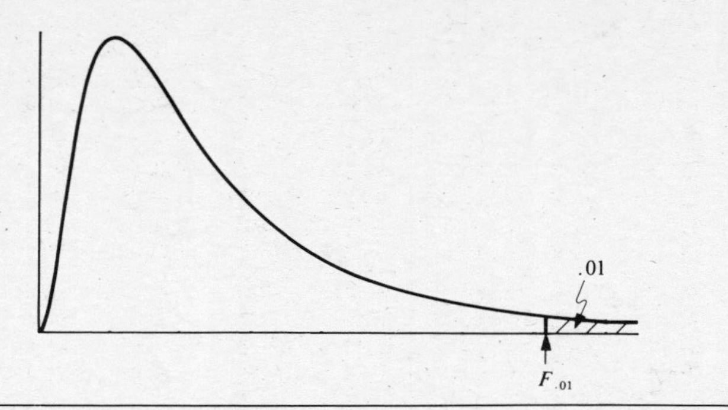

* r_1 = df for the numerator; r_2 = df for the denominator.

r_2 \ r_1	1	2	3	4	5	6	7	8	9	10	12	15	20	24	30	40	60	120	∞
1	4052	4999.5	5403	5625	5764	5859	5928	5981	6022	6056	6106	6157	6209	6235	6261	6287	6313	6339	6366
2	98.50	99.00	99.17	99.25	99.30	99.33	99.36	99.37	99.39	99.40	99.42	99.43	99.45	99.46	99.47	99.47	99.48	99.49	99.50
3	34.12	30.82	29.46	28.71	28.24	27.91	27.67	27.49	27.35	27.23	27.05	26.87	26.69	26.60	26.50	26.41	26.32	26.22	26.13
4	21.20	18.00	16.69	15.98	15.52	15.21	14.98	14.80	14.66	14.55	14.37	14.20	14.02	13.93	13.84	13.75	13.65	13.56	13.46
5	16.26	13.27	12.06	11.39	10.97	10.67	10.46	10.29	10.16	10.05	9.89	9.72	9.55	9.47	9.38	9.29	9.20	9.11	9.02
6	13.75	10.92	9.78	9.15	8.75	8.47	8.26	8.10	7.98	7.87	7.72	7.56	7.40	7.31	7.23	7.14	7.06	6.97	6.88
7	12.25	9.55	8.45	7.85	7.46	7.19	6.99	6.84	6.72	6.62	6.47	6.31	6.16	6.07	5.99	5.91	5.82	5.74	5.65
8	11.26	8.65	7.59	7.01	6.63	6.37	6.18	6.03	5.91	5.81	5.67	5.52	5.36	5.28	5.20	5.12	5.03	4.95	4.86
9	10.56	8.02	6.99	6.42	6.06	5.80	5.61	5.47	5.35	5.26	5.11	4.96	4.81	4.73	4.65	4.57	4.48	4.40	4.31
10	10.04	7.56	6.55	5.99	5.64	5.39	5.20	5.06	4.94	4.85	4.71	4.56	4.41	4.33	4.25	4.17	4.08	4.00	3.91
11	9.65	7.21	6.22	5.67	5.32	5.07	4.89	4.74	4.63	4.54	4.40	4.25	4.10	4.02	3.94	3.86	3.78	3.69	3.60
12	9.33	6.93	5.95	5.41	5.06	4.82	4.64	4.50	4.39	4.30	4.16	4.01	3.86	3.78	3.70	3.62	3.54	3.45	3.36
13	9.07	6.70	5.74	5.21	4.86	4.62	4.44	4.30	4.19	4.10	3.96	3.82	3.66	3.59	3.51	3.43	3.34	3.25	3.17
14	8.86	6.51	5.56	5.04	4.69	4.46	4.28	4.14	4.03	3.94	3.80	3.66	3.51	3.43	3.35	3.27	3.18	3.09	3.00
15	8.68	6.36	5.42	4.89	4.56	4.32	4.14	4.00	3.89	3.80	3.67	3.52	3.37	3.29	3.21	3.13	3.05	2.96	2.87
16	8.53	6.23	5.29	4.77	4.44	4.20	4.03	3.89	3.78	3.69	3.55	3.41	3.26	3.18	3.10	3.02	2.93	2.84	2.75
17	8.40	6.11	5.18	4.67	4.34	4.10	3.93	3.79	3.68	3.59	3.46	3.31	3.16	3.08	3.00	2.92	2.83	2.75	2.65
18	8.29	6.01	5.09	4.58	4.25	4.01	3.84	3.71	3.60	3.51	3.37	3.23	3.08	3.00	2.92	2.84	2.75	2.66	2.57
19	8.18	5.93	5.01	4.50	4.17	3.94	3.77	3.63	3.52	3.43	3.30	3.15	3.00	2.92	2.84	2.76	2.67	2.58	2.49
20	8.10	5.85	4.94	4.43	4.10	3.87	3.70	3.56	3.46	3.37	3.23	3.09	2.94	2.86	2.78	2.69	2.61	2.52	2.42
21	8.02	5.78	4.87	4.37	4.04	3.81	3.64	3.51	3.40	3.31	3.17	3.03	2.88	2.80	2.72	2.64	2.55	2.46	2.36
22	7.95	5.72	4.82	4.31	3.99	3.76	3.59	3.45	3.35	3.26	3.12	2.98	2.83	2.75	2.67	2.58	2.50	2.40	2.31
23	7.88	5.66	4.76	4.26	3.94	3.71	3.54	3.41	3.30	3.21	3.07	2.93	2.78	2.70	2.62	2.54	2.45	2.35	2.26
24	7.82	5.61	4.72	4.22	3.90	3.67	3.50	3.36	3.26	3.17	3.03	2.89	2.74	2.66	2.58	2.49	2.40	2.31	2.21
25	7.77	5.57	4.68	4.18	3.85	3.63	3.46	3.32	3.22	3.13	2.99	2.85	2.70	2.62	2.54	2.45	2.36	2.27	2.17
26	7.72	5.53	4.64	4.14	3.82	3.59	3.42	3.29	3.18	3.09	2.96	2.81	2.66	2.58	2.50	2.42	2.33	2.23	2.13
27	7.68	5.49	4.60	4.11	3.78	3.56	3.39	3.26	3.15	3.06	2.93	2.78	2.63	2.55	2.47	2.38	2.29	2.20	2.10
28	7.64	5.45	4.57	4.07	3.75	3.53	3.36	3.23	3.12	3.03	2.90	2.75	2.60	2.52	2.44	2.35	2.26	2.17	2.06
29	7.60	5.42	4.54	4.04	3.73	3.50	3.33	3.20	3.09	3.00	2.87	2.73	2.57	2.49	2.41	2.33	2.23	2.14	2.03
30	7.56	5.39	4.51	4.02	3.70	3.47	3.30	3.17	3.07	2.98	2.84	2.70	2.55	2.47	2.39	2.30	2.21	2.11	2.01
40	7.31	5.18	4.31	3.83	3.51	3.29	3.12	2.99	2.89	2.80	2.66	2.52	2.37	2.29	2.20	2.11	2.02	1.92	1.80
60	7.08	4.98	4.13	3.65	3.34	3.12	2.95	2.82	2.72	2.63	2.50	2.35	2.20	2.12	2.03	1.94	1.84	1.73	1.60
120	6.85	4.79	3.95	3.48	3.17	2.96	2.79	2.66	2.56	2.47	2.34	2.19	2.03	1.95	1.86	1.76	1.66	1.53	1.38
∞	6.63	4.61	3.78	3.32	3.02	2.80	2.64	2.51	2.41	2.32	2.18	2.04	1.88	1.79	1.70	1.59	1.47	1.32	1.00

TABLE 7 (*continued*)
Part B: Values of $\bar{F}_{.025}$

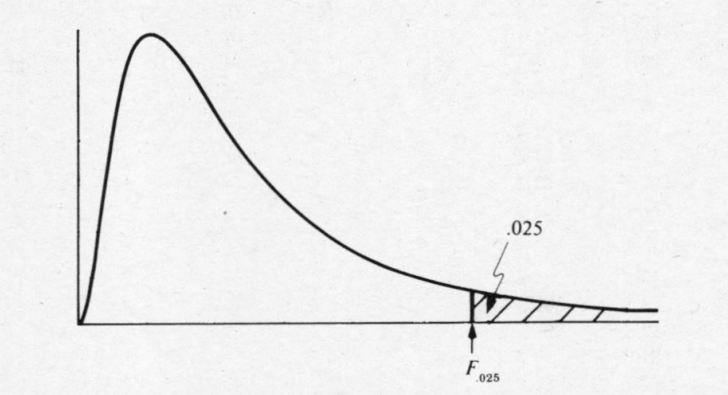

r_1 = df for the numerator; r_2 = df for the denominator.

r_2 \ r_1	1	2	3	4	5	6	7	8	9	10	12	15	20	24	30	40	60	120	∞
1	647.79	799.50	864.16	899.58	921.85	937.11	948.22	956.66	963.28	968.63	976.71	984.87	993.10	997.25	1001.4	1005.6	1009.8	1014.0	1018.3
2	38.51	39.00	39.17	39.25	39.30	39.33	39.36	39.37	39.39	39.40	39.42	39.43	39.45	39.46	39.47	39.47	39.48	39.49	39.50
3	17.44	16.04	15.44	15.10	14.89	14.74	14.62	14.54	14.47	14.42	14.34	14.25	14.17	14.12	14.08	14.04	13.99	13.95	13.90
4	12.22	10.65	9.98	9.60	9.36	9.20	9.07	8.98	8.90	8.84	8.75	8.66	8.56	8.51	8.46	8.41	8.36	8.31	8.26
5	10.00	8.43	7.76	7.39	7.15	6.98	6.85	6.76	6.68	6.62	6.52	6.43	6.33	6.28	6.23	6.18	6.12	6.07	6.02
6	8.81	7.26	6.60	6.23	5.99	5.82	5.70	5.60	5.52	5.46	5.37	5.27	5.17	5.12	5.07	5.01	4.96	4.90	4.85
7	8.07	6.54	5.89	5.52	5.29	5.12	4.99	4.90	4.82	4.76	4.67	4.57	4.47	4.42	4.36	4.31	4.25	4.20	4.14
8	7.57	6.06	5.42	5.05	4.82	4.65	4.53	4.43	4.36	4.30	4.20	4.10	4.00	3.95	3.89	3.84	3.78	3.73	3.67
9	7.21	5.71	5.08	4.72	4.48	4.32	4.20	4.10	4.03	3.96	3.87	3.77	3.67	3.61	3.56	3.51	3.45	3.39	3.33
10	6.94	5.46	4.83	4.47	4.24	4.07	3.95	3.85	3.78	3.72	3.62	3.52	3.42	3.37	3.31	3.26	3.20	3.14	3.08
11	6.72	5.26	4.63	4.28	4.04	3.88	3.76	3.66	3.59	3.53	3.43	3.33	3.23	3.17	3.12	3.06	3.00	2.94	2.88
12	6.55	5.10	4.47	4.12	3.89	3.73	3.61	3.51	3.44	3.37	3.28	3.18	3.07	3.02	2.96	2.91	2.85	2.79	2.72
13	6.41	4.97	4.35	4.00	3.77	3.60	3.48	3.39	3.31	3.25	3.15	3.05	2.95	2.89	2.84	2.78	2.72	2.66	2.60
14	6.30	4.86	4.24	3.89	3.66	3.50	3.38	3.29	3.21	3.15	3.05	2.95	2.84	2.79	2.73	2.67	2.61	2.55	2.49
15	6.20	4.77	4.15	3.80	3.58	3.41	3.29	3.20	3.12	3.06	2.96	2.86	2.76	2.70	2.64	2.59	2.52	2.46	2.40
16	6.12	4.69	4.08	3.73	3.50	3.34	3.22	3.12	3.05	2.99	2.89	2.79	2.68	2.63	2.57	2.51	2.45	2.38	2.32
17	6.04	4.62	4.01	3.66	3.44	3.28	3.16	3.06	2.98	2.92	2.82	2.72	2.62	2.56	2.50	2.44	2.38	2.32	2.25
18	5.98	4.56	3.95	3.61	3.38	3.22	3.10	3.01	2.93	2.87	2.77	2.67	2.56	2.50	2.44	2.38	2.32	2.26	2.19
19	5.92	4.51	3.90	3.56	3.33	3.17	3.05	2.96	2.88	2.82	2.72	2.62	2.51	2.45	2.39	2.33	2.27	2.20	2.13
20	5.87	4.46	3.86	3.51	3.29	3.13	3.01	2.91	2.84	2.77	2.68	2.57	2.46	2.41	2.35	2.29	2.22	2.16	2.09
21	5.83	4.42	3.82	3.48	3.25	3.09	2.97	2.87	2.80	2.73	2.64	2.53	2.42	2.37	2.31	2.25	2.18	2.11	2.04
22	5.79	4.38	3.78	3.44	3.22	3.05	2.93	2.84	2.76	2.70	2.60	2.50	2.39	2.33	2.27	2.21	2.14	2.08	2.00
23	5.75	4.35	3.75	3.41	3.18	3.02	2.90	2.81	2.73	2.67	2.57	2.47	2.36	2.30	2.24	2.18	2.11	2.04	1.97
24	5.72	4.32	3.72	3.38	3.15	2.99	2.87	2.78	2.70	2.64	2.54	2.44	2.33	2.27	2.21	2.15	2.08	2.01	1.94
25	5.69	4.29	3.69	3.35	3.13	2.97	2.85	2.75	2.68	2.61	2.51	2.41	2.30	2.24	2.18	2.12	2.05	1.98	1.91
26	5.66	4.27	3.67	3.33	3.10	2.94	2.82	2.73	2.65	2.59	2.49	2.39	2.28	2.22	2.16	2.09	2.03	1.95	1.88
27	5.63	4.24	3.65	3.31	3.08	2.92	2.80	2.71	2.63	2.57	2.47	2.36	2.25	2.19	2.13	2.07	2.00	1.93	1.85
28	5.61	4.22	3.63	3.29	3.06	2.90	2.78	2.69	2.61	2.55	2.45	2.34	2.23	2.17	2.11	2.05	1.98	1.91	1.83
29	5.59	4.20	3.61	3.27	3.04	2.88	2.76	2.67	2.59	2.53	2.43	2.32	2.21	2.15	2.09	2.03	1.96	1.89	1.81
30	5.57	4.18	3.59	3.25	3.03	2.87	2.75	2.65	2.57	2.51	2.41	2.31	2.20	2.14	2.07	2.01	1.94	1.87	1.79
40	5.42	4.05	3.46	3.13	2.90	2.74	2.62	2.53	2.45	2.39	2.29	2.18	2.07	2.01	1.94	1.88	1.80	1.72	1.64
60	5.29	3.93	3.34	3.01	2.79	2.63	2.51	2.41	2.33	2.27	2.17	2.06	1.94	1.88	1.82	1.74	1.67	1.58	1.48
120	5.15	3.80	3.23	2.89	2.67	2.52	2.39	2.30	2.22	2.16	2.05	1.95	1.82	1.76	1.69	1.61	1.53	1.43	1.31
∞	5.02	3.69	3.12	2.79	2.57	2.41	2.29	2.19	2.11	2.05	1.94	1.83	1.71	1.64	1.57	1.48	1.39	1.27	1.00

TABLE 7 ***(continued)***

Part C: Values of $F_{.05}$

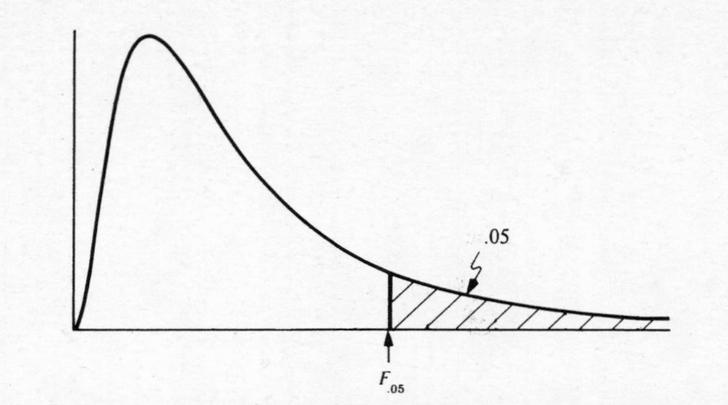

r_1 = df for the numerator; r_2 = df for the denominator.

r_2 \ r_1	1	2	3	4	5	6	7	8	9	10	12	15	20	24	30	40	60	120	∞
1	161.4	199.5	215.7	224.6	230.2	234.0	236.8	238.9	240.5	241.9	243.9	245.9	248.0	249.1	250.1	251.1	252.2	253.3	254.3
2	18.51	19.00	19.16	19.25	19.30	19.33	19.35	19.37	19.38	19.40	19.41	19.43	19.45	19.45	19.46	19.47	19.48	19.49	19.50
3	10.13	9.55	9.28	9.12	9.01	8.94	8.89	8.85	8.81	8.79	8.74	8.70	8.66	8.64	8.62	8.59	8.57	8.55	8.53
4	7.71	6.94	6.59	6.39	6.26	6.16	6.09	6.04	6.00	5.96	5.91	5.86	5.80	5.77	5.75	5.72	5.69	5.66	5.63
5	6.61	5.79	5.41	5.19	5.05	4.95	4.88	4.82	4.77	4.74	4.68	4.62	4.56	4.53	4.50	4.46	4.43	4.40	4.36
6	5.99	5.14	4.76	4.53	4.39	4.28	4.21	4.15	4.10	4.06	4.00	3.94	3.87	3.84	3.81	3.77	3.74	3.70	3.67
7	5.59	4.74	4.35	4.12	3.97	3.87	3.79	3.73	3.68	3.64	3.57	3.51	3.41	3.41	3.38	3.34	3.30	3.27	3.23
8	5.32	4.46	4.07	3.84	3.69	3.58	3.50	3.44	3.39	3.35	3.28	3.22	3.15	3.12	3.08	3.04	3.01	2.97	2.93
9	5.12	4.26	3.86	3.63	3.48	3.37	3.29	3.23	3.18	3.14	3.07	3.01	2.94	2.90	2.86	2.83	2.79	2.75	2.71
10	4.96	4.10	3.71	3.48	3.33	3.22	3.14	3.07	3.02	2.98	2.91	2.85	2.77	2.74	2.70	2.66	2.62	2.58	2.54
11	4.84	3.98	3.59	3.36	3.20	3.09	3.01	2.95	2.90	2.85	2.79	2.72	2.65	2.61	2.57	2.53	2.49	2.45	2.40
12	4.75	3.89	3.49	3.26	3.11	3.00	2.91	2.85	2.80	2.75	2.69	2.62	2.54	2.51	2.47	2.43	2.38	2.34	2.30
13	4.67	3.81	3.41	3.18	3.03	2.92	2.83	2.77	2.71	2.67	2.60	2.53	2.46	2.42	2.38	2.34	2.30	2.25	2.21
14	4.60	3.74	3.34	3.11	2.96	2.85	2.76	2.70	2.65	2.60	2.53	2.46	2.39	2.35	2.31	2.27	2.22	2.18	2.13
15	4.54	3.68	3.29	3.06	2.90	2.79	2.71	2.64	2.59	2.54	2.48	2.40	2.33	2.29	2.25	2.20	2.16	2.11	2.07
16	4.49	3.63	3.24	3.01	2.85	2.74	2.66	2.59	2.54	2.49	2.42	2.35	2.28	2.24	2.19	2.15	2.11	2.06	2.01
17	4.45	3.59	3.20	2.96	2.81	2.70	2.61	2.55	2.49	2.45	2.38	2.31	2.23	2.19	2.15	2.10	2.06	2.01	1.96
18	4.41	3.55	3.16	2.93	2.77	2.66	2.58	2.51	2.46	2.41	2.34	2.27	2.19	2.15	2.11	2.06	2.02	1.97	1.92
19	4.38	3.52	3.13	2.90	2.74	2.63	2.54	2.48	2.42	2.38	2.31	2.23	2.16	2.11	2.07	2.03	1.98	1.93	1.88
20	4.35	3.49	3.10	2.87	2.71	2.60	2.51	2.45	2.39	2.35	2.28	2.20	2.12	2.08	2.04	1.99	1.95	1.90	1.84
21	4.32	3.47	3.07	2.84	2.68	2.57	2.49	2.42	2.37	2.32	2.25	2.18	2.10	2.05	2.01	1.96	1.92	1.87	1.81
22	4.30	3.44	3.05	2.82	2.66	2.55	2.46	2.40	2.34	2.30	2.23	2.15	2.07	2.03	1.98	1.94	1.89	1.84	1.78
23	4.28	3.42	3.03	2.80	2.64	2.53	2.44	2.37	2.32	2.27	2.20	2.13	2.05	2.01	1.96	1.91	1.86	1.81	1.76
24	4.26	3.40	3.01	2.78	2.62	2.51	2.42	2.36	2.30	2.25	2.18	2.11	2.03	1.98	1.94	1.89	1.84	1.79	1.73
25	4.24	3.39	2.99	2.76	2.60	2.49	2.40	2.34	2.28	2.24	2.16	2.09	2.01	1.96	1.92	1.87	1.82	1.77	1.71
26	4.23	3.37	2.98	2.74	2.59	2.47	2.39	2.32	2.27	2.22	2.15	2.07	1.99	1.95	1.90	1.85	1.80	1.75	1.69
27	4.21	3.35	2.96	2.73	2.57	2.46	2.37	2.31	2.25	2.20	2.13	2.06	1.97	1.93	1.88	1.84	1.79	1.73	1.67
28	4.20	3.34	2.95	2.71	2.56	2.45	2.36	2.29	2.24	2.19	2.12	2.04	1.96	1.91	1.87	1.82	1.77	1.71	1.65
29	4.18	3.33	2.93	2.70	2.55	2.43	2.35	2.28	2.22	2.18	2.10	2.03	1.94	1.90	1.85	1.81	1.75	1.70	1.64
30	4.17	3.32	2.92	2.69	2.53	2.42	2.33	2.27	2.21	2.16	2.09	2.01	1.93	1.89	1.84	1.79	1.74	1.68	1.62
40	4.08	3.23	2.84	2.61	2.45	2.34	2.25	2.18	2.12	2.08	2.00	1.92	1.84	1.79	1.74	1.69	1.64	1.58	1.51
60	4.00	3.15	2.76	2.53	2.37	2.25	2.17	2.10	2.04	1.99	1.92	1.84	1.75	1.70	1.65	1.59	1.53	1.47	1.39
120	3.92	3.07	2.68	2.45	2.29	2.17	2.09	2.02	1.96	1.91	1.83	1.75	1.66	1.61	1.55	1.50	1.43	1.35	1.25
∞	3.84	3.00	2.60	2.37	2.21	2.10	2.01	1.94	1.88	1.83	1.75	1.67	1.57	1.52	1.46	1.39	1.32	1.22	1.00

TABLE 8: Critical Values for the Correlation Coefficient

	$P(R \geq r)$			
$n - 2$	.05	.025	.01	.005
1	.9877	.9969	.9995	.9999
2	.9000	.9500	.9800	.9900
3	.8053	.8783	.9343	.9587
4	.7292	.8113	.8822	.9172
5	.6694	.7544	.8329	.8745
6	.6215	.7067	.7887	.8343
7	.5822	.6664	.7497	.7977
8	.5493	.6319	.7154	.7646
9	.5214	.6020	.6850	.7348
10	.4972	.5759	.6581	.7079
11	.4761	.5529	.6338	.6835
12	.4575	.5323	.6120	.6613
13	.4408	.5139	.5922	.6411
14	.4258	.4973	.5742	.6226
15	.4123	.4821	.5577	.6054
16	.4000	.4683	.5425	.5897
17	.3887	.4555	.5285	.5750
18	.3783	.4437	.5154	.5614
19	.3687	.4328	.5033	.5487
20	.3597	.4226	.4920	.5367
25	.3282	.3808	.4450	.4869
30	.2959	.3494	.4092	.4487
35	.2746	.3246	.3809	.4182
40	.2572	.3044	.3578	.3931
45	.2428	.2875	.3383	.3721
50	.2306	.2732	.3218	.3541
60	.2108	.2500	.2948	.3248
70	.1954	.2318	.2736	.3017
80	.1829	.2172	.2565	.2829
90	.1725	.2049	.2422	.2673
100	.1638	.1946	.2300	.2540

TABLE 9: Random Numbers on the Interval (0, 1)

.18064	.85954	.03146	.95288	.93417
.52904	.26677	.33924	.13449	.25379
.81234	.08995	.72866	.06244	.31668
.83808	.67838	.02754	.55987	.61137
.12935	.77376	.97847	.40695	.13547
.65031	.18256	.74262	.31265	.69231
.84008	.30966	.79721	.49636	.30325
.85224	.88420	.13501	.35224	.39839
.72016	.23540	.43102	.96748	.42572
.34698	.75035	.87935	.02288	.72316
.63305	.78989	.54187	.64220	.47635
.57834	.68284	.39201	.70652	.21096
.40713	.04413	.10060	.70644	.83327
.14164	.85043	.32775	.81268	.42631
.74374	.12562	.56003	.72963	.83754
.95852	.71329	.15303	.99860	.11431
.19849	.66210	.68621	.65833	.27415
.21990	.35204	.63312	.13041	.58432
.83227	.23470	.41781	.89451	.10684
.09039	.08081	.17136	.80091	.76320
.87098	.85706	.80356	.60782	.91485
.51870	.37857	.10310	.71148	.84096
.14247	.78615	.93470	.03284	.28476
.91298	.41506	.77347	.40533	.97073
.67642	.82196	.34400	.16638	.40229
.41632	.37735	.01716	.20682	.58650
.15764	.16738	.08546	.50636	.76902
.55692	.92036	.00983	.27579	.06626
.41541	.39612	.13801	.76302	.83600
.64886	.86915	.87513	.92845	.19451
.31103	.61556	.39411	.32464	.90089
.98353	.29320	.40741	.30569	.66741
.75329	.01302	.79855	.17407	.35748
.07829	.75241	.30983	.58729	.23526
.62589	.13804	.69520	.42885	.81629
.53808	.38189	.94864	.75480	.49106
.65320	.99962	.61894	.21703	.23173
.93711	.03711	.28870	.89815	.29063
.16041	.84682	.13722	.70192	.47655
.04206	.46342	.90198	.74105	.82847

INDEX

Bar graph, 124
Bernoulli experiments, 140
Binomial coefficients, 141, 154
Binomial distribution, 140–44, 160
Binomial experiment, 140–141
Binomial probabilities
  normal approximation of, 141–44
  Poisson approximation of, 159–60
Binomial random variable, 141–44
Bivariate data, 67–74
Box-and-whisker diagrams, 53–54

Central limit theorem, 201–4
Central tendency, measures of, 22–27
Chebyshev's inequality, 39–40
Chi-square distribution, 208–11
Class boundaries, 2
Class limits, 2
Class mark, 2
Class width, 2
Combinations, 103–4
Complement, 97, 98–99
Compound event, 94
Conditional probability, 104–5
Continuous random variables, 171–72
Correlation coefficient, 67–70
Cumulative distribution function, 125

Data
  continuous, 2–4
  discrete, 1–2
  graphical presentation of, 1–10
  grouped, 23–25, 35–38
  ungrouped, 22–23, 33–35
Deciles, 50–51
Discrete data, 1–2
Discrete probability distributions, 121–28
Discrete random variables, 121–22
Dispersion, measures of, 33–40
Distribution
  binomial, 140–43, 160
  chi-square, 208–11
  continuous, 171–75
  discrete, 121–28
  F distribution, 214–17
  hypergeometric, 153–56
  multinomial, 161–63
  normal, 180–83
  Poisson, 156–61
  sampling, 195–217
  standard normal, 175–80
  Student's *t*, 211–14
Distribution functions, 124–25, 173–74
Distribution mean, 126
Distribution variance, 126–28

Empirical distribution function, 7–9
Empty set, 97
Events, 94–96

F distribution, 214–17
Frequency distributions, 1–4

Histograms, 4–5
Hypergeometric distribution, 153–56

Independent events, 106–7
Independent trials, 106
Intersection, 97

Least-squares regression line, 70–74
Left-tail probability, 183
Linear correlation coefficient, 67–70
Linear regression, 70–74

Mean
  distribution, 126, 174–75
  of distribution of $\bar{X}$, 197
  sample, 22–25
  sampling distribution of, 195–197, 198–204
  standard error of, 197–98
Median, 26–27
Mode, 27
Multinomial distribution, 161–63
Multiplication principle, 100
Multiplication rule for probabilities, 105–6
Mutually exclusive events, 97

Null set, 97

Ogive, 9–10
Ordered sample, 100–2

Percentiles, 50–51
Permutations, 102–3
Poisson distribution, 156–61
Probability
  binomial, 140–43
  concept of, 93–96
  conditional, 104–5
  for intersection of events, 105–7
  methods of enumeration, 100–4
  properties of, 97–99
Probability density function, 172–73
Probability functions, 122–24
Probability histogram, 143

Quartiles, 48–50, 51

Random experiment, 93–94
Random sample, 195–98
Random variables
  continuous, 171–72
  discrete, 121–22
Range, 40
Relative standing, measures of, 48–54
Right-tail probability, 182–83

Sample, random, 195–98
Sample mean, 22–25
  distribution of, 198–204
Sample space, 93–94
Sampling
  with replacement, 100–1
  without replacement, 101–2
Sampling distribution of $\bar{X}$, 195–98
Sampling distributions, 195–217
Scatter diagrams, 65–67
Simple event, 94
Standard deviation, 33–39, 127–28, 174–75
Standard error of the mean, 197
Standard normal distribution, 175–80
Statistic, 195
Stem-and-leaf diagrams, 6–7
Student's *t* distribution, 211–14

Uniform distribution, 172
Union, 97, 98–99

Variance
  distribution, 126–28, 175–75
  of distribution of $\bar{X}$, 197
  sample, 33–39

z-score, 52–53